GNU Scientific Library

Reference Manual

Third edition, for GSL Version 1.12

Mark Galassi

Los Alamos National Laboratory

Jim Davies

Department of Computer Science, Georgia Institute of Technology

James Theiler

Astrophysics and Radiation Measurements Group,
Los Alamos National Laboratory

Brian Gough

Network Theory Limited

Gerard Jungman

Theoretical Astrophysics Group, Los Alamos National Laboratory

Patrick Alken

Department of Physics, University of Colorado at Boulder

Michael Booth

Department of Physics and Astronomy, The Johns Hopkins University

Fabrice Rossi

University of Paris-Dauphine

Published by Network Theory Ltd.

A catalogue record for this book is available from the British Library.

Third edition, First printing, January 2009 for version 1.12.

Published by Network Theory Limited.

Email: info@network-theory.co.uk

ISBN: 0-9546120-7-8

This book supersedes the previous edition for version 1.8 (ISBN 0-9541617-3-4).

Original cover design by David Nicholls.

Further information about this book is available from
http://www.network-theory.co.uk/gsl/manual/

This book has an unconditional guarantee. If you are not fully satisfied with your purchase for any reason, please contact the publisher at the address above.

The texinfo source files for this manual are available from
http://www.network-theory.co.uk/gsl/manual/src/

Table of Contents

Preface

This manual documents the use of the GNU Scientific Library, a numerical library for C and C++ programmers.

The GNU Scientific Library is *free software*. The term "free software" is sometimes misunderstood—it has nothing to do with price. It is about freedom. It refers to your freedom to run, copy, distribute, study, change and improve the software. With the GNU Scientific Library you have all these freedoms.

The GNU Scientific Library is part of the GNU Project. The GNU Project was launched in 1984 to develop a complete Unix-like operating system which is free software: the GNU system. It was conceived as a way of bringing back the cooperative spirit that prevailed in the computing community in earlier days, by removing the obstacles to cooperation imposed by the owners of proprietary software.

The Free Software Foundation is a tax-exempt charity that raises funds for work on the GNU Project and is dedicated to promoting the freedom to modify and redistribute computer programs. You can support the GNU Project by becoming an associate member of the Free Software Foundation and paying regular membership dues. For more information, visit the website www.fsf.org.

Brian Gough
Publisher
December 2008

1 Introduction

The GNU Scientific Library (GSL) is a collection of routines for numerical computing. The routines have been written from scratch in C, and present a modern Applications Programming Interface (API) for C programmers, allowing wrappers to be written for very high level languages. The source code is distributed under the GNU General Public License.

1.1 Routines available in GSL

The library covers a wide range of topics in numerical computing. Routines are available for the following areas,

Complex Numbers	Roots of Polynomials
Special Functions	Vectors and Matrices
Permutations	Combinations
Sorting	BLAS Support
Linear Algebra	CBLAS Library
Fast Fourier Transforms	Eigensystems
Random Numbers	Quadrature
Random Distributions	Quasi-Random Sequences
Histograms	Statistics
Monte Carlo Integration	N-Tuples
Differential Equations	Simulated Annealing
Numerical Differentiation	Interpolation
Series Acceleration	Chebyshev Approximations
Root-Finding	Discrete Hankel Transforms
Least-Squares Fitting	Minimization
IEEE Floating-Point	Physical Constants
Basis Splines	Wavelets

The use of these routines is described in this manual. Each chapter provides detailed definitions of the functions, followed by example programs and references to the articles on which the algorithms are based.

Where possible the routines have been based on reliable public-domain packages such as FFTPACK and QUADPACK, which the developers of GSL have reimplemented in C with modern coding conventions.

1.2 GSL is Free Software

The subroutines in the GNU Scientific Library are "free software"; this means that everyone is free to use them, and to redistribute them in other free programs. The library is not in the public domain; it is copyrighted and there are conditions on its distribution. These conditions are designed to permit everything that a good cooperating citizen would want to do. What is not allowed is to try to prevent others from further sharing any version of the software that they might get from you.

Specifically, we want to make sure that you have the right to share copies of programs that you are given which use the GNU Scientific Library, that you receive their source code or else can get it if you want it, that you can change these programs or use pieces of them in new free programs, and that you know you can do these things.

To make sure that everyone has such rights, we have to forbid you to deprive anyone else of these rights. For example, if you distribute copies of any code which uses the GNU Scientific Library, you must give the recipients all the rights that you have received. You must make sure that they, too, receive or can get the source code, both to the library and the code which uses it. And you must tell them their rights. This means that the library should not be redistributed in proprietary programs.

Also, for our own protection, we must make certain that everyone finds out that there is no warranty for the GNU Scientific Library. If these programs are modified by someone else and passed on, we want their recipients to know that what they have is not what we distributed, so that any problems introduced by others will not reflect on our reputation.

The precise conditions for the distribution of software related to the GNU Scientific Library are found in the GNU General Public License (see [GNU General Public License], page 523). Further information about this license is available from the GNU Project webpage *Frequently Asked Questions* about *the GNU GPL*,

>   http://www.gnu.org/copyleft/gpl-faq.html

The Free Software Foundation also operates a license consulting service for commercial users (contact details available from http://www.fsf.org/).

1.3 Obtaining GSL

The source code for the library can be obtained in different ways, by copying it from a friend, purchasing it on CDROM or downloading it from the internet. A list of public ftp servers which carry the source code can be found on the GNU website,

>   http://www.gnu.org/software/gsl/

The preferred platform for the library is a GNU system, which allows it to take advantage of additional features in the GNU C compiler and GNU C library. However, the library is fully portable and should compile on most systems with a C compiler.

Announcements of new releases, updates and other relevant events are made on the `info-gsl@gnu.org` mailing list. To subscribe to this low-volume list, send an email of the following form:

```
To: info-gsl-request@gnu.org
Subject: subscribe
```

You will receive a response asking you to reply in order to confirm your subscription.

1.4 No Warranty

The software described in this manual has no warranty, it is provided "as is". It is your responsibility to validate the behavior of the routines and their accuracy using the source code provided, or to purchase support and warranties from commercial redistributors. Consult the GNU General Public license for further details (see [GNU General Public License], page 523).

1.5 Reporting Bugs

A list of known bugs can be found in the 'BUGS' file included in the GSL distribution or online in the GSL bug tracker.[1] Details of compilation problems can be found in the 'INSTALL' file.

If you find a bug which is not listed in these files, please report it to `bug-gsl@gnu.org`.

All bug reports should include:

- The version number of GSL
- The hardware and operating system
- The compiler used, including version number and compilation options
- A description of the bug behavior
- A short program which exercises the bug

It is useful if you can check whether the same problem occurs when the library is compiled without optimization. Thank you.

Any errors or omissions in this manual can also be reported to the same address.

---

[1] `http://savannah.gnu.org/bugs/?group=gsl`

1.6 Further Information

Additional information, including online copies of this manual, links to related projects, and mailing list archives are available from the website mentioned above.

Any questions about the use and installation of the library can be asked on the mailing list `help-gsl@gnu.org`. To subscribe to this list, send an email of the following form:

```
To: help-gsl-request@gnu.org
Subject: subscribe
```

This mailing list can be used to ask questions not covered by this manual, and to contact the developers of the library.

If you would like to refer to the GNU Scientific Library in a journal article, the recommended way is to cite this reference manual, e.g. *M. Galassi et al, GNU Scientific Library Reference Manual (3rd Ed.), ISBN 0-9546120-7-8.*

If you want to give a url, use "`http://www.gnu.org/software/gsl/`".

1.7 Conventions used in this manual

This manual contains many examples which can be typed at the keyboard. A command entered at the terminal is shown like this,

```
$ command
```

The first character on the line is the terminal prompt, and should not be typed. The dollar sign '$' is used as the standard prompt in this manual, although some systems may use a different character.

The examples assume the use of the GNU operating system. There may be minor differences in the output on other systems. The commands for setting environment variables use the Bourne shell syntax of the standard GNU shell (bash).

2 Using the library

This chapter describes how to compile programs that use GSL, and introduces its conventions.

2.1 An Example Program

The following short program demonstrates the use of the library by computing the value of the Bessel function $J_0(x)$ for $x = 5$,

```
#include <stdio.h>
#include <gsl/gsl_sf_bessel.h>

int
main (void)
{
  double x = 5.0;
  double y = gsl_sf_bessel_J0 (x);
  printf ("J0(%g) = %.18e\n", x, y);
  return 0;
}
```

The output is shown below, and should be correct to double-precision accuracy,[1]

```
J0(5) = -1.775967713143382920e-01
```

The steps needed to compile this program are described in the following sections.

2.2 Compiling and Linking

The library header files are installed in their own 'gsl' directory. You should write any preprocessor include statements with a 'gsl/' directory prefix thus,

```
#include <gsl/gsl_math.h>
```

If the directory is not installed on the standard search path of your compiler you will also need to provide its location to the preprocessor as a command line flag. The default location of the 'gsl' directory is '/usr/local/include/gsl'. A typical compilation command for a source file 'example.c' with the GNU C compiler gcc is,

```
$ gcc -Wall -I/usr/local/include -c example.c
```

This results in an object file 'example.o'. The default include path for gcc searches '/usr/local/include' automatically so the -I option can actually be omitted when GSL is installed in its default location.

---

[1] The last few digits may vary slightly depending on the compiler and platform used—this is normal.

2.2.1 Linking programs with the library

The library is installed as a single file, 'libgsl.a'. A shared version of the library 'libgsl.so' is also installed on systems that support shared libraries. The default location of these files is '/usr/local/lib'. If this directory is not on the standard search path of your linker you will also need to provide its location as a command line flag.

To link against the library you need to specify both the main library and a supporting CBLAS library, which provides standard basic linear algebra subroutines. A suitable CBLAS implementation is provided in the library 'libgslcblas.a' if your system does not provide one. The following example shows how to link an application with the library,

```
$ gcc -L/usr/local/lib example.o -lgsl -lgslcblas -lm
```

The default library path for gcc searches '/usr/local/lib' automatically so the -L option can be omitted when GSL is installed in its default location.

2.2.2 Linking with an alternative BLAS library

The following command line shows how you would link the same application with an alternative CBLAS library called 'libcblas',

```
$ gcc example.o -lgsl -lcblas -lm
```

For the best performance an optimized platform-specific CBLAS library should be used for -lcblas. The library must conform to the CBLAS standard. The ATLAS package provides a portable high-performance BLAS library with a CBLAS interface. It is free software and should be installed for any work requiring fast vector and matrix operations. The following command line will link with the ATLAS library and its CBLAS interface,

```
$ gcc example.o -lgsl -lcblas -latlas -lm
```

If the ATLAS library is installed in a non-standard directory use the -L option to add it to the search path, as described above.

For more information about BLAS functions see Chapter 12 [BLAS Support], page 137.

2.3 Shared Libraries

To run a program linked with the shared version of the library the operating system must be able to locate the corresponding '.so' file at runtime. If the library cannot be found, the following error will occur:

```
$ ./a.out
./a.out: error while loading shared libraries:
libgsl.so.0: cannot open shared object file: No such
file or directory
```

To avoid this error, either modify the system dynamic linker configuration[2] or define the shell variable LD_LIBRARY_PATH to include the directory where the library is installed.

---

[2] '/etc/ld.so.conf' on GNU/Linux systems.

For example, in the Bourne shell (/bin/sh or /bin/bash), the library search path can be set with the following commands:

```
$ LD_LIBRARY_PATH=/usr/local/lib
$ export LD_LIBRARY_PATH
$ ./example
```

In the C-shell (/bin/csh or /bin/tcsh) the equivalent command is,

```
% setenv LD_LIBRARY_PATH /usr/local/lib
```

The standard prompt for the C-shell in the example above is the percent character '%', and should not be typed as part of the command.

To save retyping these commands each session they can be placed in an individual or system-wide login file.

To compile a statically linked version of the program, use the -static flag in gcc,

```
$ gcc -static example.o -lgsl -lgslcblas -lm
```

2.4 ANSI C Compliance

The library is written in ANSI C and is intended to conform to the ANSI C standard (C89). It should be portable to any system with a working ANSI C compiler.

The library does not rely on any non-ANSI extensions in the interface it exports to the user. Programs you write using GSL can be ANSI compliant. Extensions which can be used in a way compatible with pure ANSI C are supported, however, via conditional compilation. This allows the library to take advantage of compiler extensions on those platforms which support them.

When an ANSI C feature is known to be broken on a particular system the library will exclude any related functions at compile-time. This should make it impossible to link a program that would use these functions and give incorrect results.

To avoid namespace conflicts all exported function names and variables have the prefix gsl_, while exported macros have the prefix GSL_.

2.5 Inline functions

The inline keyword is not part of the original ANSI C standard (C89) so the library does not export any inline function definitions by default. Inline functions were introduced officially in the newer C99 standard but most C89 compilers have also included inline as an extension for a long time.

To allow the use of inline functions, the library provides optional inline versions of performance-critical routines by conditional compilation in the exported header files. The inline versions of these functions can be included by defining the macro HAVE_INLINE when compiling an application,

```
$ gcc -Wall -c -DHAVE_INLINE example.c
```

If you use autoconf this macro can be defined automatically. If you do not define the macro HAVE_INLINE then the slower non-inlined versions of the functions will be used instead.

By default, the actual form of the inline keyword is extern inline, which is a gcc extension that eliminates unnecessary function definitions. If the form extern inline causes problems with other compilers a stricter autoconf test can be used, see Appendix C [Autoconf Macros], page 501.

When compiling with gcc in C99 mode (gcc -std=c99) the header files automatically switch to C99-compatible inline function declarations instead of extern inline. With other C99 compilers, define the macro GSL_C99_INLINE to use these declarations.

2.6 Long double

In general, the algorithms in the library are written for double precision only. The long double type is not supported for actual computation.

One reason for this choice is that the precision of long double is platform dependent. The IEEE standard only specifies the minimum precision of extended precision numbers, while the precision of double is the same on all platforms.

However, it is sometimes necessary to interact with external data in long-double format, so the vector and matrix datatypes include long-double versions.

It should be noted that in some system libraries the stdio.h formatted input/output functions printf and scanf are not implemented correctly for long double. Undefined or incorrect results are avoided by testing these functions during the configure stage of library compilation and eliminating certain GSL functions which depend on them if necessary. The corresponding line in the configure output looks like this,

```
checking whether printf works with long double... no
```

Consequently when long double formatted input/output does not work on a given system it should be impossible to link a program which uses GSL functions dependent on this.

If it is necessary to work on a system which does not support formatted long double input/output then the options are to use binary formats or to convert long double results into double for reading and writing.

2.7 Portability functions

To help in writing portable applications GSL provides some implementations of functions that are found in other libraries, such as the BSD math library. You can write your application to use the native versions of these functions, and substitute the GSL versions via a preprocessor macro if they are unavailable on another platform.

For example, after determining whether the BSD function hypot is available you can include the following macro definitions in a file 'config.h' with your application,

```
/* Substitute gsl_hypot for missing system hypot */

#ifndef HAVE_HYPOT
#define hypot gsl_hypot
#endif
```

The application source files can then use the include command #include <config.h> to replace each occurrence of hypot by gsl_hypot when hypot is not available. This substitution can be made automatically if you use autoconf, see Appendix C [Autoconf Macros], page 501.

In most circumstances the best strategy is to use the native versions of these functions when available, and fall back to GSL versions otherwise, since this allows your application to take advantage of any platform-specific optimizations in the system library. This is the strategy used within GSL itself.

2.8 Alternative optimized functions

The main implementation of some functions in the library will not be optimal on all architectures. For example, there are several ways to compute a Gaussian random variate and their relative speeds are platform-dependent. In cases like this the library provides alternative implementations of these functions with the same interface. If you write your application using calls to the standard implementation you can select an alternative version later via a preprocessor definition. It is also possible to introduce your own optimized functions this way while retaining portability. The following lines demonstrate the use of a platform-dependent choice of methods for sampling from the Gaussian distribution,

```
#ifdef SPARC
#define gsl_ran_gaussian gsl_ran_gaussian_ratio_method
#elif INTEL
#define gsl_ran_gaussian my_gaussian
#endif
```

These lines would be placed in the configuration header file 'config.h' of the application, which should then be included by all the source files. Note that the alternative implementations will not produce bit-for-bit identical results, and in the case of random number distributions will produce an entirely different stream of random variates.

2.9 Support for different numeric types

Many functions in the library are defined for different numeric types. This feature is implemented by varying the name of the function with a type-related modifier—a primitive form of C++ templates. The modifier is inserted into the function name after the initial module prefix. The following table shows the function names defined for all the numeric types of an imaginary module `gsl_foo` with function `fn`,

`gsl_foo_fn`	double
`gsl_foo_long_double_fn`	long double
`gsl_foo_float_fn`	float
`gsl_foo_long_fn`	long
`gsl_foo_ulong_fn`	unsigned long
`gsl_foo_int_fn`	int
`gsl_foo_uint_fn`	unsigned int
`gsl_foo_short_fn`	short
`gsl_foo_ushort_fn`	unsigned short
`gsl_foo_char_fn`	char
`gsl_foo_uchar_fn`	unsigned char

The normal numeric precision double is considered the default and does not require a suffix. For example, the function `gsl_stats_mean` computes the mean of double precision numbers, while the function `gsl_stats_int_mean` computes the mean of integers.

A corresponding scheme is used for library defined types, such as `gsl_vector` and `gsl_matrix`. In this case the modifier is appended to the type name. For example, if a module defines a new type-dependent struct or typedef `gsl_foo` it is modified for other types in the following way,

`gsl_foo`	double
`gsl_foo_long_double`	long double
`gsl_foo_float`	float
`gsl_foo_long`	long
`gsl_foo_ulong`	unsigned long
`gsl_foo_int`	int
`gsl_foo_uint`	unsigned int
`gsl_foo_short`	short
`gsl_foo_ushort`	unsigned short
`gsl_foo_char`	char
`gsl_foo_uchar`	unsigned char

When a module contains type-dependent definitions the library provides individual header files for each type. The filenames are modified as shown in the below. For convenience the default header includes the definitions for all the types. To include only the double precision header file, or any other specific type, use its individual filename.

`#include <gsl/gsl_foo.h>`	All types
`#include <gsl/gsl_foo_double.h>`	double
`#include <gsl/gsl_foo_long_double.h>`	long double
`#include <gsl/gsl_foo_float.h>`	float
`#include <gsl/gsl_foo_long.h>`	long

`#include <gsl/gsl_foo_ulong.h>`	unsigned long
`#include <gsl/gsl_foo_int.h>`	int
`#include <gsl/gsl_foo_uint.h>`	unsigned int
`#include <gsl/gsl_foo_short.h>`	short
`#include <gsl/gsl_foo_ushort.h>`	unsigned short
`#include <gsl/gsl_foo_char.h>`	char
`#include <gsl/gsl_foo_uchar.h>`	unsigned char

2.10 Compatibility with C++

The library header files automatically define functions to have `extern "C"` linkage when included in C++ programs. This allows the functions to be called directly from C++.

To use C++ exception handling within user-defined functions passed to the library as parameters, the library must be built with the additional CFLAGS compilation option '`-fexceptions`'.

2.11 Aliasing of arrays

The library assumes that arrays, vectors and matrices passed as modifiable arguments are not aliased and do not overlap with each other. This removes the need for the library to handle overlapping memory regions as a special case, and allows additional optimizations to be used. If overlapping memory regions are passed as modifiable arguments then the results of such functions will be undefined. If the arguments will not be modified (for example, if a function prototype declares them as const arguments) then overlapping or aliased memory regions can be safely used.

2.12 Thread-safety

The library can be used in multi-threaded programs. All the functions are thread-safe, in the sense that they do not use static variables. Memory is always associated with objects and not with functions. For functions which use *workspace* objects as temporary storage the workspaces should be allocated on a per-thread basis. For functions which use *table* objects as read-only memory the tables can be used by multiple threads simultaneously. Table arguments are always declared const in function prototypes, to indicate that they may be safely accessed by different threads.

There are a small number of static global variables which are used to control the overall behavior of the library (e.g. whether to use range-checking, the function to call on fatal error, etc). These variables are set directly by the user, so they should be initialized once at program startup and not modified by different threads.

2.13 Deprecated Functions

From time to time, it may be necessary for the definitions of some functions
to be altered or removed from the library. In these circumstances the functions
will first be declared *deprecated* and then removed from subsequent versions
of the library. Functions that are deprecated can be disabled in the current
release by setting the preprocessor definition GSL_DISABLE_DEPRECATED. This
allows existing code to be tested for forwards compatibility.

2.14 Code Reuse

Where possible the routines in the library have been written to avoid de-
pendencies between modules and files. This should make it possible to extract
individual functions for use in your own applications, without needing to have
the whole library installed. You may need to define certain macros such as GSL_
ERROR and remove some #include statements in order to compile the files as
standalone units. Reuse of the library code in this way is encouraged, subject
to the terms of the GNU General Public License.

3 Error Handling

This chapter describes the way that GSL functions report and handle errors. By examining the status information returned by every function you can determine whether it succeeded or failed, and if it failed you can find out what the precise cause of failure was. You can also define your own error handling functions to modify the default behavior of the library.

The functions described in this section are declared in the header file 'gsl_errno.h'.

3.1 Error Reporting

The library follows the thread-safe error reporting conventions of the POSIX Threads library. Functions return a non-zero error code to indicate an error and 0 to indicate success.

```
int status = gsl_function (...)

if (status) { /* an error occurred */
  .....
  /* status value specifies the type of error */
}
```

The routines report an error whenever they cannot perform the task requested of them. For example, a root-finding function would return a non-zero error code if could not converge to the requested accuracy, or exceeded a limit on the number of iterations. Situations like this are a normal occurrence when using any mathematical library and you should check the return status of the functions that you call.

Whenever a routine reports an error the return value specifies the type of error. The return value is analogous to the value of the variable errno in the C library. The caller can examine the return code and decide what action to take, including ignoring the error if it is not considered serious.

In addition to reporting errors by return codes the library also has an error handler function gsl_error. This function is called by other library functions when they report an error, just before they return to the caller. The default behavior of the error handler is to print a message and abort the program,

```
gsl: file.c:67: ERROR: invalid argument supplied by user
Default GSL error handler invoked.
Aborted
```

The purpose of the gsl_error handler is to provide a function where a breakpoint can be set that will catch library errors when running under the debugger. It is not intended for use in production programs, which should handle any errors using the return codes.

3.2 Error Codes

The error code numbers returned by library functions are defined in the file 'gsl_errno.h'. They all have the prefix GSL_ and expand to non-zero constant integer values. Error codes above 1024 are reserved for applications, and are not used by the library. Many of the error codes use the same base name as the corresponding error code in the C library. Here are some of the most common error codes,

int GSL_EDOM   Macro
    Domain error; used by mathematical functions when an argument value does not fall into the domain over which the function is defined (like EDOM in the C library)

int GSL_ERANGE   Macro
    Range error; used by mathematical functions when the result value is not representable because of overflow or underflow (like ERANGE in the C library)

int GSL_ENOMEM   Macro
    No memory available. The system cannot allocate more virtual memory because its capacity is full (like ENOMEM in the C library). This error is reported when a GSL routine encounters problems when trying to allocate memory with malloc.

int GSL_EINVAL   Macro
    Invalid argument. This is used to indicate various kinds of problems with passing the wrong argument to a library function (like EINVAL in the C library).

The error codes can be converted into an error message using the function gsl_strerror.

const char * gsl_strerror (const int *gsl_errno*)   Function
    This function returns a pointer to a string describing the error code *gsl_errno*. For example,

```
      printf ("error: %s\n", gsl_strerror (status));
```

    would print an error message like error: output range error for a status value of GSL_ERANGE.

3.3 Error Handlers

The default behavior of the GSL error handler is to print a short message and call abort. When this default is in use programs will stop with a core-dump whenever a library routine reports an error. This is intended as a fail-safe default for programs which do not check the return status of library routines (we don't encourage you to write programs this way).

If you turn off the default error handler it is your responsibility to check the return values of routines and handle them yourself. You can also customize the error behavior by providing a new error handler. For example, an alternative error handler could log all errors to a file, ignore certain error conditions (such as underflows), or start the debugger and attach it to the current process when an error occurs.

All GSL error handlers have the type gsl_error_handler_t, which is defined in 'gsl_errno.h',

gsl_error_handler_t   Data Type

This is the type of GSL error handler functions. An error handler will be passed four arguments which specify the reason for the error (a string), the name of the source file in which it occurred (also a string), the line number in that file (an integer) and the error number (an integer). The source file and line number are set at compile time using the __FILE__ and __LINE__ directives in the preprocessor. An error handler function returns type void. Error handler functions should be defined like this,

```
void handler (const char * reason,
              const char * file,
              int line,
              int gsl_errno)
```

To request the use of your own error handler you need to call the function gsl_set_error_handler which is also declared in 'gsl_errno.h',

gsl_error_handler_t * gsl_set_error_handler   Function
        (gsl_error_handler_t * new_handler)

This function sets a new error handler, new_handler, for the GSL library routines. The previous handler is returned (so that you can restore it later). Note that the pointer to a user defined error handler function is stored in a static variable, so there can be only one error handler per program. This function should be not be used in multi-threaded programs except to set up a program-wide error handler from a master thread. The following example shows how to set and restore a new error handler,

```
/* save original handler, install new handler */
old_handler = gsl_set_error_handler (&my_handler);

/* code uses new handler */
.....

/* restore original handler */
gsl_set_error_handler (old_handler);
```

To use the default behavior (abort on error) set the error handler to NULL,

```
old_handler = gsl_set_error_handler (NULL);
```

`gsl_error_handler_t * gsl_set_error_handler_off ()`   *Function*
    This function turns off the error handler by defining an error handler which
    does nothing. This will cause the program to continue after any error, so
    the return values from any library routines must be checked. This is the
    recommended behavior for production programs. The previous handler is
    returned (so that you can restore it later).

The error behavior can be changed for specific applications by recompiling
the library with a customized definition of the GSL_ERROR macro in the file
'gsl_errno.h'.

3.4 Using GSL error reporting in your own functions

If you are writing numerical functions in a program which also uses GSL code
you may find it convenient to adopt the same error reporting conventions as in
the library.

To report an error you need to call the function gsl_error with a string
describing the error and then return an appropriate error code from gsl_errno.
h, or a special value, such as NaN. For convenience the file 'gsl_errno.h' defines
two macros which carry out these steps:

`GSL_ERROR (reason, gsl_errno)`   *Macro*
    This macro reports an error using the GSL conventions and returns a status
    value of gsl_errno. It expands to the following code fragment,

```
gsl_error (reason, __FILE__, __LINE__, gsl_errno);
return gsl_errno;
```

    The macro definition in 'gsl_errno.h' actually wraps the code in a do { ...
    } while (0) block to prevent possible parsing problems.

Here is an example of how the macro could be used to report that a routine
did not achieve a requested tolerance. To report the error the routine needs to
return the error code GSL_ETOL.

```
if (residual > tolerance)
  {
    GSL_ERROR("residual exceeds tolerance", GSL_ETOL);
  }
```

`GSL_ERROR_VAL (reason, gsl_errno, value)`   *Macro*
    This macro is the same as GSL_ERROR but returns a user-defined value of
    value instead of an error code. It can be used for mathematical functions
    that return a floating point value.

The following example shows how to return a NaN at a mathematical singularity using the GSL_ERROR_VAL macro,

```
if (x == 0)
  {
    GSL_ERROR_VAL("argument lies on singularity",
                  GSL_ERANGE, GSL_NAN);
  }
```

3.5 Examples

Here is an example of some code which checks the return value of a function where an error might be reported,

```
#include <stdio.h>
#include <gsl/gsl_errno.h>
#include <gsl/gsl_fft_complex.h>

...
  int status;
  size_t n = 37;

  gsl_set_error_handler_off();

  status = gsl_fft_complex_radix2_forward (data, stride, n);

  if (status) {
    if (status == GSL_EINVAL) {
       fprintf (stderr, "invalid argument, n=%d\n", n);
    } else {
       fprintf (stderr, "failed, gsl_errno=%d\n",
                        status);
    }
    exit (-1);
  }
  ...
```

The function gsl_fft_complex_radix2 only accepts integer lengths which are a power of two. If the variable n is not a power of two then the call to the library function will return GSL_EINVAL, indicating that the length argument is invalid. The function call to gsl_set_error_handler_off stops the default error handler from aborting the program. The else clause catches any other possible errors.

4 Mathematical Functions

This chapter describes basic mathematical functions. Some of these functions are present in system libraries, but the alternative versions given here can be used as a substitute when the system functions are not available.

The functions and macros described in this chapter are defined in the header file 'gsl_math.h'.

4.1 Mathematical Constants

The library ensures that the standard BSD mathematical constants are defined. For reference, here is a list of the constants:

M_E
> The base of exponentials, e

M_LOG2E
> The base-2 logarithm of e, $\log_2(e)$

M_LOG10E
> The base-10 logarithm of e, $\log_{10}(e)$

M_SQRT2
> The square root of two, $\sqrt{2}$

M_SQRT1_2
> The square root of one-half, $\sqrt{1/2}$

M_SQRT3
> The square root of three, $\sqrt{3}$

M_PI
> The constant pi, π

M_PI_2
> Pi divided by two, $\pi/2$

M_PI_4
> Pi divided by four, $\pi/4$

M_SQRTPI
> The square root of pi, $\sqrt{\pi}$

M_2_SQRTPI
> Two divided by the square root of pi, $2/\sqrt{\pi}$

M_1_PI
> The reciprocal of pi, $1/\pi$

M_2_PI
> Twice the reciprocal of pi, $2/\pi$

M_LN10

    The natural logarithm of ten, ln(10)

M_LN2

    The natural logarithm of two, ln(2)

M_LNPI

    The natural logarithm of pi, ln(π)

M_EULER

    Euler's constant, γ

4.2 Infinities and Not-a-number

GSL_POSINF   Macro

    This macro contains the IEEE representation of positive infinity, $+\infty$. It is
    computed from the expression +1.0/0.0.

GSL_NEGINF   Macro

    This macro contains the IEEE representation of negative infinity, $-\infty$. It is
    computed from the expression -1.0/0.0.

GSL_NAN   Macro

    This macro contains the IEEE representation of the Not-a-Number symbol,
    NaN. It is computed from the ratio 0.0/0.0.

int gsl_isnan (const double x)   Function

    This function returns 1 if x is not-a-number.

int gsl_isinf (const double x)   Function

    This function returns $+1$ if x is positive infinity, -1 if x is negative infinity
    and 0 otherwise.[1]

int gsl_finite (const double x)   Function

    This function returns 1 if x is a real number, and 0 if it is infinite or not-a-
    number.

---

[1] Note that the C99 standard only requires the system isinf function to return a
non-zero value, without the sign of the infinity. The implementation in some earlier
versions of GSL used the system isinf function and may have this behavior on
some platforms. Therefore, it is advisable to test the sign of x separately, if needed,
rather than relying the sign of the return value from gsl_isinf().

4.3 Elementary Functions

The following routines provide portable implementations of functions found in the BSD math library. When native versions are not available the functions described here can be used instead. The substitution can be made automatically if you use autoconf to compile your application (see Section 2.7 [Portability functions], page 11).

double **gsl_log1p** (const double x)   *Function*
> This function computes the value of $\log(1 + x)$ in a way that is accurate for small x. It provides an alternative to the BSD math function log1p(x).

double **gsl_expm1** (const double x)   *Function*
> This function computes the value of $\exp(x) - 1$ in a way that is accurate for small x. It provides an alternative to the BSD math function expm1(x).

double **gsl_hypot** (const double x, const double y)   *Function*
> This function computes the value of $\sqrt{x^2 + y^2}$ in a way that avoids overflow. It provides an alternative to the BSD math function hypot(x,y).

double **gsl_hypot3** (const double x, const double y, const double z)   *Function*
> This function computes the value of $\sqrt{x^2 + y^2 + z^2}$ in a way that avoids overflow.

double **gsl_acosh** (const double x)   *Function*
> This function computes the value of $\operatorname{arccosh}(x)$. It provides an alternative to the standard math function acosh(x).

double **gsl_asinh** (const double x)   *Function*
> This function computes the value of $\operatorname{arcsinh}(x)$. It provides an alternative to the standard math function asinh(x).

double **gsl_atanh** (const double x)   *Function*
> This function computes the value of $\operatorname{arctanh}(x)$. It provides an alternative to the standard math function atanh(x).

double **gsl_ldexp** (double x, int e)   *Function*
> This function computes the value of $x * 2^e$. It provides an alternative to the standard math function ldexp(x,e).

double **gsl_frexp** (double x, int * e)   *Function*
> This function splits the number x into its normalized fraction f and exponent e, such that $x = f * 2^e$ and $0.5 \leq f < 1$. The function returns f and stores the exponent in e. If x is zero, both f and e are set to zero. This function provides an alternative to the standard math function frexp(x, e).

4.4 Small integer powers

A common complaint about the standard C library is its lack of a function for calculating (small) integer powers. GSL provides some simple functions to fill this gap. For reasons of efficiency, these functions do not check for overflow or underflow conditions.

double gsl_pow_int (double x, int n)   Function

> This routine computes the power x^n for integer n. The power is computed efficiently—for example, x^8 is computed as $((x^2)^2)^2$, requiring only 3 multiplications. A version of this function which also computes the numerical error in the result is available as gsl_sf_pow_int_e.

double gsl_pow_2 (const double x)   Function
double gsl_pow_3 (const double x)   Function
double gsl_pow_4 (const double x)   Function
double gsl_pow_5 (const double x)   Function
double gsl_pow_6 (const double x)   Function
double gsl_pow_7 (const double x)   Function
double gsl_pow_8 (const double x)   Function
double gsl_pow_9 (const double x)   Function

> These functions can be used to compute small integer powers x^2, x^3, etc. efficiently. The functions will be inlined when HAVE_INLINE is defined, so that use of these functions should be as efficient as explicitly writing the corresponding product expression.

```
#include <gsl/gsl_math.h>
double y = gsl_pow_4 (3.141)   /* compute 3.141**4 */
```

4.5 Testing the Sign of Numbers

GSL_SIGN (x)   Macro

> This macro returns the sign of x. It is defined as ((x) >= 0 ? 1 : -1). Note that with this definition the sign of zero is positive (regardless of its IEEE sign bit).

4.6 Testing for Odd and Even Numbers

GSL_IS_ODD (n)   Macro

> This macro evaluates to 1 if n is odd and 0 if n is even. The argument n must be of integer type.

GSL_IS_EVEN (n)   Macro

> This macro is the opposite of GSL_IS_ODD(n). It evaluates to 1 if n is even and 0 if n is odd. The argument n must be of integer type.

4.7 Maximum and Minimum functions

Note that the following macros perform multiple evaluations of their arguments, so they should not be used with arguments that have side effects (such as a call to a random number generator).

GSL_MAX (a, b)   Macro
    This macro returns the maximum of a and b. It is defined as ((a) > (b) ? (a):(b)).

GSL_MIN (a, b)   Macro
    This macro returns the minimum of a and b. It is defined as ((a) < (b) ? (a):(b)).

extern inline double GSL_MAX_DBL (double a, double b)   Function
    This function returns the maximum of the double precision numbers a and b using an inline function. The use of a function allows for type checking of the arguments as an extra safety feature. On platforms where inline functions are not available the macro GSL_MAX will be automatically substituted.

extern inline double GSL_MIN_DBL (double a, double b)   Function
    This function returns the minimum of the double precision numbers a and b using an inline function. The use of a function allows for type checking of the arguments as an extra safety feature. On platforms where inline functions are not available the macro GSL_MIN will be automatically substituted.

extern inline int GSL_MAX_INT (int a, int b)   Function
extern inline int GSL_MIN_INT (int a, int b)   Function
    These functions return the maximum or minimum of the integers a and b using an inline function. On platforms where inline functions are not available the macros GSL_MAX or GSL_MIN will be automatically substituted.

extern inline long double GSL_MAX_LDBL (long double a, long   Function
        double b)
extern inline long double GSL_MIN_LDBL (long double a, long   Function
        double b)
    These functions return the maximum or minimum of the long doubles a and b using an inline function. On platforms where inline functions are not available the macros GSL_MAX or GSL_MIN will be automatically substituted.

4.8 Approximate Comparison of Floating Point Numbers

It is sometimes useful to be able to compare two floating point numbers approximately, to allow for rounding and truncation errors. The following function implements the approximate floating-point comparison algorithm proposed by D.E. Knuth in Section 4.2.2 of *Seminumerical Algorithms* (3rd edition).

int gsl_fcmp (double x, double y, double *epsilon*)   Function
  This function determines whether x and y are approximately equal to a relative accuracy *epsilon*.

  The relative accuracy is measured using an interval of size 2δ, where $\delta = 2^k \epsilon$ and k is the maximum base-2 exponent of x and y as computed by the function frexp.

  If x and y lie within this interval, they are considered approximately equal and the function returns 0. Otherwise if $x < y$, the function returns -1, or if $x > y$, the function returns $+1$.

  Note that x and y are compared to relative accuracy, so this function is not suitable for testing whether a value is approximately zero.

  The implementation is based on the package fcmp by T.C. Belding.

5 Complex Numbers

The functions described in this chapter provide support for complex numbers. The algorithms take care to avoid unnecessary intermediate underflows and overflows, allowing the functions to be evaluated over as much of the complex plane as possible.

For multiple-valued functions the branch cuts have been chosen to follow the conventions of Abramowitz and Stegun in the *Handbook of Mathematical Functions*. The functions return principal values which are the same as those in GNU Calc, which in turn are the same as those in *Common Lisp, The Language (Second Edition)*[1] and the HP-28/48 series of calculators.

The complex types are defined in the header file 'gsl_complex.h', while the corresponding complex functions and arithmetic operations are defined in 'gsl_complex_math.h'.

5.1 Representation of complex numbers

Complex numbers are represented using the type gsl_complex. The internal representation of this type may vary across platforms and should not be accessed directly. The functions and macros described below allow complex numbers to be manipulated in a portable way.

For reference, the default form of the gsl_complex type is given by the following struct,

```
typedef struct
{
  double dat[2];
} gsl_complex;
```

The real and imaginary part are stored in contiguous elements of a two element array. This eliminates any padding between the real and imaginary parts, dat[0] and dat[1], allowing the struct to be mapped correctly onto packed complex arrays.

gsl_complex gsl_complex_rect (double x, double y)   *Function*
  This function uses the rectangular cartesian components (x,y) to return the complex number $z = x + iy$. An inline version of this function is used when HAVE_INLINE is defined.

gsl_complex gsl_complex_polar (double r, double *theta*)   *Function*
  This function returns the complex number $z = r \exp(i\theta) = r(\cos(\theta) + i\sin(\theta))$ from the polar representation (r,*theta*).

GSL_REAL (z)   *Macro*
GSL_IMAG (z)   *Macro*
  These macros return the real and imaginary parts of the complex number z.

---

[1] Note that the first edition uses different definitions.

GSL_SET_COMPLEX (*zp*, *x*, *y*)   Macro
    This macro uses the cartesian components (x,y) to set the real and imaginary
    parts of the complex number pointed to by *zp*. For example,

        GSL_SET_COMPLEX(&z, 3, 4)

    sets *z* to be $3 + 4i$.

GSL_SET_REAL (*zp*,*x*)   Macro
GSL_SET_IMAG (*zp*,*y*)   Macro
    These macros allow the real and imaginary parts of the complex number
    pointed to by *zp* to be set independently.

5.2 Properties of complex numbers

double gsl_complex_arg (gsl_complex *z*)   Function
    This function returns the argument of the complex number z, $\arg(z)$, where
    $-\pi < \arg(z) \le \pi$.

double gsl_complex_abs (gsl_complex *z*)   Function
    This function returns the magnitude of the complex number z, $|z|$.

double gsl_complex_abs2 (gsl_complex *z*)   Function
    This function returns the squared magnitude of the complex number z, $|z|^2$.

double gsl_complex_logabs (gsl_complex *z*)   Function
    This function returns the natural logarithm of the magnitude of the complex
    number z, $\log|z|$. It allows an accurate evaluation of $\log|z|$ when $|z|$ is close
    to one. The direct evaluation of log(gsl_complex_abs(z)) would lead to a
    loss of precision in this case.

5.3 Complex arithmetic operators

gsl_complex gsl_complex_add (gsl_complex *a*, gsl_complex *b*)   Function
    This function returns the sum of the complex numbers a and b, $z = a + b$.

gsl_complex gsl_complex_sub (gsl_complex *a*, gsl_complex *b*)   Function
    This function returns the difference of the complex numbers a and b, $z =
    a - b$.

gsl_complex gsl_complex_mul (gsl_complex *a*, gsl_complex *b*)   Function
    This function returns the product of the complex numbers a and b, $z = ab$.

gsl_complex gsl_complex_div (gsl_complex *a*, gsl_complex *b*)   Function
    This function returns the quotient of the complex numbers a and b, $z = a/b$.

gsl_complex gsl_complex_add_real (gsl_complex *a*, double *x*)   Function
    This function returns the sum of the complex number a and the real number
    x, $z = a + x$.

gsl_complex gsl_complex_sub_real (gsl_complex a, double x)   Function
This function returns the difference of the complex number a and the real
number x, $z = a - x$.

gsl_complex gsl_complex_mul_real (gsl_complex a, double x)   Function
This function returns the product of the complex number a and the real
number x, $z = ax$.

gsl_complex gsl_complex_div_real (gsl_complex a, double x)   Function
This function returns the quotient of the complex number a and the real
number x, $z = a/x$.

gsl_complex gsl_complex_add_imag (gsl_complex a, double y)   Function
This function returns the sum of the complex number a and the imaginary
number iy, $z = a + iy$.

gsl_complex gsl_complex_sub_imag (gsl_complex a, double y)   Function
This function returns the difference of the complex number a and the imaginary number iy, $z = a - iy$.

gsl_complex gsl_complex_mul_imag (gsl_complex a, double y)   Function
This function returns the product of the complex number a and the imaginary number iy, $z = a * (iy)$.

gsl_complex gsl_complex_div_imag (gsl_complex a, double y)   Function
This function returns the quotient of the complex number a and the imaginary number iy, $z = a/(iy)$.

gsl_complex gsl_complex_conjugate (gsl_complex z)   Function
This function returns the complex conjugate of the complex number z, $z^* = x - iy$.

gsl_complex gsl_complex_inverse (gsl_complex z)   Function
This function returns the inverse, or reciprocal, of the complex number z,
$1/z = (x - iy)/(x^2 + y^2)$.

gsl_complex gsl_complex_negative (gsl_complex z)   Function
This function returns the negative of the complex number z, $-z = (-x) + i(-y)$.

5.4 Elementary Complex Functions

`gsl_complex gsl_complex_sqrt (gsl_complex z)`   Function
   This function returns the square root of the complex number z, $\sqrt{z}$. The
   branch cut is the negative real axis. The result always lies in the right half
   of the complex plane.

`gsl_complex gsl_complex_sqrt_real (double x)`   Function
   This function returns the complex square root of the real number x, where
   x may be negative.

`gsl_complex gsl_complex_pow (gsl_complex z, gsl_complex a)`   Function
   The function returns the complex number z raised to the complex power
   a, z^a. This is computed as $\exp(\log(z) * a)$ using complex logarithms and
   complex exponentials.

`gsl_complex gsl_complex_pow_real (gsl_complex z, double x)`   Function
   This function returns the complex number z raised to the real power x, z^x.

`gsl_complex gsl_complex_exp (gsl_complex z)`   Function
   This function returns the complex exponential of the complex number z,
   $\exp(z)$.

`gsl_complex gsl_complex_log (gsl_complex z)`   Function
   This function returns the complex natural logarithm (base e) of the complex
   number z, $\log(z)$. The branch cut is the negative real axis.

`gsl_complex gsl_complex_log10 (gsl_complex z)`   Function
   This function returns the complex base-10 logarithm of the complex number
   z, $\log_{10}(z)$.

`gsl_complex gsl_complex_log_b (gsl_complex z, gsl_complex b)`   Function
   This function returns the complex base-b logarithm of the complex number
   z, $\log_b(z)$. This quantity is computed as the ratio $\log(z)/\log(b)$.

5.5 Complex Trigonometric Functions

`gsl_complex gsl_complex_sin (gsl_complex z)`   Function
   This function returns the complex sine of the complex number z, $\sin(z) =$
   $(\exp(iz) - \exp(-iz))/(2i)$.

`gsl_complex gsl_complex_cos (gsl_complex z)`   Function
   This function returns the complex cosine of the complex number z, $\cos(z) =$
   $(\exp(iz) + \exp(-iz))/2$.

`gsl_complex gsl_complex_tan (gsl_complex z)`   Function
   This function returns the complex tangent of the complex number z,
   $\tan(z) = \sin(z)/\cos(z)$.

`gsl_complex gsl_complex_sec (gsl_complex z)`   *Function*
 This function returns the complex secant of the complex number z, $\sec(z) = 1/\cos(z)$.

`gsl_complex gsl_complex_csc (gsl_complex z)`   *Function*
 This function returns the complex cosecant of the complex number z, $\csc(z) = 1/\sin(z)$.

`gsl_complex gsl_complex_cot (gsl_complex z)`   *Function*
 This function returns the complex cotangent of the complex number z, $\cot(z) = 1/\tan(z)$.

5.6 Inverse Complex Trigonometric Functions

`gsl_complex gsl_complex_arcsin (gsl_complex z)`   *Function*
 This function returns the complex arcsine of the complex number z, $\arcsin(z)$. The branch cuts are on the real axis, less than -1 and greater than 1.

`gsl_complex gsl_complex_arcsin_real (double z)`   *Function*
 This function returns the complex arcsine of the real number z, $\arcsin(z)$. For z between -1 and 1, the function returns a real value in the range $[-\pi/2, \pi/2]$. For z less than -1 the result has a real part of $-\pi/2$ and a positive imaginary part. For z greater than 1 the result has a real part of $\pi/2$ and a negative imaginary part.

`gsl_complex gsl_complex_arccos (gsl_complex z)`   *Function*
 This function returns the complex arccosine of the complex number z, $\arccos(z)$. The branch cuts are on the real axis, less than -1 and greater than 1.

`gsl_complex gsl_complex_arccos_real (double z)`   *Function*
 This function returns the complex arccosine of the real number z, $\arccos(z)$. For z between -1 and 1, the function returns a real value in the range $[0, \pi]$. For z less than -1 the result has a real part of π and a negative imaginary part. For z greater than 1 the result is purely imaginary and positive.

`gsl_complex gsl_complex_arctan (gsl_complex z)`   *Function*
 This function returns the complex arctangent of the complex number z, $\arctan(z)$. The branch cuts are on the imaginary axis, below $-i$ and above i.

`gsl_complex gsl_complex_arcsec (gsl_complex z)`   *Function*
 This function returns the complex arcsecant of the complex number z, $\text{arcsec}(z) = \arccos(1/z)$.

`gsl_complex gsl_complex_arcsec_real (double z)`   *Function*
 This function returns the complex arcsecant of the real number z, $\text{arcsec}(z) = \arccos(1/z)$.

gsl_complex gsl_complex_arccsc (gsl_complex z)   Function
    This function returns the complex arccosecant of the complex number z,
    $\text{arccsc}(z) = \arcsin(1/z)$.

gsl_complex gsl_complex_arccsc_real (double z)   Function
    This function returns the complex arccosecant of the real number z,
    $\text{arccsc}(z) = \arcsin(1/z)$.

gsl_complex gsl_complex_arccot (gsl_complex z)   Function
    This function returns the complex arccotangent of the complex number z,
    $\text{arccot}(z) = \arctan(1/z)$.

5.7 Complex Hyperbolic Functions

gsl_complex gsl_complex_sinh (gsl_complex z)   Function
    This function returns the complex hyperbolic sine of the complex number z,
    $\sinh(z) = (\exp(z) - \exp(-z))/2$.

gsl_complex gsl_complex_cosh (gsl_complex z)   Function
    This function returns the complex hyperbolic cosine of the complex number
    z, $\cosh(z) = (\exp(z) + \exp(-z))/2$.

gsl_complex gsl_complex_tanh (gsl_complex z)   Function
    This function returns the complex hyperbolic tangent of the complex number
    z, $\tanh(z) = \sinh(z)/\cosh(z)$.

gsl_complex gsl_complex_sech (gsl_complex z)   Function
    This function returns the complex hyperbolic secant of the complex number
    z, $\text{sech}(z) = 1/\cosh(z)$.

gsl_complex gsl_complex_csch (gsl_complex z)   Function
    This function returns the complex hyperbolic cosecant of the complex num-
    ber z, $\text{csch}(z) = 1/\sinh(z)$.

gsl_complex gsl_complex_coth (gsl_complex z)   Function
    This function returns the complex hyperbolic cotangent of the complex num-
    ber z, $\coth(z) = 1/\tanh(z)$.

5.8 Inverse Complex Hyperbolic Functions

gsl_complex gsl_complex_arcsinh (gsl_complex z)   Function
This function returns the complex hyperbolic arcsine of the complex number
z, arcsinh(z). The branch cuts are on the imaginary axis, below $-i$ and above
i.

gsl_complex gsl_complex_arccosh (gsl_complex z)   Function
This function returns the complex hyperbolic arccosine of the complex number z, arccosh(z). The branch cut is on the real axis, less than 1. Note that
in this case we use the negative square root in formula 4.6.21 of Abramowitz
& Stegun giving $\mathrm{arccosh}(z) = \log(z - \sqrt{z^2 - 1})$.

gsl_complex gsl_complex_arccosh_real (double z)   Function
This function returns the complex hyperbolic arccosine of the real number
z, arccosh(z).

gsl_complex gsl_complex_arctanh (gsl_complex z)   Function
This function returns the complex hyperbolic arctangent of the complex
number z, arctanh(z). The branch cuts are on the real axis, less than -1
and greater than 1.

gsl_complex gsl_complex_arctanh_real (double z)   Function
This function returns the complex hyperbolic arctangent of the real number
z, arctanh(z).

gsl_complex gsl_complex_arcsech (gsl_complex z)   Function
This function returns the complex hyperbolic arcsecant of the complex number z, $\mathrm{arcsech}(z) = \mathrm{arccosh}(1/z)$.

gsl_complex gsl_complex_arccsch (gsl_complex z)   Function
This function returns the complex hyperbolic arccosecant of the complex
number z, $\mathrm{arccsch}(z) = \mathrm{arcsin}(1/z)$.

gsl_complex gsl_complex_arccoth (gsl_complex z)   Function
This function returns the complex hyperbolic arccotangent of the complex
number z, $\mathrm{arccoth}(z) = \mathrm{arctanh}(1/z)$.

5.9 References and Further Reading

The implementations of the elementary and trigonometric functions are based on the following papers,

T. E. Hull, Thomas F. Fairgrieve, Ping Tak Peter Tang, "Implementing Complex Elementary Functions Using Exception Handling", *ACM Transactions on Mathematical Software*, Volume 20 (1994), pp 215–244, Corrigenda, p553

T. E. Hull, Thomas F. Fairgrieve, Ping Tak Peter Tang, "Implementing the complex arcsin and arccosine functions using exception handling", *ACM Transactions on Mathematical Software*, Volume 23 (1997) pp 299–335

The general formulas and details of branch cuts can be found in the following books,

Abramowitz and Stegun, *Handbook of Mathematical Functions*, "Circular Functions in Terms of Real and Imaginary Parts", Formulas 4.3.55–58, "Inverse Circular Functions in Terms of Real and Imaginary Parts", Formulas 4.4.37–39, "Hyperbolic Functions in Terms of Real and Imaginary Parts", Formulas 4.5.49–52, "Inverse Hyperbolic Functions—relation to Inverse Circular Functions", Formulas 4.6.14–19.

Dave Gillespie, *Calc Manual*, Free Software Foundation, ISBN 1-882114-18-3

6 Polynomials

This chapter describes functions for evaluating and solving polynomials. There are routines for finding real and complex roots of quadratic and cubic equations using analytic methods. An iterative polynomial solver is also available for finding the roots of general polynomials with real coefficients (of any order). The functions are declared in the header file 'gsl_poly.h'.

6.1 Polynomial Evaluation

The functions described here evaluate the polynomial $c[0] + c[1]x + c[2]x^2 + \ldots + c[len-1]x^{len-1}$ using Horner's method for stability. Inline versions of these functions are used when HAVE_INLINE is defined.

double gsl_poly_eval (const double c[], const int *len*, const   Function
        double x)

This function evaluates a polynomial with real coefficients for the real variable x.

gsl_complex gsl_poly_complex_eval (const double c[], const   Function
        int *len*, const gsl_complex z)

This function evaluates a polynomial with real coefficients for the complex variable z.

gsl_complex gsl_complex_poly_complex_eval (const gsl_complex   Function
        c[], const int *len*, const gsl_complex z)

This function evaluates a polynomial with complex coefficients for the complex variable z.

6.2 Divided Difference Representation of Polynomials

The functions described here manipulate polynomials stored in Newton's divided-difference representation. The use of divided-differences is described in Abramowitz & Stegun sections 25.1.4 and 25.2.26.

int gsl_poly_dd_init (double dd[], const double xa[], const   Function
        double ya[], size_t *size*)

This function computes a divided-difference representation of the interpolating polynomial for the points (xa, ya) stored in the arrays xa and ya of length *size*. On output the divided-differences of (xa,ya) are stored in the array dd, also of length *size*.

double gsl_poly_dd_eval (const double dd[], const double   Function
        xa[], const size_t *size*, const double x)

This function evaluates the polynomial stored in divided-difference form in the arrays dd and xa of length *size* at the point x. An inline version of this function is used when HAVE_INLINE is defined.

int gsl_poly_dd_taylor (double c[], double xp, const double   Function
        dd[], const double xa[], size_t size, double w[])
This function converts the divided-difference representation of a polynomial
to a Taylor expansion. The divided-difference representation is supplied in
the arrays dd and xa of length size. On output the Taylor coefficients of the
polynomial expanded about the point xp are stored in the array c also of
length size. A workspace of length size must be provided in the array w.

6.3 Quadratic Equations

int gsl_poly_solve_quadratic (double a, double b, double c,   Function
        double * x0, double * x1)
This function finds the real roots of the quadratic equation,

$$ax^2 + bx + c = 0$$

The number of real roots (either zero, one or two) is returned, and their
locations are stored in $x0$ and $x1$. If no real roots are found then $x0$ and $x1$
are not modified. If one real root is found (i.e. if $a = 0$) then it is stored in
$x0$. When two real roots are found they are stored in $x0$ and $x1$ in ascending
order. The case of coincident roots is not considered special. For example
$(x - 1)^2 = 0$ will have two roots, which happen to have exactly equal values.
The number of roots found depends on the sign of the discriminant $b^2 - 4ac$.
This will be subject to rounding and cancellation errors when computed in
double precision, and will also be subject to errors if the coefficients of the
polynomial are inexact. These errors may cause a discrete change in the
number of roots. However, for polynomials with small integer coefficients
the discriminant can always be computed exactly.

int gsl_poly_complex_solve_quadratic (double a, double b,   Function
        double c, gsl_complex * z0, gsl_complex * z1)
This function finds the complex roots of the quadratic equation,

$$az^2 + bz + c = 0$$

The number of complex roots is returned (either one or two) and the locations of the roots are stored in $z0$ and $z1$. The roots are returned in
ascending order, sorted first by their real components and then by their
imaginary components. If only one real root is found (i.e. if $a = 0$) then it
is stored in $z0$.

6.4 Cubic Equations

int gsl_poly_solve_cubic (double a, double b, double c, double   *Function*
    * x0, double * x1, double * x2)

This function finds the real roots of the cubic equation,

$$x^3 + ax^2 + bx + c = 0$$

with a leading coefficient of unity. The number of real roots (either one or
three) is returned, and their locations are stored in $x0$, $x1$ and $x2$. If one
real root is found then only $x0$ is modified. When three real roots are found
they are stored in $x0$, $x1$ and $x2$ in ascending order. The case of coincident
roots is not considered special. For example, the equation $(x - 1)^3 = 0$ will
have three roots with exactly equal values. As in the quadratic case, finite
precision may cause equal or closely-spaced real roots to move off the real
axis into the complex plane, leading to a discrete change in the number of
real roots.

int gsl_poly_complex_solve_cubic (double a, double b, double   *Function*
        c, gsl_complex * z0, gsl_complex * z1, gsl_complex * z2)

This function finds the complex roots of the cubic equation,

$$z^3 + az^2 + bz + c = 0$$

The number of complex roots is returned (always three) and the locations
of the roots are stored in $z0$, $z1$ and $z2$. The roots are returned in ascending
order, sorted first by their real components and then by their imaginary
components.

6.5 General Polynomial Equations

The roots of polynomial equations cannot be found analytically beyond the
special cases of the quadratic, cubic and quartic equation. The algorithm described in this section uses an iterative method to find the approximate locations
of roots of higher order polynomials.

gsl_poly_complex_workspace *   *Function*
        gsl_poly_complex_workspace_alloc (size_t n)

This function allocates space for a gsl_poly_complex_workspace struct and
a workspace suitable for solving a polynomial with n coefficients using the
routine gsl_poly_complex_solve.

The function returns a pointer to the newly allocated gsl_poly_complex_
workspace if no errors were detected, and a null pointer in the case of error.

void gsl_poly_complex_workspace_free   *Function*
        (gsl_poly_complex_workspace * w)

This function frees all the memory associated with the workspace w.

int gsl_poly_complex_solve (const double * a, size_t n,   Function
         gsl_poly_complex_workspace * w, gsl_complex_packed_ptr z)

This function computes the roots of the general polynomial $P(x) = a_0 + a_1 x + a_2 x^2 + ... + a_{n-1} x^{n-1}$ using balanced-QR reduction of the companion matrix. The parameter n specifies the length of the coefficient array. The coefficient of the highest order term must be non-zero. The function requires a workspace w of the appropriate size. The $n - 1$ roots are returned in the packed complex array z of length $2(n - 1)$, alternating real and imaginary parts.

The function returns GSL_SUCCESS if all the roots are found. If the QR reduction does not converge, the error handler is invoked with an error code of GSL_EFAILED. Note that due to finite precision, roots of higher multiplicity are returned as a cluster of simple roots with reduced accuracy. The solution of polynomials with higher-order roots requires specialized algorithms that take the multiplicity structure into account (see e.g. Z. Zeng, Algorithm 835, ACM Transactions on Mathematical Software, Volume 30, Issue 2 (2004), pp 218–236).

6.6 Examples

To demonstrate the use of the general polynomial solver we will take the polynomial $P(x) = x^5 - 1$ which has the following roots,

$$1, e^{2\pi i/5}, e^{4\pi i/5}, e^{6\pi i/5}, e^{8\pi i/5}$$

The following program will find these roots.

```
#include <stdio.h>
#include <gsl/gsl_poly.h>

int
main (void)
{
  int i;
  /* coefficients of P(x) =  -1 + x^5  */
  double a[6] = { -1, 0, 0, 0, 0, 1 };
  double z[10];

  gsl_poly_complex_workspace * w
      = gsl_poly_complex_workspace_alloc (6);

  gsl_poly_complex_solve (a, 6, w, z);

  gsl_poly_complex_workspace_free (w);

  for (i = 0; i < 5; i++)
    {
      printf ("z%d = %+.18f %+.18f\n",
              i, z[2*i], z[2*i+1]);
```

```
      }

    return 0;
  }
```
The output of the program is,
```
  $ ./a.out
    z0 = -0.809016994374947451 +0.587785252292473137
    z1 = -0.809016994374947451 -0.587785252292473137
    z2 = +0.309016994374947451 +0.951056516295153642
    z3 = +0.309016994374947451 -0.951056516295153642
    z4 = +1.000000000000000000 +0.000000000000000000
```
which agrees with the analytic result, $z_n = \exp(2\pi n i/5)$.

6.7 References and Further Reading

The balanced-QR method and its error analysis are described in the following papers,

R.S. Martin, G. Peters and J.H. Wilkinson, "The QR Algorithm for Real Hessenberg Matrices", *Numerische Mathematik*, 14 (1970), 219–231.

B.N. Parlett and C. Reinsch, "Balancing a Matrix for Calculation of Eigenvalues and Eigenvectors", *Numerische Mathematik*, 13 (1969), 293–304.

A. Edelman and H. Murakami, "Polynomial roots from companion matrix eigenvalues", *Mathematics of Computation*, Vol. 64, No. 210 (1995), 763–776.

The formulas for divided differences are given in Abramowitz and Stegun,

Abramowitz and Stegun, *Handbook of Mathematical Functions*, Sections 25.1.4 and 25.2.26.

7 Special Functions

This chapter describes the GSL special function library. The library includes routines for calculating the values of Airy functions, Bessel functions, Clausen functions, Coulomb wave functions, Coupling coefficients, the Dawson function, Debye functions, Dilogarithms, Elliptic integrals, Jacobi elliptic functions, Error functions, Exponential integrals, Fermi-Dirac functions, Gamma functions, Gegenbauer functions, Hypergeometric functions, Laguerre functions, Legendre functions and Spherical Harmonics, the Psi (Digamma) Function, Synchrotron functions, Transport functions, Trigonometric functions and Zeta functions. Each routine also computes an estimate of the numerical error in the calculated value of the function.

The functions in this chapter are declared in individual header files, such as 'gsl_sf_airy.h', 'gsl_sf_bessel.h', etc. The complete set of header files can be included using the file 'gsl_sf.h'.

7.1 Usage

The special functions are available in two calling conventions, a *natural form* which returns the numerical value of the function and an *error-handling form* which returns an error code. The two types of function provide alternative ways of accessing the same underlying code.

The *natural form* returns only the value of the function and can be used directly in mathematical expressions. For example, the following function call will compute the value of the Bessel function $J_0(x)$,

```
double y = gsl_sf_bessel_J0 (x);
```

There is no way to access an error code or to estimate the error using this method. To allow access to this information the alternative error-handling form stores the value and error in a modifiable argument,

```
gsl_sf_result result;
int status = gsl_sf_bessel_J0_e (x, &result);
```

The error-handling functions have the suffix _e. The returned status value indicates error conditions such as overflow, underflow or loss of precision. If there are no errors the error-handling functions return GSL_SUCCESS.

7.2 The gsl_sf_result struct

The error handling form of the special functions always calculate an error estimate along with the value of the result. Therefore, structures are provided for amalgamating a value and error estimate. These structures are declared in the header file 'gsl_sf_result.h'.

The gsl_sf_result struct contains value and error fields.

```
typedef struct
{
  double val;
  double err;
} gsl_sf_result;
```

The field *val* contains the value and the field *err* contains an estimate of the absolute error in the value.

In some cases, an overflow or underflow can be detected and handled by a function. In this case, it may be possible to return a scaling exponent as well as an error/value pair in order to save the result from exceeding the dynamic range of the built-in types. The gsl_sf_result_e10 struct contains value and error fields as well as an exponent field such that the actual result is obtained as result * 10^(e10).

```
typedef struct
{
  double val;
  double err;
  int    e10;
} gsl_sf_result_e10;
```

7.3 Modes

The goal of the library is to achieve double precision accuracy wherever possible. However the cost of evaluating some special functions to double precision can be significant, particularly where very high order terms are required. In these cases a mode argument allows the accuracy of the function to be reduced in order to improve performance. The following precision levels are available for the mode argument,

GSL_PREC_DOUBLE

> Double-precision, a relative accuracy of approximately 2×10^{-16}.

GSL_PREC_SINGLE

> Single-precision, a relative accuracy of approximately 1×10^{-7}.

GSL_PREC_APPROX

> Approximate values, a relative accuracy of approximately 5×10^{-4}.

The approximate mode provides the fastest evaluation at the lowest accuracy.

7.4 Airy Functions and Derivatives

The Airy functions $Ai(x)$ and $Bi(x)$ are defined by the integral representations,

$$Ai(x) = \frac{1}{\pi} \int_0^\infty \cos(t^3/3 + xt)\, dt,$$

$$Bi(x) = \frac{1}{\pi} \int_0^\infty (e^{-t^3/3} + \sin(t^3/3 + xt))\, dt.$$

For further information see Abramowitz & Stegun, Section 10.4. The Airy functions are defined in the header file 'gsl_sf_airy.h'.

7.4.1 Airy Functions

double gsl_sf_airy_Ai (double x, gsl_mode_t *mode*)   Function
int gsl_sf_airy_Ai_e (double x, gsl_mode_t *mode*,   Function
        gsl_sf_result * *result*)
    These routines compute the Airy function $Ai(x)$ with an accuracy specified by *mode*.

double gsl_sf_airy_Bi (double x, gsl_mode_t *mode*)   Function
int gsl_sf_airy_Bi_e (double x, gsl_mode_t *mode*,   Function
        gsl_sf_result * *result*)
    These routines compute the Airy function $Bi(x)$ with an accuracy specified by *mode*.

double gsl_sf_airy_Ai_scaled (double x, gsl_mode_t *mode*)   Function
int gsl_sf_airy_Ai_scaled_e (double x, gsl_mode_t *mode*,   Function
        gsl_sf_result * *result*)
    These routines compute a scaled version of the Airy function $S_A(x)Ai(x)$.
    For $x > 0$ the scaling factor $S_A(x)$ is $\exp(+(2/3)x^{3/2})$, and is 1 for $x < 0$.

double gsl_sf_airy_Bi_scaled (double x, gsl_mode_t *mode*)   Function
int gsl_sf_airy_Bi_scaled_e (double x, gsl_mode_t *mode*,   Function
        gsl_sf_result * *result*)
    These routines compute a scaled version of the Airy function $S_B(x)Bi(x)$.
    For $x > 0$ the scaling factor $S_B(x)$ is $\exp(-(2/3)x^{3/2})$, and is 1 for $x < 0$.

7.4.2 Derivatives of Airy Functions

double gsl_sf_airy_Ai_deriv (double x, gsl_mode_t *mode*)   Function
int gsl_sf_airy_Ai_deriv_e (double x, gsl_mode_t *mode*,   Function
        gsl_sf_result * *result*)
    These routines compute the Airy function derivative $Ai'(x)$ with an accuracy specified by *mode*.

double `gsl_sf_airy_Bi_deriv` (double x, gsl_mode_t *mode*)   Function
int `gsl_sf_airy_Bi_deriv_e` (double x, gsl_mode_t *mode*,   Function
          `gsl_sf_result * *result*`)

These routines compute the Airy function derivative $Bi'(x)$ with an accuracy
specified by *mode*.

double `gsl_sf_airy_Ai_deriv_scaled` (double x, gsl_mode_t   Function
          mode)
int `gsl_sf_airy_Ai_deriv_scaled_e` (double x, gsl_mode_t   Function
          mode, gsl_sf_result * *result*)

These routines compute the scaled Airy function derivative $S_A(x)Ai'(x)$.
For $x > 0$ the scaling factor $S_A(x)$ is $\exp(+(2/3)x^{3/2})$, and is 1 for $x < 0$.

double `gsl_sf_airy_Bi_deriv_scaled` (double x, gsl_mode_t   Function
          mode)
int `gsl_sf_airy_Bi_deriv_scaled_e` (double x, gsl_mode_t   Function
          mode, gsl_sf_result * *result*)

These routines compute the scaled Airy function derivative $S_B(x)Bi'(x)$.
For $x > 0$ the scaling factor $S_B(x)$ is $\exp(-(2/3)x^{3/2})$, and is 1 for $x < 0$.

7.4.3 Zeros of Airy Functions

double `gsl_sf_airy_zero_Ai` (unsigned int s)   Function
int `gsl_sf_airy_zero_Ai_e` (unsigned int s, gsl_sf_result *   Function
          result)

These routines compute the location of the s-th zero of the Airy function
$Ai(x)$.

double `gsl_sf_airy_zero_Bi` (unsigned int s)   Function
int `gsl_sf_airy_zero_Bi_e` (unsigned int s, gsl_sf_result *   Function
          result)

These routines compute the location of the s-th zero of the Airy function
$Bi(x)$.

7.4.4 Zeros of Derivatives of Airy Functions

double `gsl_sf_airy_zero_Ai_deriv` (unsigned int s)   Function
int `gsl_sf_airy_zero_Ai_deriv_e` (unsigned int s,   Function
          `gsl_sf_result * *result*`)

These routines compute the location of the s-th zero of the Airy function
derivative $Ai'(x)$.

double `gsl_sf_airy_zero_Bi_deriv` (unsigned int s)   Function
int `gsl_sf_airy_zero_Bi_deriv_e` (unsigned int s,   Function
          `gsl_sf_result * *result*`)

These routines compute the location of the s-th zero of the Airy function
derivative $Bi'(x)$.

7.5 Bessel Functions

The routines described in this section compute the Cylindrical Bessel functions $J_n(x)$, $Y_n(x)$, Modified cylindrical Bessel functions $I_n(x)$, $K_n(x)$, Spherical Bessel functions $j_l(x)$, $y_l(x)$, and Modified Spherical Bessel functions $i_l(x)$, $k_l(x)$. For more information see Abramowitz & Stegun, Chapters 9 and 10. The Bessel functions are defined in the header file 'gsl_sf_bessel.h'.

7.5.1 Regular Cylindrical Bessel Functions

double gsl_sf_bessel_J0 (double x)   Function
int gsl_sf_bessel_J0_e (double x, gsl_sf_result * result)   Function
  These routines compute the regular cylindrical Bessel function of zeroth
  order, $J_0(x)$.

double gsl_sf_bessel_J1 (double x)   Function
int gsl_sf_bessel_J1_e (double x, gsl_sf_result * result)   Function
  These routines compute the regular cylindrical Bessel function of first order,
  $J_1(x)$.

double gsl_sf_bessel_Jn (int n, double x)   Function
int gsl_sf_bessel_Jn_e (int n, double x, gsl_sf_result *   Function
        result)
  These routines compute the regular cylindrical Bessel function of order n,
  $J_n(x)$.

int gsl_sf_bessel_Jn_array (int nmin, int nmax, double x,   Function
        double result_array[])
  This routine computes the values of the regular cylindrical Bessel functions
  $J_n(x)$ for n from nmin to nmax inclusive, storing the results in the array result_array. The values are computed using recurrence relations for efficiency,
  and therefore may differ slightly from the exact values.

7.5.2 Irregular Cylindrical Bessel Functions

double gsl_sf_bessel_Y0 (double x)   Function
int gsl_sf_bessel_Y0_e (double x, gsl_sf_result * result)   Function
  These routines compute the irregular cylindrical Bessel function of zeroth
  order, $Y_0(x)$, for $x > 0$.

double gsl_sf_bessel_Y1 (double x)   Function
int gsl_sf_bessel_Y1_e (double x, gsl_sf_result * result)   Function
  These routines compute the irregular cylindrical Bessel function of first order, $Y_1(x)$, for $x > 0$.

double gsl_sf_bessel_Yn (int n,double x)   Function
int gsl_sf_bessel_Yn_e (int n,double x, gsl_sf_result * result)   Function
  These routines compute the irregular cylindrical Bessel function of order n,
  $Y_n(x)$, for $x > 0$.

int gsl_sf_bessel_Yn_array (int *nmin*, int *nmax*, double x,   *Function*
           double *result_array* [])
   This routine computes the values of the irregular cylindrical Bessel functions
   $Y_n(x)$ for n from *nmin* to *nmax* inclusive, storing the results in the array
   result_array. The domain of the function is $x > 0$. The values are computed
   using recurrence relations for efficiency, and therefore may differ slightly
   from the exact values.

7.5.3 Regular Modified Cylindrical Bessel Functions

double gsl_sf_bessel_I0 (double x)   *Function*
int gsl_sf_bessel_I0_e (double x, gsl_sf_result * *result*)   *Function*
   These routines compute the regular modified cylindrical Bessel function of
   zeroth order, $I_0(x)$.

double gsl_sf_bessel_I1 (double x)   *Function*
int gsl_sf_bessel_I1_e (double x, gsl_sf_result * *result*)   *Function*
   These routines compute the regular modified cylindrical Bessel function of
   first order, $I_1(x)$.

double gsl_sf_bessel_In (int n, double x)   *Function*
int gsl_sf_bessel_In_e (int n, double x, gsl_sf_result *   *Function*
           result)
   These routines compute the regular modified cylindrical Bessel function of
   order n, $I_n(x)$.

int gsl_sf_bessel_In_array (int *nmin*, int *nmax*, double x,   *Function*
           double *result_array* [])
   This routine computes the values of the regular modified cylindrical Bessel
   functions $I_n(x)$ for n from *nmin* to *nmax* inclusive, storing the results in
   the array *result_array*. The start of the range *nmin* must be positive or
   zero. The values are computed using recurrence relations for efficiency, and
   therefore may differ slightly from the exact values.

double gsl_sf_bessel_I0_scaled (double x)   *Function*
int gsl_sf_bessel_I0_scaled_e (double x, gsl_sf_result *   *Function*
           result)
   These routines compute the scaled regular modified cylindrical Bessel func-
   tion of zeroth order $\exp(-|x|)I_0(x)$.

double gsl_sf_bessel_I1_scaled (double x)   *Function*
int gsl_sf_bessel_I1_scaled_e (double x, gsl_sf_result *   *Function*
           result)
   These routines compute the scaled regular modified cylindrical Bessel func-
   tion of first order $\exp(-|x|)I_1(x)$.

double gsl_sf_bessel_In_scaled (int n, double x)   *Function*
int gsl_sf_bessel_In_scaled_e (int n, double x, gsl_sf_result   *Function*
           * *result*)
   These routines compute the scaled regular modified cylindrical Bessel func-
   tion of order n, $\exp(-|x|)I_n(x)$

int gsl_sf_bessel_In_scaled_array (int *nmin*, int *nmax*,   Function
      double x, double *result_array*[])
> This routine computes the values of the scaled regular cylindrical Bessel
> functions $\exp(-|x|)I_n(x)$ for n from *nmin* to *nmax* inclusive, storing the
> results in the array *result_array*. The start of the range *nmin* must be
> positive or zero. The values are computed using recurrence relations for
> efficiency, and therefore may differ slightly from the exact values.

7.5.4 Irregular Modified Cylindrical Bessel Functions

double gsl_sf_bessel_K0 (double x)   Function
int gsl_sf_bessel_K0_e (double x, gsl_sf_result * *result*)   Function
> These routines compute the irregular modified cylindrical Bessel function of
> zeroth order, $K_0(x)$, for $x > 0$.

double gsl_sf_bessel_K1 (double x)   Function
int gsl_sf_bessel_K1_e (double x, gsl_sf_result * *result*)   Function
> These routines compute the irregular modified cylindrical Bessel function of
> first order, $K_1(x)$, for $x > 0$.

double gsl_sf_bessel_Kn (int n, double x)   Function
int gsl_sf_bessel_Kn_e (int n, double x, gsl_sf_result *   Function
      result)
> These routines compute the irregular modified cylindrical Bessel function of
> order n, $K_n(x)$, for $x > 0$.

int gsl_sf_bessel_Kn_array (int *nmin*, int *nmax*, double x,   Function
      double *result_array*[])
> This routine computes the values of the irregular modified cylindrical Bessel
> functions $K_n(x)$ for n from *nmin* to *nmax* inclusive, storing the results in
> the array *result_array*. The start of the range *nmin* must be positive or
> zero. The domain of the function is $x > 0$. The values are computed using
> recurrence relations for efficiency, and therefore may differ slightly from the
> exact values.

double gsl_sf_bessel_K0_scaled (double x)   Function
int gsl_sf_bessel_K0_scaled_e (double x, gsl_sf_result *   Function
      result)
> These routines compute the scaled irregular modified cylindrical Bessel func-
> tion of zeroth order $\exp(x)K_0(x)$ for $x > 0$.

double gsl_sf_bessel_K1_scaled (double x)   Function
int gsl_sf_bessel_K1_scaled_e (double x, gsl_sf_result *   Function
      result)
> These routines compute the scaled irregular modified cylindrical Bessel func-
> tion of first order $\exp(x)K_1(x)$ for $x > 0$.

double gsl_sf_bessel_Kn_scaled (int n, double x)   Function
int gsl_sf_bessel_Kn_scaled_e (int n, double x, gsl_sf_result   Function
            * result)

These routines compute the scaled irregular modified cylindrical Bessel function of order n, $\exp(x)K_n(x)$, for $x > 0$.

int gsl_sf_bessel_Kn_scaled_array (int nmin, int nmax,   Function
            double x, double result_array[])

This routine computes the values of the scaled irregular cylindrical Bessel functions $\exp(x)K_n(x)$ for n from nmin to nmax inclusive, storing the results in the array result_array. The start of the range nmin must be positive or zero. The domain of the function is $x > 0$. The values are computed using recurrence relations for efficiency, and therefore may differ slightly from the exact values.

7.5.5 Regular Spherical Bessel Functions

double gsl_sf_bessel_j0 (double x)   Function
int gsl_sf_bessel_j0_e (double x, gsl_sf_result * result)   Function

These routines compute the regular spherical Bessel function of zeroth order, $j_0(x) = \sin(x)/x$.

double gsl_sf_bessel_j1 (double x)   Function
int gsl_sf_bessel_j1_e (double x, gsl_sf_result * result)   Function

These routines compute the regular spherical Bessel function of first order, $j_1(x) = (\sin(x)/x - \cos(x))/x$.

double gsl_sf_bessel_j2 (double x)   Function
int gsl_sf_bessel_j2_e (double x, gsl_sf_result * result)   Function

These routines compute the regular spherical Bessel function of second order, $j_2(x) = ((3/x^2 - 1)\sin(x) - 3\cos(x)/x)/x$.

double gsl_sf_bessel_jl (int l, double x)   Function
int gsl_sf_bessel_jl_e (int l, double x, gsl_sf_result * result) Function

These routines compute the regular spherical Bessel function of order l, $j_l(x)$, for $l \geq 0$ and $x \geq 0$.

int gsl_sf_bessel_jl_array (int lmax, double x, double   Function
            result_array[])

This routine computes the values of the regular spherical Bessel functions $j_l(x)$ for l from 0 to lmax inclusive for $lmax \geq 0$ and $x \geq 0$, storing the results in the array result_array. The values are computed using recurrence relations for efficiency, and therefore may differ slightly from the exact values.

int gsl_sf_bessel_jl_steed_array (int *lmax*, double x, double   Function
            * *jl_x_array*)

This routine uses Steed's method to compute the values of the regular spherical Bessel functions $j_l(x)$ for l from 0 to *lmax* inclusive for $lmax \geq 0$ and $x \geq 0$, storing the results in the array *result_array*. The Steed/Barnett algorithm is described in *Comp. Phys. Comm.* 21, 297 (1981). Steed's method is more stable than the recurrence used in the other functions but is also slower.

7.5.6 Irregular Spherical Bessel Functions

double gsl_sf_bessel_y0 (double x)   Function
int gsl_sf_bessel_y0_e (double x, gsl_sf_result * *result*)   Function

These routines compute the irregular spherical Bessel function of zeroth order, $y_0(x) = -\cos(x)/x$.

double gsl_sf_bessel_y1 (double x)   Function
int gsl_sf_bessel_y1_e (double x, gsl_sf_result * *result*)   Function

These routines compute the irregular spherical Bessel function of first order, $y_1(x) = -(\cos(x)/x + \sin(x))/x$.

double gsl_sf_bessel_y2 (double x)   Function
int gsl_sf_bessel_y2_e (double x, gsl_sf_result * *result*)   Function

These routines compute the irregular spherical Bessel function of second order, $y_2(x) = (-3/x^3 + 1/x)\cos(x) - (3/x^2)\sin(x)$.

double gsl_sf_bessel_yl (int *l*, double x)   Function
int gsl_sf_bessel_yl_e (int *l*, double x, gsl_sf_result * *result*)   Function

These routines compute the irregular spherical Bessel function of order l, $y_l(x)$, for $l \geq 0$.

int gsl_sf_bessel_yl_array (int *lmax*, double x, double   Function
            result_array[])

This routine computes the values of the irregular spherical Bessel functions $y_l(x)$ for l from 0 to *lmax* inclusive for $lmax \geq 0$, storing the results in the array *result_array*. The values are computed using recurrence relations for efficiency, and therefore may differ slightly from the exact values.

7.5.7 Regular Modified Spherical Bessel Functions

The regular modified spherical Bessel functions $i_l(x)$ are related to the modified Bessel functions of fractional order, $i_l(x) = \sqrt{\pi/(2x)}I_{l+1/2}(x)$

double gsl_sf_bessel_i0_scaled (double x)   Function
int gsl_sf_bessel_i0_scaled_e (double x, gsl_sf_result *   Function
            result)

These routines compute the scaled regular modified spherical Bessel function of zeroth order, $\exp(-|x|)i_0(x)$.

double gsl_sf_bessel_i1_scaled (double x)   Function
int gsl_sf_bessel_i1_scaled_e (double x, gsl_sf_result *   Function
      result)
    These routines compute the scaled regular modified spherical Bessel function
    of first order, $\exp(-|x|)i_1(x)$.

double gsl_sf_bessel_i2_scaled (double x)   Function
int gsl_sf_bessel_i2_scaled_e (double x, gsl_sf_result *   Function
      result)
    These routines compute the scaled regular modified spherical Bessel function
    of second order, $\exp(-|x|)i_2(x)$

double gsl_sf_bessel_il_scaled (int l, double x)   Function
int gsl_sf_bessel_il_scaled_e (int l, double x, gsl_sf_result   Function
      * result)
    These routines compute the scaled regular modified spherical Bessel function
    of order l, $\exp(-|x|)i_l(x)$

int gsl_sf_bessel_il_scaled_array (int lmax, double x, double   Function
      result_array [])
    This routine computes the values of the scaled regular modified cylindrical
    Bessel functions $\exp(-|x|)i_l(x)$ for l from 0 to lmax inclusive for $lmax \geq 0$,
    storing the results in the array result_array. The values are computed using
    recurrence relations for efficiency, and therefore may differ slightly from the
    exact values.

7.5.8 Irregular Modified Spherical Bessel Functions

    The irregular modified spherical Bessel functions, denoted by $k_l(x)$, are re-
lated to the irregular modified Bessel functions of fractional order, $k_l(x) = \sqrt{\pi/(2x)}K_{l+1/2}(x)$.

double gsl_sf_bessel_k0_scaled (double x)   Function
int gsl_sf_bessel_k0_scaled_e (double x, gsl_sf_result *   Function
      result)
    These routines compute the scaled irregular modified spherical Bessel func-
    tion of zeroth order, $\exp(x)k_0(x)$, for $x > 0$.

double gsl_sf_bessel_k1_scaled (double x)   Function
int gsl_sf_bessel_k1_scaled_e (double x, gsl_sf_result *   Function
      result)
    These routines compute the scaled irregular modified spherical Bessel func-
    tion of first order, $\exp(x)k_1(x)$, for $x > 0$.

double gsl_sf_bessel_k2_scaled (double x)   Function
int gsl_sf_bessel_k2_scaled_e (double x, gsl_sf_result *   Function
      result)
    These routines compute the scaled irregular modified spherical Bessel func-
    tion of second order, $\exp(x)k_2(x)$, for $x > 0$.

```
double gsl_sf_bessel_kl_scaled (int l, double x)                    Function
int gsl_sf_bessel_kl_scaled_e (int l, double x, gsl_sf_result       Function
        * result)
```
These routines compute the scaled irregular modified spherical Bessel function of order l, $\exp(x)k_l(x)$, for $x > 0$.

```
int gsl_sf_bessel_kl_scaled_array (int lmax, double x, double       Function
        result_array[])
```
This routine computes the values of the scaled irregular modified spherical Bessel functions $\exp(x)k_l(x)$ for l from 0 to $lmax$ inclusive for $lmax \geq 0$ and $x > 0$, storing the results in the array result_array. The values are computed using recurrence relations for efficiency, and therefore may differ slightly from the exact values.

7.5.9 Regular Bessel Function—Fractional Order

```
double gsl_sf_bessel_Jnu (double nu, double x)                      Function
int gsl_sf_bessel_Jnu_e (double nu, double x, gsl_sf_result *       Function
        result)
```
These routines compute the regular cylindrical Bessel function of fractional order ν, $J_\nu(x)$.

```
int gsl_sf_bessel_sequence_Jnu_e (double nu, gsl_mode_t            Function
        mode, size_t size, double v[])
```
This function computes the regular cylindrical Bessel function of fractional order ν, $J_\nu(x)$, evaluated at a series of x values. The array v of length size contains the x values. They are assumed to be strictly ordered and positive. The array is over-written with the values of $J_\nu(x_i)$.

7.5.10 Irregular Bessel Functions—Fractional Order

```
double gsl_sf_bessel_Ynu (double nu, double x)                      Function
int gsl_sf_bessel_Ynu_e (double nu, double x, gsl_sf_result *       Function
        result)
```
These routines compute the irregular cylindrical Bessel function of fractional order ν, $Y_\nu(x)$.

7.5.11 Regular Modified Bessel Functions—Fractional Order

```
double gsl_sf_bessel_Inu (double nu, double x)                      Function
int gsl_sf_bessel_Inu_e (double nu, double x, gsl_sf_result *       Function
        result)
```
These routines compute the regular modified Bessel function of fractional order ν, $I_\nu(x)$ for $x > 0$, $\nu > 0$.

double gsl_sf_bessel_Inu_scaled (double *nu*, double *x*)   Function
int gsl_sf_bessel_Inu_scaled_e (double *nu*, double *x*,   Function
        gsl_sf_result * *result*)

These routines compute the scaled regular modified Bessel function of fractional order ν, $\exp(-|x|)I_\nu(x)$ for $x > 0$, $\nu > 0$.

7.5.12 Irregular Modified Bessel Functions—Fractional Order

double gsl_sf_bessel_Knu (double *nu*, double *x*)   Function
int gsl_sf_bessel_Knu_e (double *nu*, double *x*, gsl_sf_result *   Function
        result)

These routines compute the irregular modified Bessel function of fractional order ν, $K_\nu(x)$ for $x > 0$, $\nu > 0$.

double gsl_sf_bessel_lnKnu (double *nu*, double *x*)   Function
int gsl_sf_bessel_lnKnu_e (double *nu*, double *x*, gsl_sf_result   Function
        * *result*)

These routines compute the logarithm of the irregular modified Bessel function of fractional order ν, $\ln(K_\nu(x))$ for $x > 0$, $\nu > 0$.

double gsl_sf_bessel_Knu_scaled (double *nu*, double *x*)   Function
int gsl_sf_bessel_Knu_scaled_e (double *nu*, double *x*,   Function
        gsl_sf_result * *result*)

These routines compute the scaled irregular modified Bessel function of fractional order ν, $\exp(+|x|)K_\nu(x)$ for $x > 0$, $\nu > 0$.

7.5.13 Zeros of Regular Bessel Functions

double gsl_sf_bessel_zero_J0 (unsigned int *s*)   Function
int gsl_sf_bessel_zero_J0_e (unsigned int *s*, gsl_sf_result *   Function
        result)

These routines compute the location of the s-th positive zero of the Bessel function $J_0(x)$.

double gsl_sf_bessel_zero_J1 (unsigned int *s*)   Function
int gsl_sf_bessel_zero_J1_e (unsigned int *s*, gsl_sf_result *   Function
        result)

These routines compute the location of the s-th positive zero of the Bessel function $J_1(x)$.

double gsl_sf_bessel_zero_Jnu (double *nu*, unsigned int *s*)   Function
int gsl_sf_bessel_zero_Jnu_e (double *nu*, unsigned int *s*,   Function
        gsl_sf_result * *result*)

These routines compute the location of the s-th positive zero of the Bessel function $J_\nu(x)$. The current implementation does not support negative values of *nu*.

7.6 Clausen Functions

The Clausen function is defined by the following integral,

$$Cl_2(x) = -\int_0^x dt \log(2\sin(t/2))$$

It is related to the dilogarithm by $Cl_2(\theta) = \mathrm{Im}(Li_2(e^{i\theta}))$. The Clausen functions are declared in the header file 'gsl_sf_clausen.h'.

double gsl_sf_clausen (double x)   Function
int gsl_sf_clausen_e (double x, gsl_sf_result * result)   Function
    These routines compute the Clausen integral $Cl_2(x)$.

7.7 Coulomb Functions

The prototypes of the Coulomb functions are declared in the header file 'gsl_sf_coulomb.h'. Both bound state and scattering solutions are available.

7.7.1 Normalized Hydrogenic Bound States

double gsl_sf_hydrogenicR_1 (double Z, double r)   Function
int gsl_sf_hydrogenicR_1_e (double Z, double r, gsl_sf_result   Function
    * result)
    These routines compute the lowest-order normalized hydrogenic bound state
    radial wavefunction $R_1 := 2Z\sqrt{Z}\exp(-Zr)$.

double gsl_sf_hydrogenicR (int n, int l, double Z, double r)   Function
int gsl_sf_hydrogenicR_e (int n, int l, double Z, double r,   Function
    gsl_sf_result * result)
    These routines compute the n-th normalized hydrogenic bound state radial
    wavefunction,

$$R_n := \frac{2Z^{3/2}}{n^2}\left(\frac{2Zr}{n}\right)^l \sqrt{\frac{(n-l-1)!}{(n+l)!}}\exp(-Zr/n)L_{n-l-1}^{2l+1}(2Zr/n).$$

where $L_b^a(x)$ is the generalized Laguerre polynomial (see Section 7.22 [Laguerre Functions], page 74). The normalization is chosen such that the wavefunction ψ is given by $\psi(n,l,r) = R_n Y_{lm}$.

7.7.2 Coulomb Wave Functions

The Coulomb wave functions $F_L(\eta, x)$, $G_L(\eta, x)$ are described in Abramowitz & Stegun, Chapter 14. Because there can be a large dynamic range of values for these functions, overflows are handled gracefully. If an overflow occurs, GSL_EOVRFLW is signalled and exponent(s) are returned through the modifiable parameters exp_F, exp_G. The full solution can be reconstructed from the following relations,

$$F_L(\eta, x) = fc[k_L] * \exp(exp_F)$$
$$G_L(\eta, x) = gc[k_L] * \exp(exp_G)$$

$$F_L'(\eta, x) = fcp[k_L] * \exp(exp_F)$$
$$G_L'(\eta, x) = gcp[k_L] * \exp(exp_G)$$

int gsl_sf_coulomb_wave_FG_e (double eta, double x, double   Function
        L_F, int k, gsl_sf_result * F, gsl_sf_result * Fp,
        gsl_sf_result * G, gsl_sf_result * Gp, double * exp_F,
        double * exp_G)

This function computes the Coulomb wave functions $F_L(\eta, x)$, $G_{L-k}(\eta, x)$ and their derivatives $F_L'(\eta, x)$, $G_{L-k}'(\eta, x)$ with respect to x. The parameters are restricted to $L, L - k > -1/2$, $x > 0$ and integer k. Note that L itself is not restricted to being an integer. The results are stored in the parameters F, G for the function values and Fp, Gp for the derivative values. If an overflow occurs, GSL_EOVRFLW is returned and scaling exponents are stored in the modifiable parameters exp_F, exp_G.

int gsl_sf_coulomb_wave_F_array (double L_min, int kmax,   Function
        double eta, double x, double fc_array[], double * F_exponent)

This function computes the Coulomb wave function $F_L(\eta, x)$ for $L = Lmin \ldots Lmin + kmax$, storing the results in fc_array. In the case of overflow the exponent is stored in $F\_exponent$.

int gsl_sf_coulomb_wave_FG_array (double L_min, int kmax,   Function
        double eta, double x, double fc_array[], double gc_array[],
        double * F_exponent, double * G_exponent)

This function computes the functions $F_L(\eta, x)$, $G_L(\eta, x)$ for the range $L = Lmin \ldots Lmin + kmax$ storing the results in fc_array and gc_array. In the case of overflow the exponents are stored in $F\_exponent$ and $G\_exponent$.

int gsl_sf_coulomb_wave_FGp_array (double L_min, int kmax,   Function
        double eta, double x, double fc_array[], double fcp_array[],
        double gc_array[], double gcp_array[], double * F_exponent,
        double * G_exponent)

This function computes the functions $F_L(\eta, x)$, $G_L(\eta, x)$ and their derivatives $F_L'(\eta, x)$, $G_L'(\eta, x)$ for $L = Lmin \ldots Lmin + kmax$ storing the results in fc_array, gc_array, fcp_array and gcp_array. In the case of overflow the exponents are stored in $F\_exponent$ and $G\_exponent$.

int gsl_sf_coulomb_wave_sphF_array (double L_min, int $kmax$,   Function
          double eta, double x, double fc_array[], double
          $F\_exponent$[])

This function computes the Coulomb wave function divided by the argument $F_L(\eta, x)/x$ for $L = Lmin \ldots Lmin + kmax$, storing the results in fc_array. In the case of overflow the exponent is stored in $F\_exponent$. This function reduces to spherical Bessel functions in the limit $\eta \to 0$.

7.7.3 Coulomb Wave Function Normalization Constant

The Coulomb wave function normalization constant is defined in Abramowitz 14.1.7.

int gsl_sf_coulomb_CL_e (double L, double eta, gsl_sf_result *   Function
          $result$)

This function computes the Coulomb wave function normalization constant $C_L(\eta)$ for $L > -1$.

int gsl_sf_coulomb_CL_array (double $Lmin$, int $kmax$, double   Function
          eta, double cl[])

This function computes the Coulomb wave function normalization constant $C_L(\eta)$ for $L = Lmin \ldots Lmin + kmax$, $Lmin > -1$.

7.8 Coupling Coefficients

The Wigner 3-j, 6-j and 9-j symbols give the coupling coefficients for combined angular momentum vectors. Since the arguments of the standard coupling coefficient functions are integer or half-integer, the arguments of the following functions are, by convention, integers equal to twice the actual spin value. For information on the 3-j coefficients see Abramowitz & Stegun, Section 27.9. The functions described in this section are declared in the header file 'gsl_sf_coupling.h'.

7.8.1 3-j Symbols

double gsl_sf_coupling_3j (int two_ja, int two_jb, int two_jc,   Function
          int two_ma, int two_mb, int two_mc)

int gsl_sf_coupling_3j_e (int two_ja, int two_jb, int two_jc,   Function
          int two_ma, int two_mb, int two_mc, gsl_sf_result * $result$)

These routines compute the Wigner 3-j coefficient,

$$\begin{pmatrix} ja & jb & jc \\ ma & mb & mc \end{pmatrix}$$

where the arguments are given in half-integer units, $ja = two\_ja/2$, $ma = two\_ma/2$, etc.

7.8.2 6-j Symbols

double gsl_sf_coupling_6j (int *two_ja*, int *two_jb*, int *two_jc*,   Function
        int *two_jd*, int *two_je*, int *two_jf*)
int gsl_sf_coupling_6j_e (int *two_ja*, int *two_jb*, int *two_jc*,   Function
        int *two_jd*, int *two_je*, int *two_jf*, gsl_sf_result * *result*)
These routines compute the Wigner 6-j coefficient,

$$\begin{Bmatrix} ja & jb & jc \\ jd & je & jf \end{Bmatrix}$$

where the arguments are given in half-integer units, $ja = two\_ja/2$, $ma = two\_ma/2$, etc.

7.8.3 9-j Symbols

double gsl_sf_coupling_9j (int *two_ja*, int *two_jb*, int *two_jc*,   Function
        int *two_jd*, int *two_je*, int *two_jf*, int *two_jg*, int *two_jh*, int
        two_ji)
int gsl_sf_coupling_9j_e (int *two_ja*, int *two_jb*, int *two_jc*,   Function
        int *two_jd*, int *two_je*, int *two_jf*, int *two_jg*, int *two_jh*, int
        two_ji, gsl_sf_result * *result*)
These routines compute the Wigner 9-j coefficient,

$$\begin{Bmatrix} ja & jb & jc \\ jd & je & jf \\ jg & jh & ji \end{Bmatrix}$$

where the arguments are given in half-integer units, $ja = two\_ja/2$, $ma = two\_ma/2$, etc.

7.9 Dawson Function

The Dawson integral is defined by $\exp(-x^2) \int_0^x dt \exp(t^2)$. A table of Dawson's integral can be found in Abramowitz & Stegun, Table 7.5. The Dawson functions are declared in the header file 'gsl_sf_dawson.h'.

double gsl_sf_dawson (double x)   Function
int gsl_sf_dawson_e (double x, gsl_sf_result * *result*)   Function
These routines compute the value of Dawson's integral for x.

7.10 Debye Functions

The Debye functions $D_n(x)$ are defined by the following integral,

$$D_n(x) = \frac{n}{x^n} \int_0^x dt \frac{t^n}{e^t - 1}$$

For further information see Abramowitz & Stegun, Section 27.1. The Debye functions are declared in the header file 'gsl_sf_debye.h'.

double gsl_sf_debye_1 (double x)   Function
int gsl_sf_debye_1_e (double x, gsl_sf_result * result)   Function
  These routines compute the first-order Debye function, which is defined as
  $D_1(x) = (1/x) \int_0^x dt(t/(e^t - 1))$.

double gsl_sf_debye_2 (double x)   Function
int gsl_sf_debye_2_e (double x, gsl_sf_result * result)   Function
  These routines compute the second-order Debye function, which is defined
  as $D_2(x) = (2/x^2) \int_0^x dt(t^2/(e^t - 1))$.

double gsl_sf_debye_3 (double x)   Function
int gsl_sf_debye_3_e (double x, gsl_sf_result * result)   Function
  These routines compute the third-order Debye function, which is defined as
  $D_3(x) = (3/x^3) \int_0^x dt(t^3/(e^t - 1))$.

double gsl_sf_debye_4 (double x)   Function
int gsl_sf_debye_4_e (double x, gsl_sf_result * result)   Function
  These routines compute the fourth-order Debye function, which is defined
  as $D_4(x) = (4/x^4) \int_0^x dt(t^4/(e^t - 1))$.

double gsl_sf_debye_5 (double x)   Function
int gsl_sf_debye_5_e (double x, gsl_sf_result * result)   Function
  These routines compute the fifth-order Debye function, which is defined as
  $D_5(x) = (5/x^5) \int_0^x dt(t^5/(e^t - 1))$.

double gsl_sf_debye_6 (double x)   Function
int gsl_sf_debye_6_e (double x, gsl_sf_result * result)   Function
  These routines compute the sixth-order Debye function, which is defined as
  $D_6(x) = (6/x^6) \int_0^x dt(t^6/(e^t - 1))$.

7.11 Dilogarithm

The functions described in this section are declared in the header file 'gsl_sf_dilog.h'.

7.11.1 Real Argument

double gsl_sf_dilog (double x)   *Function*

int gsl_sf_dilog_e (double x, gsl_sf_result * result)   *Function*

These routines compute the dilogarithm for a real argument. In Lewin's notation this is $Li_2(x)$, the real part of the dilogarithm of a real x. It is defined by the integral representation $Li_2(x) = -\mathrm{Re} \int_0^x ds \log(1-s)/s$. Note that $\mathrm{Im}(Li_2(x)) = 0$ for $x \le 1$, and $-\pi \log(x)$ for $x > 1$.

Note that Abramowitz & Stegun refer to the Spence integral $S(x) = Li_2(1 - x)$ as the dilogarithm rather than $Li_2(x)$.

7.11.2 Complex Argument

int gsl_sf_complex_dilog_e (double r, double *theta*,   *Function*
          gsl_sf_result * *result_re*, gsl_sf_result * *result_im*)

This function computes the full complex-valued dilogarithm for the complex argument $z = r \exp(i\theta)$. The real and imaginary parts of the result are returned in *result_re*, *result_im*.

7.12 Elementary Operations

The following functions allow for the propagation of errors when combining quantities by multiplication. The functions are declared in the header file 'gsl_sf_elementary.h'.

int gsl_sf_multiply_e (double x, double y, gsl_sf_result *   *Function*
          result)

This function multiplies x and y storing the product and its associated error in *result*.

int gsl_sf_multiply_err_e (double x, double dx, double y,   *Function*
          double dy, gsl_sf_result * result)

This function multiplies x and y with associated absolute errors dx and dy. The product $xy \pm xy \sqrt{(dx/x)^2 + (dy/y)^2}$ is stored in *result*.

7.13 Elliptic Integrals

The functions described in this section are declared in the header file
'gsl_sf_ellint.h'. Further information about the elliptic integrals can be
found in Abramowitz & Stegun, Chapter 17.

7.13.1 Definition of Legendre Forms

The Legendre forms of elliptic integrals $F(\phi, k)$, $E(\phi, k)$ and $\Pi(\phi, k, n)$ are
defined by,

$$F(\phi, k) = \int_0^\phi dt \frac{1}{\sqrt{(1 - k^2 \sin^2(t))}}$$

$$E(\phi, k) = \int_0^\phi dt \sqrt{(1 - k^2 \sin^2(t))}$$

$$\Pi(\phi, k, n) = \int_0^\phi dt \frac{1}{(1 + n \sin^2(t))\sqrt{1 - k^2 \sin^2(t)}}$$

The complete Legendre forms are denoted by $K(k) = F(\pi/2, k)$ and $E(k) =
E(\pi/2, k)$.

The notation used here is based on Carlson, *Numerische Mathematik* 33
(1979) 1 and differs slightly from that used by Abramowitz & Stegun, where
the functions are given in terms of the parameter $m = k^2$ and n is replaced by
$-n$.

7.13.2 Definition of Carlson Forms

The Carlson symmetric forms of elliptical integrals $RC(x, y)$, $RD(x, y, z)$,
$RF(x, y, z)$ and $RJ(x, y, z, p)$ are defined by,

$$RC(x, y) = 1/2 \int_0^\infty dt (t + x)^{-1/2} (t + y)^{-1}$$

$$RD(x, y, z) = 3/2 \int_0^\infty dt (t + x)^{-1/2} (t + y)^{-1/2} (t + z)^{-3/2}$$

$$RF(x, y, z) = 1/2 \int_0^\infty dt (t + x)^{-1/2} (t + y)^{-1/2} (t + z)^{-1/2}$$

$$RJ(x, y, z, p) = 3/2 \int_0^\infty dt (t + x)^{-1/2} (t + y)^{-1/2} (t + z)^{-1/2} (t + p)^{-1}$$

7.13.3 Legendre Form of Complete Elliptic Integrals

double gsl_sf_ellint_Kcomp (double k, gsl_mode_t *mode*)   Function
int gsl_sf_ellint_Kcomp_e (double k, gsl_mode_t *mode*,   Function
      gsl_sf_result * *result*)

These routines compute the complete elliptic integral $K(k)$ to the accuracy specified by the mode variable *mode*. Note that Abramowitz & Stegun define this function in terms of the parameter $m = k^2$.

double gsl_sf_ellint_Ecomp (double k, gsl_mode_t *mode*)   Function
int gsl_sf_ellint_Ecomp_e (double k, gsl_mode_t *mode*,   Function
      gsl_sf_result * *result*)

These routines compute the complete elliptic integral $E(k)$ to the accuracy specified by the mode variable *mode*. Note that Abramowitz & Stegun define this function in terms of the parameter $m = k^2$.

double gsl_sf_ellint_Pcomp (double k, double n, gsl_mode_t   Function
      mode)
int gsl_sf_ellint_Pcomp_e (double k, double n, gsl_mode_t   Function
      mode, gsl_sf_result * *result*)

These routines compute the complete elliptic integral $\Pi(k, n)$ to the accuracy specified by the mode variable *mode*. Note that Abramowitz & Stegun define this function in terms of the parameters $m = k^2$ and $\sin^2(\alpha) = k^2$, with the change of sign $n \to -n$.

7.13.4 Legendre Form of Incomplete Elliptic Integrals

double gsl_sf_ellint_F (double *phi*, double k, gsl_mode_t   Function
      mode)
int gsl_sf_ellint_F_e (double *phi*, double k, gsl_mode_t *mode*,   Function
      gsl_sf_result * *result*)

These routines compute the incomplete elliptic integral $F(\phi, k)$ to the accuracy specified by the mode variable *mode*. Note that Abramowitz & Stegun define this function in terms of the parameter $m = k^2$.

double gsl_sf_ellint_E (double *phi*, double k, gsl_mode_t   Function
      mode)
int gsl_sf_ellint_E_e (double *phi*, double k, gsl_mode_t *mode*,   Function
      gsl_sf_result * *result*)

These routines compute the incomplete elliptic integral $E(\phi, k)$ to the accuracy specified by the mode variable *mode*. Note that Abramowitz & Stegun define this function in terms of the parameter $m = k^2$.

double gsl_sf_ellint_P (double *phi*, double k, double n,   Function
      gsl_mode_t *mode*)
int gsl_sf_ellint_P_e (double *phi*, double k, double n,   Function
      gsl_mode_t *mode*, gsl_sf_result * *result*)

These routines compute the incomplete elliptic integral $\Pi(\phi, k, n)$ to the accuracy specified by the mode variable *mode*. Note that Abramowitz &

Stegun define this function in terms of the parameters $m = k^2$ and $\sin^2(\alpha) = k^2$, with the change of sign $n \to -n$.

double gsl_sf_ellint_D (double *phi*, double *k*, double *n*,   *Function*
        gsl_mode_t *mode*)
int gsl_sf_ellint_D_e (double *phi*, double *k*, double *n*,   *Function*
        gsl_mode_t *mode*, gsl_sf_result * *result*)
These functions compute the incomplete elliptic integral $D(\phi, k)$ which is defined through the Carlson form $RD(x, y, z)$ by the following relation,

$$D(\phi, k, n) = \frac{1}{3}(\sin \phi)^3 RD(1 - \sin^2(\phi), 1 - k^2 \sin^2(\phi), 1).$$

The argument n is not used and will be removed in a future release.

7.13.5 Carlson Forms

double gsl_sf_ellint_RC (double *x*, double *y*, gsl_mode_t *mode*)   *Function*
int gsl_sf_ellint_RC_e (double *x*, double *y*, gsl_mode_t *mode*,   *Function*
        gsl_sf_result * *result*)
These routines compute the incomplete elliptic integral $RC(x, y)$ to the accuracy specified by the mode variable *mode*.

double gsl_sf_ellint_RD (double *x*, double *y*, double *z*,   *Function*
        gsl_mode_t *mode*)
int gsl_sf_ellint_RD_e (double *x*, double *y*, double *z*,   *Function*
        gsl_mode_t *mode*, gsl_sf_result * *result*)
These routines compute the incomplete elliptic integral $RD(x, y, z)$ to the accuracy specified by the mode variable *mode*.

double gsl_sf_ellint_RF (double *x*, double *y*, double *z*,   *Function*
        gsl_mode_t *mode*)
int gsl_sf_ellint_RF_e (double *x*, double *y*, double *z*,   *Function*
        gsl_mode_t *mode*, gsl_sf_result * *result*)
These routines compute the incomplete elliptic integral $RF(x, y, z)$ to the accuracy specified by the mode variable *mode*.

double gsl_sf_ellint_RJ (double *x*, double *y*, double *z*, double   *Function*
        p, gsl_mode_t *mode*)
int gsl_sf_ellint_RJ_e (double *x*, double *y*, double *z*, double   *Function*
        p, gsl_mode_t *mode*, gsl_sf_result * *result*)
These routines compute the incomplete elliptic integral $RJ(x, y, z, p)$ to the accuracy specified by the mode variable *mode*.

7.14 Elliptic Functions (Jacobi)

The Jacobian Elliptic functions are defined in Abramowitz & Stegun, Chapter 16. The functions are declared in the header file 'gsl_sf_elljac.h'.

int gsl_sf_elljac_e (double u, double m, double * sn, double *   *Function*
      cn, double * dn)

   This function computes the Jacobian elliptic functions $sn(u|m)$, $cn(u|m)$, $dn(u|m)$ by descending Landen transformations.

7.15 Error Functions

The error function is described in Abramowitz & Stegun, Chapter 7. The functions in this section are declared in the header file 'gsl_sf_erf.h'.

7.15.1 Error Function

double gsl_sf_erf (double x)   *Function*
int gsl_sf_erf_e (double x, gsl_sf_result * $result$)   *Function*

   These routines compute the Gaussian error function erf(x), where erf(x) $= (2/\sqrt{\pi}) \int_0^x dt \exp(-t^2)$.

7.15.2 Complementary Error Function

double gsl_sf_erfc (double x)   *Function*
int gsl_sf_erfc_e (double x, gsl_sf_result * $result$)   *Function*

   These routines compute the complementary error function erfc(x) $= 1 -$ erf(x) $= (2/\sqrt{\pi}) \int_x^\infty \exp(-t^2)$.

7.15.3 Log Complementary Error Function

double gsl_sf_log_erfc (double x)   *Function*
int gsl_sf_log_erfc_e (double x, gsl_sf_result * $result$)   *Function*

   These routines compute the logarithm of the complementary error function log(erfc(x)).

7.15.4 Probability functions

The probability functions for the Normal or Gaussian distribution are described in Abramowitz & Stegun, Section 26.2.

double gsl_sf_erf_Z (double x)   *Function*
int gsl_sf_erf_Z_e (double x, gsl_sf_result * $result$)   *Function*

   These routines compute the Gaussian probability density function $Z(x) = (1/\sqrt{2\pi}) \exp(-x^2/2)$.

double gsl_sf_erf_Q (double x)   *Function*
int gsl_sf_erf_Q_e (double x, gsl_sf_result * $result$)   *Function*

   These routines compute the upper tail of the Gaussian probability function $Q(x) = (1/\sqrt{2\pi}) \int_x^\infty dt \exp(-t^2/2)$.

The *hazard function* for the normal distribution, also known as the inverse Mill's ratio, is defined as,

$$h(x) = \frac{Z(x)}{Q(x)} = \sqrt{\frac{2}{\pi}} \frac{\exp(-x^2/2)}{\operatorname{erfc}(x/\sqrt{2})}$$

It decreases rapidly as x approaches $-\infty$ and asymptotes to $h(x) \sim x$ as x approaches $+\infty$.

double gsl_sf_hazard (double x)   *Function*
int gsl_sf_hazard_e (double x, gsl_sf_result * *result*)   *Function*
  These routines compute the hazard function for the normal distribution.

7.16 Exponential Functions

The functions described in this section are declared in the header file 'gsl_sf_exp.h'.

7.16.1 Exponential Function

double gsl_sf_exp (double x)   *Function*
int gsl_sf_exp_e (double x, gsl_sf_result * *result*)   *Function*
  These routines provide an exponential function $\exp(x)$ using GSL semantics and error checking.

int gsl_sf_exp_e10_e (double x, gsl_sf_result_e10 * *result*)   *Function*
  This function computes the exponential $\exp(x)$ using the gsl_sf_result_e10 type to return a result with extended range. This function may be useful if the value of $\exp(x)$ would overflow the numeric range of double.

double gsl_sf_exp_mult (double x, double y)   *Function*
int gsl_sf_exp_mult_e (double x, double y, gsl_sf_result *   *Function*
          result)
  These routines exponentiate x and multiply by the factor y to return the product $y \exp(x)$.

int gsl_sf_exp_mult_e10_e (const double x, const double y,   *Function*
          gsl_sf_result_e10 * *result*)
  This function computes the product $y \exp(x)$ using the gsl_sf_result_e10 type to return a result with extended numeric range.

7.16.2 Relative Exponential Functions

double gsl_sf_expm1 (double x)   Function
int gsl_sf_expm1_e (double x, gsl_sf_result * *result*)   Function
    These routines compute the quantity $\exp(x) - 1$ using an algorithm that is
    accurate for small x.

double gsl_sf_exprel (double x)   Function
int gsl_sf_exprel_e (double x, gsl_sf_result * *result*)   Function
    These routines compute the quantity $(\exp(x) - 1)/x$ using an algorithm that
    is accurate for small x. For small x the algorithm is based on the expansion
    $(\exp(x) - 1)/x = 1 + x/2 + x^2/(2*3) + x^3/(2*3*4) + \ldots$.

double gsl_sf_exprel_2 (double x)   Function
int gsl_sf_exprel_2_e (double x, gsl_sf_result * *result*)   Function
    These routines compute the quantity $2(\exp(x)-1-x)/x^2$ using an algorithm
    that is accurate for small x. For small x the algorithm is based on the
    expansion $2(\exp(x) - 1 - x)/x^2 = 1 + x/3 + x^2/(3*4) + x^3/(3*4*5) + \ldots$.

double gsl_sf_exprel_n (int n, double x)   Function
int gsl_sf_exprel_n_e (int n, double x, gsl_sf_result * *result*)   Function
    These routines compute the N-relative exponential, which is the n-th gen-
    eralization of the functions gsl_sf_exprel and gsl_sf_exprel2. The N-
    relative exponential is given by,

$$\mathrm{exprel}_N(x) = N!/x^N \left(\exp(x) - \sum_{k=0}^{N-1} x^k/k! \right)$$
$$= 1 + x/(N+1) + x^2/((N+1)(N+2)) + \cdots$$
$$= {}_1F_1(1, 1+N, x)$$

7.16.3 Exponentiation With Error Estimate

int gsl_sf_exp_err_e (double x, double dx, gsl_sf_result *   Function
                  result)
    This function exponentiates x with an associated absolute error dx.

int gsl_sf_exp_err_e10_e (double x, double dx,   Function
                  gsl_sf_result_e10 * *result*)
    This function exponentiates a quantity x with an associated absolute er-
    ror dx using the gsl_sf_result_e10 type to return a result with extended
    range.

int gsl_sf_exp_mult_err_e (double x, double dx, double y,   Function
                  double dy, gsl_sf_result * *result*)
    This routine computes the product $y \exp(x)$ for the quantities x, y with
    associated absolute errors dx, dy.

```
int gsl_sf_exp_mult_err_e10_e (double x, double dx, double y,        Function
          double dy, gsl_sf_result_e10 * result)
```
This routine computes the product $y\exp(x)$ for the quantities x, y with associated absolute errors dx, dy using the `gsl_sf_result_e10` type to return a result with extended range.

7.17 Exponential Integrals

Information on the exponential integrals can be found in Abramowitz & Stegun, Chapter 5. These functions are declared in the header file 'gsl_sf_expint.h'.

7.17.1 Exponential Integral

```
double gsl_sf_expint_E1 (double x)                                   Function
int gsl_sf_expint_E1_e (double x, gsl_sf_result * result)           Function
```
These routines compute the exponential integral $E_1(x)$,

$$E_1(x) := \mathrm{Re} \int_1^\infty dt\, \exp(-xt)/t.$$

```
double gsl_sf_expint_E2 (double x)                                   Function
int gsl_sf_expint_E2_e (double x, gsl_sf_result * result)           Function
```
These routines compute the second-order exponential integral $E_2(x)$,

$$E_2(x) := \mathrm{Re} \int_1^\infty dt\, \exp(-xt)/t^2.$$

```
double gsl_sf_expint_En (int n, double x)                            Function
int gsl_sf_expint_En_e (int n, double x, gsl_sf_result *            Function
          result)
```
These routines compute the exponential integral $E_n(x)$ of order n,

$$E_n(x) := \mathrm{Re} \int_1^\infty dt\, \exp(-xt)/t^n.$$

7.17.2 Ei(x)

```
double gsl_sf_expint_Ei (double x)                                   Function
int gsl_sf_expint_Ei_e (double x, gsl_sf_result * result)           Function
```
These routines compute the exponential integral $Ei(x)$,

$$Ei(x) := -PV\left(\int_{-x}^\infty dt\, \exp(-t)/t \right)$$

where PV denotes the principal value of the integral.

7.17.3 Hyperbolic Integrals

double gsl_sf_Shi (double x)   Function
int gsl_sf_Shi_e (double x, gsl_sf_result * result)   Function
   These routines compute the integral $Shi(x) = \int_0^x dt \, \sinh(t)/t$.

double gsl_sf_Chi (double x)   Function
int gsl_sf_Chi_e (double x, gsl_sf_result * result)   Function
   These routines compute the integral $Chi(x) := \mathrm{Re}[\gamma_E + \log(x) + \int_0^x dt(\cosh[t] - 1)/t]$, where γ_E is the Euler constant (available as the macro M_EULER).

7.17.4 Ei_3(x)

double gsl_sf_expint_3 (double x)   Function
int gsl_sf_expint_3_e (double x, gsl_sf_result * result)   Function
   These routines compute the third-order exponential integral $Ei_3(x) = \int_0^x dt \, \exp(-t^3)$ for $x \geq 0$.

7.17.5 Trigonometric Integrals

double gsl_sf_Si (const double x)   Function
int gsl_sf_Si_e (double x, gsl_sf_result * result)   Function
   These routines compute the Sine integral $Si(x) = \int_0^x dt \, \sin(t)/t$.

double gsl_sf_Ci (const double x)   Function
int gsl_sf_Ci_e (double x, gsl_sf_result * result)   Function
   These routines compute the Cosine integral $Ci(x) = -\int_x^\infty dt \, \cos(t)/t$ for $x > 0$.

7.17.6 Arctangent Integral

double gsl_sf_atanint (double x)   Function
int gsl_sf_atanint_e (double x, gsl_sf_result * result)   Function
   These routines compute the Arctangent integral, which is defined as $AtanInt(x) = \int_0^x dt \, \arctan(t)/t$.

7.18 Fermi-Dirac Function

The functions described in this section are declared in the header file 'gsl_sf_fermi_dirac.h'.

7.18.1 Complete Fermi-Dirac Integrals

The complete Fermi-Dirac integral $F_j(x)$ is given by,

$$F_j(x) := \frac{1}{\Gamma(j+1)} \int_0^\infty dt \frac{t^j}{(\exp(t-x)+1)}$$

Note that the Fermi-Dirac integral is sometimes defined without the normalisation factor in other texts.

double **gsl_sf_fermi_dirac_m1** (double x)   *Function*
int **gsl_sf_fermi_dirac_m1_e** (double x, gsl_sf_result * *result*)   *Function*
> These routines compute the complete Fermi-Dirac integral with an index of -1. This integral is given by $F_{-1}(x) = e^x/(1+e^x)$.

double **gsl_sf_fermi_dirac_0** (double x)   *Function*
int **gsl_sf_fermi_dirac_0_e** (double x, gsl_sf_result * *result*)   *Function*
> These routines compute the complete Fermi-Dirac integral with an index of 0. This integral is given by $F_0(x) = \ln(1+e^x)$.

double **gsl_sf_fermi_dirac_1** (double x)   *Function*
int **gsl_sf_fermi_dirac_1_e** (double x, gsl_sf_result * *result*)   *Function*
> These routines compute the complete Fermi-Dirac integral with an index of 1, $F_1(x) = \int_0^\infty dt(t/(\exp(t-x)+1))$.

double **gsl_sf_fermi_dirac_2** (double x)   *Function*
int **gsl_sf_fermi_dirac_2_e** (double x, gsl_sf_result * *result*)   *Function*
> These routines compute the complete Fermi-Dirac integral with an index of 2, $F_2(x) = (1/2)\int_0^\infty dt(t^2/(\exp(t-x)+1))$.

double **gsl_sf_fermi_dirac_int** (int j, double x)   *Function*
int **gsl_sf_fermi_dirac_int_e** (int j, double x, gsl_sf_result *   *Function*
>   *result*)
> These routines compute the complete Fermi-Dirac integral with an integer index of j, $F_j(x) = (1/\Gamma(j+1))\int_0^\infty dt(t^j/(\exp(t-x)+1))$.

double **gsl_sf_fermi_dirac_mhalf** (double x)   *Function*
int **gsl_sf_fermi_dirac_mhalf_e** (double x, gsl_sf_result *   *Function*
>   *result*)
> These routines compute the complete Fermi-Dirac integral $F_{-1/2}(x)$.

double **gsl_sf_fermi_dirac_half** (double x)   *Function*
int **gsl_sf_fermi_dirac_half_e** (double x, gsl_sf_result *   *Function*
>   *result*)
> These routines compute the complete Fermi-Dirac integral $F_{1/2}(x)$.

double gsl_sf_fermi_dirac_3half (double x)   Function
int gsl_sf_fermi_dirac_3half_e (double x, gsl_sf_result *   Function
        result)
    These routines compute the complete Fermi-Dirac integral $F_{3/2}(x)$.

7.18.2 Incomplete Fermi-Dirac Integrals

The incomplete Fermi-Dirac integral $F_j(x, b)$ is given by,

$$F_j(x, b) := \frac{1}{\Gamma(j+1)} \int_b^\infty dt \frac{t^j}{(\exp(t-x)+1)}$$

double gsl_sf_fermi_dirac_inc_0 (double x, double b)   Function
int gsl_sf_fermi_dirac_inc_0_e (double x, double b,   Function
        gsl_sf_result * result)
    These routines compute the incomplete Fermi-Dirac integral with an index
    of zero, $F_0(x, b) = \ln(1 + e^{b-x}) - (b - x)$.

7.19 Gamma and Beta Functions

The functions described in this section are declared in the header file
'gsl_sf_gamma.h'.

7.19.1 Gamma Functions

The Gamma function is defined by the following integral,

$$\Gamma(x) = \int_0^\infty dt\, t^{x-1} \exp(-t)$$

It is related to the factorial function by $\Gamma(n) = (n - 1)!$ for positive integer n.
Further information on the Gamma function can be found in Abramowitz &
Stegun, Chapter 6. The functions described in this section are declared in the
header file 'gsl_sf_gamma.h'.

double gsl_sf_gamma (double x)   Function
int gsl_sf_gamma_e (double x, gsl_sf_result * result)   Function
    These routines compute the Gamma function $\Gamma(x)$, subject to x not being
    a negative integer or zero. The function is computed using the real Lanczos
    method. The maximum value of x such that $\Gamma(x)$ is not considered an
    overflow is given by the macro GSL_SF_GAMMA_XMAX and is 171.0.

double gsl_sf_lngamma (double x)   Function
int gsl_sf_lngamma_e (double x, gsl_sf_result * result)   Function
    These routines compute the logarithm of the Gamma function, $\log(\Gamma(x))$,
    subject to x not being a negative integer or zero. For $x < 0$ the real part
    of $\log(\Gamma(x))$ is returned, which is equivalent to $\log(|\Gamma(x)|)$. The function is
    computed using the real Lanczos method.

int gsl_sf_lngamma_sgn_e (double x, gsl_sf_result * result_lg,   Function
          double * sgn)
This routine computes the sign of the gamma function and the logarithm of
its magnitude, subject to x not being a negative integer or zero. The function is computed using the real Lanczos method. The value of the gamma
function can be reconstructed using the relation $\Gamma(x) = sgn * \exp(result\_lg)$.

double gsl_sf_gammastar (double x)   Function
int gsl_sf_gammastar_e (double x, gsl_sf_result * result)   Function
These routines compute the regulated Gamma Function $\Gamma^*(x)$ for $x > 0$.
The regulated gamma function is given by,

$$\Gamma^*(x) = \Gamma(x)/(\sqrt{2\pi}x^{(x-1/2)}\exp(-x))$$
$$= \left(1 + \frac{1}{12x} + ...\right) \quad \text{for } x \to \infty$$

and is a useful suggestion of Temme.

double gsl_sf_gammainv (double x)   Function
int gsl_sf_gammainv_e (double x, gsl_sf_result * result)   Function
These routines compute the reciprocal of the gamma function, $1/\Gamma(x)$ using
the real Lanczos method.

int gsl_sf_lngamma_complex_e (double zr, double zi,   Function
          gsl_sf_result * lnr, gsl_sf_result * arg)
This routine computes $\log(\Gamma(z))$ for complex $z = z_r + iz_i$ and z not a
negative integer or zero, using the complex Lanczos method. The returned
parameters are $lnr = \log|\Gamma(z)|$ and $arg = \arg(\Gamma(z))$ in $(-\pi, \pi]$. Note that
the phase part (arg) is not well-determined when $|z|$ is very large, due to
inevitable roundoff in restricting to $(-\pi, \pi]$. This will result in a GSL_ELOSS
error when it occurs. The absolute value part (lnr), however, never suffers
from loss of precision.

7.19.2 Factorials

Although factorials can be computed from the Gamma function, using the
relation $n! = \Gamma(n + 1)$ for non-negative integer n, it is usually more efficient
to call the functions in this section, particularly for small values of n, whose
factorial values are maintained in hardcoded tables.

double gsl_sf_fact (unsigned int n)   Function
int gsl_sf_fact_e (unsigned int n, gsl_sf_result * result)   Function
These routines compute the factorial $n!$. The factorial is related to the
Gamma function by $n! = \Gamma(n + 1)$. The maximum value of n such that $n!$
is not considered an overflow is given by the macro GSL_SF_FACT_NMAX and
is 170.

double gsl_sf_doublefact (unsigned int n)   Function
int gsl_sf_doublefact_e (unsigned int n, gsl_sf_result *   Function
        result)
    These routines compute the double factorial $n!! = n(n-2)(n-4)\ldots$. The
    maximum value of n such that $n!!$ is not considered an overflow is given by
    the macro GSL_SF_DOUBLEFACT_NMAX and is 297.

double gsl_sf_lnfact (unsigned int n)   Function
int gsl_sf_lnfact_e (unsigned int n, gsl_sf_result * result)   Function
    These routines compute the logarithm of the factorial of n, $\log(n!)$. The
    algorithm is faster than computing $\ln(\Gamma(n+1))$ via gsl_sf_lngamma for
    $n < 170$, but defers for larger n.

double gsl_sf_lndoublefact (unsigned int n)   Function
int gsl_sf_lndoublefact_e (unsigned int n, gsl_sf_result *   Function
        result)
    These routines compute the logarithm of the double factorial of n, $\log(n!!)$.

double gsl_sf_choose (unsigned int n, unsigned int m)   Function
int gsl_sf_choose_e (unsigned int n, unsigned int m,   Function
        gsl_sf_result * result)
    These routines compute the combinatorial factor n choose m $= n!/(m!(n-m)!)$

double gsl_sf_lnchoose (unsigned int n, unsigned int m)   Function
int gsl_sf_lnchoose_e (unsigned int n, unsigned int m,   Function
        gsl_sf_result * result)
    These routines compute the logarithm of n choose m. This is equivalent to
    the sum $\log(n!) - \log(m!) - \log((n-m)!)$.

double gsl_sf_taylorcoeff (int n, double x)   Function
int gsl_sf_taylorcoeff_e (int n, double x, gsl_sf_result *   Function
        result)
    These routines compute the Taylor coefficient $x^n/n!$ for $x \geq 0$, $n \geq 0$.

7.19.3 Pochhammer Symbol

double gsl_sf_poch (double a, double x)   Function
int gsl_sf_poch_e (double a, double x, gsl_sf_result * result)   Function
    These routines compute the Pochhammer symbol $(a)_x = \Gamma(a+x)/\Gamma(a)$,
    subject to a and $a+x$ not being negative integers or zero. The Pochhammer
    symbol is also known as the Apell symbol and sometimes written as (a, x).

double gsl_sf_lnpoch (double a, double x)   Function
int gsl_sf_lnpoch_e (double a, double x, gsl_sf_result *   Function
        result)
    These routines compute the logarithm of the Pochhammer symbol,
    $\log((a)_x) = \log(\Gamma(a+x)/\Gamma(a))$ for $a > 0$, $a + x > 0$.

int gsl_sf_lnpoch_sgn_e (double a, double x, gsl_sf_result *   Function
        result, double * sgn)

These routines compute the sign of the Pochhammer symbol and the loga-
rithm of its magnitude. The computed parameters are $result = \log(|(a)_x|)$
and $sgn = \text{sgn}((a)_x)$ where $(a)_x = \Gamma(a + x)/\Gamma(a)$, subject to a, $a + x$ not
being negative integers or zero.

double gsl_sf_pochrel (double a, double x)   Function
int gsl_sf_pochrel_e (double a, double x, gsl_sf_result *   Function
        result)

These routines compute the relative Pochhammer symbol $((a)_x - 1)/x$ where
$(a)_x = \Gamma(a + x)/\Gamma(a)$.

7.19.4 Incomplete Gamma Functions

double gsl_sf_gamma_inc (double a, double x)   Function
int gsl_sf_gamma_inc_e (double a, double x, gsl_sf_result *   Function
        result)

These functions compute the unnormalized incomplete Gamma Function
$\Gamma(a, x) = \int_x^\infty dt\, t^{(a-1)} \exp(-t)$ for a real and $x \geq 0$.

double gsl_sf_gamma_inc_Q (double a, double x)   Function
int gsl_sf_gamma_inc_Q_e (double a, double x, gsl_sf_result *   Function
        result)

These routines compute the normalized incomplete Gamma Function
$Q(a, x) = 1/\Gamma(a) \int_x^\infty dt\, t^{(a-1)} \exp(-t)$ for $a > 0$, $x \geq 0$.

double gsl_sf_gamma_inc_P (double a, double x)   Function
int gsl_sf_gamma_inc_P_e (double a, double x, gsl_sf_result *   Function
        result)

These routines compute the complementary normalized incomplete Gamma
Function $P(a, x) = 1 - Q(a, x) = 1/\Gamma(a) \int_0^x dt\, t^{(a-1)} \exp(-t)$ for $a > 0$,
$x \geq 0$.

Note that Abramowitz & Stegun call $P(a, x)$ the incomplete gamma function
(section 6.5).

7.19.5 Beta Functions

double gsl_sf_beta (double a, double b)   Function
int gsl_sf_beta_e (double a, double b, gsl_sf_result * result)   Function

These routines compute the Beta Function, $B(a, b) = \Gamma(a)\Gamma(b)/\Gamma(a + b)$
subject to a and b not being negative integers.

double gsl_sf_lnbeta (double a, double b)   Function
int gsl_sf_lnbeta_e (double a, double b, gsl_sf_result *   Function
        result)

These routines compute the logarithm of the Beta Function, $\log(B(a, b))$
subject to a and b not being negative integers.

7.19.6 Incomplete Beta Function

double gsl_sf_beta_inc (double a, double b, double x)   *Function*
int gsl_sf_beta_inc_e (double a, double b, double x,   *Function*
        gsl_sf_result * result)

    These routines compute the normalized incomplete Beta function $I_x(a,b) = B_x(a,b)/B(a,b)$ where $B_x(a,b) = \int_0^x t^{a-1}(1-t)^{b-1}dt$ for $0 \leq x \leq 1$. For $a > 0$, $b > 0$ the value is computed using a continued fraction expansion. For all other values it is computed using the relation $I_x(a,b,x) = (1/a)x^a\,{}_2F_1(a,1-b,a+1,x)/B(a,b)$.

7.20 Gegenbauer Functions

    The Gegenbauer polynomials are defined in Abramowitz & Stegun, Chapter 22, where they are known as Ultraspherical polynomials. The functions described in this section are declared in the header file 'gsl_sf_gegenbauer.h'.

double gsl_sf_gegenpoly_1 (double lambda, double x)   *Function*
double gsl_sf_gegenpoly_2 (double lambda, double x)   *Function*
double gsl_sf_gegenpoly_3 (double lambda, double x)   *Function*
int gsl_sf_gegenpoly_1_e (double lambda, double x,   *Function*
        gsl_sf_result * result)
int gsl_sf_gegenpoly_2_e (double lambda, double x,   *Function*
        gsl_sf_result * result)
int gsl_sf_gegenpoly_3_e (double lambda, double x,   *Function*
        gsl_sf_result * result)

    These functions evaluate the Gegenbauer polynomials $C_n^{(\lambda)}(x)$ using explicit representations for $n = 1, 2, 3$.

double gsl_sf_gegenpoly_n (int n, double lambda, double x)   *Function*
int gsl_sf_gegenpoly_n_e (int n, double lambda, double x,   *Function*
        gsl_sf_result * result)

    These functions evaluate the Gegenbauer polynomial $C_n^{(\lambda)}(x)$ for a specific value of n, lambda, x subject to $\lambda > -1/2$, $n \geq 0$.

int gsl_sf_gegenpoly_array (int nmax, double lambda, double   *Function*
        x, double result_array[])

    This function computes an array of Gegenbauer polynomials $C_n^{(\lambda)}(x)$ for $n = 0, 1, 2, \ldots, nmax$, subject to $\lambda > -1/2$, $nmax \geq 0$.

7.21 Hypergeometric Functions

Hypergeometric functions are described in Abramowitz & Stegun, Chapters 13 and 15. These functions are declared in the header file 'gsl_sf_hyperg.h'.

double gsl_sf_hyperg_0F1 (double c, double x)   Function
int gsl_sf_hyperg_0F1_e (double c, double x, gsl_sf_result *   Function
     result)
    These routines compute the hypergeometric function $_0F_1(c, x)$.

double gsl_sf_hyperg_1F1_int (int m, int n, double x)   Function
int gsl_sf_hyperg_1F1_int_e (int m, int n, double x,   Function
     gsl_sf_result * result)
    These routines compute the confluent hypergeometric function of the first kind $_1F_1(m, n, x) = M(m, n, x)$ for integer parameters m, n.

double gsl_sf_hyperg_1F1 (double a, double b, double x)   Function
int gsl_sf_hyperg_1F1_e (double a, double b, double x,   Function
     gsl_sf_result * result)
    These routines compute the confluent hypergeometric function $_1F_1(a, b, x) = M(a, b, x)$ for general parameters a, b.

double gsl_sf_hyperg_U_int (int m, int n, double x)   Function
int gsl_sf_hyperg_U_int_e (int m, int n, double x,   Function
     gsl_sf_result * result)
    These routines compute the confluent hypergeometric function $U(m, n, x)$ for integer parameters m, n.

int gsl_sf_hyperg_U_int_e10_e (int m, int n, double x,   Function
     gsl_sf_result_e10 * result)
    This routine computes the confluent hypergeometric function $U(m, n, x)$ for integer parameters m, n using the gsl_sf_result_e10 type to return a result with extended range.

double gsl_sf_hyperg_U (double a, double b, double x)   Function
int gsl_sf_hyperg_U_e (double a, double b, double x,   Function
     gsl_sf_result * result)
    These routines compute the confluent hypergeometric function $U(a, b, x)$.

int gsl_sf_hyperg_U_e10_e (double a, double b, double x,   Function
     gsl_sf_result_e10 * result)
    This routine computes the confluent hypergeometric function $U(a, b, x)$ using the gsl_sf_result_e10 type to return a result with extended range.

double gsl_sf_hyperg_2F1 (double a, double b, double c, double Function
     x)
int gsl_sf_hyperg_2F1_e (double a, double b, double c, double Function
     x, gsl_sf_result * result)
    These routines compute the Gauss hypergeometric function $_2F_1(a, b, c, x) = F(a, b, c, x)$ for $|x| < 1$.

If the arguments (a, b, c, x) are too close to a singularity then the function can return the error code GSL_EMAXITER when the series approximation converges too slowly. This occurs in the region of $x = 1$, $c - a - b = m$ for integer m.

double gsl_sf_hyperg_2F1_conj (double aR, double aI, double c,   Function
            double x)
int gsl_sf_hyperg_2F1_conj_e (double aR, double aI, double c,   Function
            double x, gsl_sf_result * result)
These routines compute the Gauss hypergeometric function $_2F_1(a_R + ia_I, a_R - ia_I, c, x)$ with complex parameters for $|x| < 1$. exceptions:

double gsl_sf_hyperg_2F1_renorm (double a, double b, double c,   Function
            double x)
int gsl_sf_hyperg_2F1_renorm_e (double a, double b, double c,   Function
            double x, gsl_sf_result * result)
These routines compute the renormalized Gauss hypergeometric function $_2F_1(a, b, c, x)/\Gamma(c)$ for $|x| < 1$.

double gsl_sf_hyperg_2F1_conj_renorm (double aR, double aI,   Function
            double c, double x)
int gsl_sf_hyperg_2F1_conj_renorm_e (double aR, double aI,   Function
            double c, double x, gsl_sf_result * result)
These routines compute the renormalized Gauss hypergeometric function $_2F_1(a_R + ia_I, a_R - ia_I, c, x)/\Gamma(c)$ for $|x| < 1$.

double gsl_sf_hyperg_2F0 (double a, double b, double x)   Function
int gsl_sf_hyperg_2F0_e (double a, double b, double x,   Function
            gsl_sf_result * result)
These routines compute the hypergeometric function $_2F_0(a, b, x)$. The series representation is a divergent hypergeometric series. However, for $x < 0$ we have $_2F_0(a, b, x) = (-1/x)^a U(a, 1 + a - b, -1/x)$

7.22 Laguerre Functions

The generalized Laguerre polynomials are defined in terms of confluent hypergeometric functions as $L_n^a(x) = ((a + 1)_n/n!)_1F_1(-n, a + 1, x)$, and are sometimes referred to as the associated Laguerre polynomials. They are related to the plain Laguerre polynomials $L_n(x)$ by $L_n^0(x) = L_n(x)$ and $L_n^k(x) = (-1)^k(d^k/dx^k)L_{(n+k)}(x)$. For more information see Abramowitz & Stegun, Chapter 22.

The functions described in this section are declared in the header file 'gsl_sf_laguerre.h'.

```
double gsl_sf_laguerre_1 (double a, double x)              Function
double gsl_sf_laguerre_2 (double a, double x)              Function
double gsl_sf_laguerre_3 (double a, double x)              Function
int gsl_sf_laguerre_1_e (double a, double x, gsl_sf_result *   Function
        result)
int gsl_sf_laguerre_2_e (double a, double x, gsl_sf_result *   Function
        result)
int gsl_sf_laguerre_3_e (double a, double x, gsl_sf_result *   Function
        result)
```
These routines evaluate the generalized Laguerre polynomials $L_1^a(x)$, $L_2^a(x)$, $L_3^a(x)$ using explicit representations.

```
double gsl_sf_laguerre_n (const int n, const double a, const   Function
        double x)
int gsl_sf_laguerre_n_e (int n, double a, double x,           Function
        gsl_sf_result * result)
```
These routines evaluate the generalized Laguerre polynomials $L_n^a(x)$ for $a > -1$, $n \geq 0$.

7.23 Lambert W Functions

Lambert's W functions, $W(x)$, are defined to be solutions of the equation $W(x)\exp(W(x)) = x$. This function has multiple branches for $x < 0$; however, it has only two real-valued branches. We define $W_0(x)$ to be the principal branch, where $W > -1$ for $x < 0$, and $W_{-1}(x)$ to be the other real branch, where $W < -1$ for $x < 0$. The Lambert functions are declared in the header file 'gsl_sf_lambert.h'.

```
double gsl_sf_lambert_W0 (double x)                        Function
int gsl_sf_lambert_W0_e (double x, gsl_sf_result * result) Function
```
These compute the principal branch of the Lambert W function, $W_0(x)$.

```
double gsl_sf_lambert_Wm1 (double x)                       Function
int gsl_sf_lambert_Wm1_e (double x, gsl_sf_result * result) Function
```
These compute the secondary real-valued branch of the Lambert W function, $W_{-1}(x)$.

7.24 Legendre Functions and Spherical Harmonics

The Legendre Functions and Legendre Polynomials are described in Abramowitz & Stegun, Chapter 8. These functions are declared in the header file 'gsl_sf_legendre.h'.

7.24.1 Legendre Polynomials

double gsl_sf_legendre_P1 (double x)	Function
double gsl_sf_legendre_P2 (double x)	Function
double gsl_sf_legendre_P3 (double x)	Function
int gsl_sf_legendre_P1_e (double x, gsl_sf_result * result)	Function
int gsl_sf_legendre_P2_e (double x, gsl_sf_result * result)	Function
int gsl_sf_legendre_P3_e (double x, gsl_sf_result * result)	Function

These functions evaluate the Legendre polynomials $P_l(x)$ using explicit representations for $l = 1, 2, 3$.

double gsl_sf_legendre_Pl (int l, double x)	Function
int gsl_sf_legendre_Pl_e (int l, double x, gsl_sf_result * result)	Function

These functions evaluate the Legendre polynomial $P_l(x)$ for a specific value of l, x subject to $l \geq 0$, $|x| \leq 1$

int gsl_sf_legendre_Pl_array (int lmax, double x, double result_array[])	Function
int gsl_sf_legendre_Pl_deriv_array (int lmax, double x, double result_array[], double result_deriv_array[])	Function

These functions compute an array of Legendre polynomials $P_l(x)$, and optionally their derivatives $dP_l(x)/dx$, for $l = 0, \ldots, lmax$, $|x| \leq 1$

double gsl_sf_legendre_Q0 (double x)	Function
int gsl_sf_legendre_Q0_e (double x, gsl_sf_result * result)	Function

These routines compute the Legendre function $Q_0(x)$ for $x > -1$, $x \neq 1$.

double gsl_sf_legendre_Q1 (double x)	Function
int gsl_sf_legendre_Q1_e (double x, gsl_sf_result * result)	Function

These routines compute the Legendre function $Q_1(x)$ for $x > -1$, $x \neq 1$.

double gsl_sf_legendre_Ql (int l, double x)	Function
int gsl_sf_legendre_Ql_e (int l, double x, gsl_sf_result * result)	Function

These routines compute the Legendre function $Q_l(x)$ for $x > -1$, $x \neq 1$ and $l \geq 0$.

7.24.2 Associated Legendre Polynomials and Spherical Harmonics

The following functions compute the associated Legendre Polynomials $P_l^m(x)$. Note that this function grows combinatorially with l and can overflow for l larger than about 150. There is no trouble for small m, but overflow occurs when m and l are both large. Rather than allow overflows, these functions refuse to calculate $P_l^m(x)$ and return GSL_EOVRFLW when they can sense that l and m are too big.

If you want to calculate a spherical harmonic, then *do not* use these functions. Instead use gsl_sf_legendre_sphPlm below, which uses a similar recursion, but with the normalized functions.

```
double gsl_sf_legendre_Plm (int l, int m, double x)                Function
int gsl_sf_legendre_Plm_e (int l, int m, double x,                 Function
          gsl_sf_result * result)
```
These routines compute the associated Legendre polynomial $P_l^m(x)$ for $m \geq 0$, $l \geq m$, $|x| \leq 1$.

```
int gsl_sf_legendre_Plm_array (int lmax, int m, double x,          Function
          double result_array[])
int gsl_sf_legendre_Plm_deriv_array (int lmax, int m, double       Function
          x, double result_array[], double result_deriv_array[])
```
These functions compute an array of Legendre polynomials $P_l^m(x)$, and optionally their derivatives $dP_l^m(x)/dx$, for $m \geq 0$, $l = |m|, \ldots, lmax$, $|x| \leq 1$.

```
double gsl_sf_legendre_sphPlm (int l, int m, double x)             Function
int gsl_sf_legendre_sphPlm_e (int l, int m, double x,              Function
          gsl_sf_result * result)
```
These routines compute the normalized associated Legendre polynomial $\sqrt{(2l+1)/(4\pi)}\sqrt{(l-m)!/(l+m)!}P_l^m(x)$ suitable for use in spherical harmonics. The parameters must satisfy $m \geq 0$, $l \geq m$, $|x| \leq 1$. Theses routines avoid the overflows that occur for the standard normalization of $P_l^m(x)$.

```
int gsl_sf_legendre_sphPlm_array (int lmax, int m, double x,       Function
          double result_array[])
int gsl_sf_legendre_sphPlm_deriv_array (int lmax, int m,           Function
          double x, double result_array[], double result_deriv_array[])
```
These functions compute an array of normalized associated Legendre functions $\sqrt{(2l+1)/(4\pi)}\sqrt{(l-m)!/(l+m)!}P_l^m(x)$, and optionally their derivatives, for $m \geq 0$, $l = |m|, \ldots, lmax$, $|x| \leq 1$

```
int gsl_sf_legendre_array_size (const int lmax, const int m)       Function
```
This function returns the size of result_array[] needed for the array versions of $P_l^m(x)$, $lmax - m + 1$. An inline version of this function is used when HAVE_INLINE is defined.

7.24.3 Conical Functions

The Conical Functions $P^\mu_{-(1/2)+i\lambda}(x)$ and $Q^\mu_{-(1/2)+i\lambda}$ are described in Abramowitz & Stegun, Section 8.12.

```
double gsl_sf_conicalP_half (double lambda, double x)              Function
int gsl_sf_conicalP_half_e (double lambda, double x,               Function
          gsl_sf_result * result)
```
These routines compute the value of the irregular Spherical Conical Function $P^{1/2}_{-1/2+i\lambda}(x)$ for $x > -1$.

double gsl_sf_conicalP_mhalf (double *lambda*, double x)   Function
int gsl_sf_conicalP_mhalf_e (double *lambda*, double x,   Function
        gsl_sf_result * *result*)

These routines compute the regular Spherical Conical Function $P_{-1/2+i\lambda}^{-1/2}(x)$ for $x > -1$.

double gsl_sf_conicalP_0 (double *lambda*, double x)   Function
int gsl_sf_conicalP_0_e (double *lambda*, double x,   Function
        gsl_sf_result * *result*)

These routines compute the conical function $P_{-1/2+i\lambda}^0(x)$ for $x > -1$.

double gsl_sf_conicalP_1 (double *lambda*, double x)   Function
int gsl_sf_conicalP_1_e (double *lambda*, double x,   Function
        gsl_sf_result * *result*)

These routines compute the conical function $P_{-1/2+i\lambda}^1(x)$ for $x > -1$.

double gsl_sf_conicalP_sph_reg (int *l*, double *lambda*, double   Function
        x)
int gsl_sf_conicalP_sph_reg_e (int *l*, double *lambda*, double x,   Function
        gsl_sf_result * *result*)

These routines compute the value of the Regular Spherical Conical Function $P_{-1/2+i\lambda}^{-1/2-l}(x)$ for $x > -1$, $l \geq -1$.

double gsl_sf_conicalP_cyl_reg (int *m*, double *lambda*, double   Function
        x)
int gsl_sf_conicalP_cyl_reg_e (int *m*, double *lambda*, double   Function
        x, gsl_sf_result * *result*)

These routines compute the value of the Regular Cylindrical Conical Function $P_{-1/2+i\lambda}^{-m}(x)$ for $x > -1$, $m \geq -1$.

7.24.4 Radial Functions for Hyperbolic Space

The following spherical functions are specializations of Legendre functions which give the regular eigenfunctions of the Laplacian on a 3-dimensional hyperbolic space $H3d$. Of particular interest is the flat limit, $\lambda \to \infty$, $\eta \to 0$, $\lambda\eta$ fixed.

double gsl_sf_legendre_H3d_0 (double *lambda*, double eta)   Function
int gsl_sf_legendre_H3d_0_e (double *lambda*, double eta,   Function
        gsl_sf_result * *result*)

These routines compute the zeroth radial eigenfunction of the Laplacian on the 3-dimensional hyperbolic space,

$$L_0^{H3d}(\lambda, \eta) := \frac{\sin(\lambda\eta)}{\lambda \sinh(\eta)}$$

for $\eta \geq 0$. In the flat limit this takes the form $L_0^{H3d}(\lambda, \eta) = j_0(\lambda\eta)$.

```
double gsl_sf_legendre_H3d_1 (double lambda, double eta)          Function
int gsl_sf_legendre_H3d_1_e (double lambda, double eta,           Function
        gsl_sf_result * result)
```
These routines compute the first radial eigenfunction of the Laplacian on the 3-dimensional hyperbolic space,

$$L_1^{H3d}(\lambda, \eta) := \frac{1}{\sqrt{\lambda^2 + 1}} \left(\frac{\sin(\lambda\eta)}{\lambda \sinh(\eta)} \right) (\coth(\eta) - \lambda \cot(\lambda\eta))$$

for $\eta \geq 0$. In the flat limit this takes the form $L_1^{H3d}(\lambda, \eta) = j_1(\lambda\eta)$.

```
double gsl_sf_legendre_H3d (int l, double lambda, double eta)     Function
int gsl_sf_legendre_H3d_e (int l, double lambda, double eta,      Function
        gsl_sf_result * result)
```
These routines compute the l-th radial eigenfunction of the Laplacian on the 3-dimensional hyperbolic space $\eta \geq 0$, $l \geq 0$. In the flat limit this takes the form $L_l^{H3d}(\lambda, \eta) = j_l(\lambda\eta)$.

```
int gsl_sf_legendre_H3d_array (int lmax, double lambda,           Function
        double eta, double result_array[])
```
This function computes an array of radial eigenfunctions $L_l^{H3d}(\lambda, \eta)$ for $0 \leq l \leq lmax$.

7.25 Logarithm and Related Functions

Information on the properties of the Logarithm function can be found in Abramowitz & Stegun, Chapter 4. The functions described in this section are declared in the header file 'gsl_sf_log.h'.

```
double gsl_sf_log (double x)                                      Function
int gsl_sf_log_e (double x, gsl_sf_result * result)              Function
```
These routines compute the logarithm of x, $\log(x)$, for $x > 0$.

```
double gsl_sf_log_abs (double x)                                  Function
int gsl_sf_log_abs_e (double x, gsl_sf_result * result)          Function
```
These routines compute the logarithm of the magnitude of x, $\log(|x|)$, for $x \neq 0$.

```
int gsl_sf_complex_log_e (double zr, double zi, gsl_sf_result *   Function
        lnr, gsl_sf_result * theta)
```
This routine computes the complex logarithm of $z = z_r + iz_i$. The results are returned as lnr, theta such that $\exp(lnr + i\theta) = z_r + iz_i$, where θ lies in the range $[-\pi, \pi]$.

```
double gsl_sf_log_1plusx (double x)                               Function
int gsl_sf_log_1plusx_e (double x, gsl_sf_result * result)       Function
```
These routines compute $\log(1 + x)$ for $x > -1$ using an algorithm that is accurate for small x.

double gsl_sf_log_1plusx_mx (double x)   Function
int gsl_sf_log_1plusx_mx_e (double x, gsl_sf_result * result)   Function
    These routines compute $\log(1 + x) - x$ for $x > -1$ using an algorithm that
    is accurate for small x.

7.26 Mathieu Functions

The routines described in this section compute the angular and radial Mathieu
functions, and their characteristic values. Mathieu functions are the solutions
of the following two differential equations:

$$\frac{d^2 y}{dv^2} + (a - 2q \cos 2v)y = 0,$$

$$\frac{d^2 f}{du^2} - (a - 2q \cosh 2u)f = 0.$$

The angular Mathieu functions $ce_r(x, q)$, $se_r(x, q)$ are the even and odd periodic
solutions of the first equation, which is known as Mathieu's equation. These
exist only for the discrete sequence of characteristic values $a = a_r(q)$ (even-
periodic) and $a = b_r(q)$ (odd-periodic).

The radial Mathieu functions $Mc_r^{(j)}(z, q)$, $Ms_r^{(j)}(z, q)$ are the solutions of
the second equation, which is referred to as Mathieu's modified equation. The
radial Mathieu functions of the first, second, third and fourth kind are denoted
by the parameter j, which takes the value 1, 2, 3 or 4.

For more information on the Mathieu functions, see Abramowitz and Stegun,
Chapter 20. These functions are defined in the header file 'gsl_sf_mathieu.h'.

7.26.1 Mathieu Function Workspace

The Mathieu functions can be computed for a single order or for multiple
orders, using array-based routines. The array-based routines require a preallocated workspace.

gsl_sf_mathieu_workspace * gsl_sf_mathieu_alloc (size_t n,   Function
            double qmax)
    This function returns a workspace for the array versions of the Mathieu
    routines. The arguments n and qmax specify the maximum order and q-
    value of Mathieu functions which can be computed with this workspace.

void gsl_sf_mathieu_free (gsl_sf_mathieu_workspace *work)   Function
    This function frees the workspace work.

7.26.2 Mathieu Function Characteristic Values

int gsl_sf_mathieu_a (int n, double q, gsl_sf_result *result)   Function
int gsl_sf_mathieu_b (int n, double q, gsl_sf_result *result)   Function
  These routines compute the characteristic values $a_n(q)$, $b_n(q)$ of the Mathieu
  functions $ce_n(q, x)$ and $se_n(q, x)$, respectively.

int gsl_sf_mathieu_a_array (int order_min, int order_max,   Function
            double q, gsl_sf_mathieu_workspace *work, double
            result_array[])
int gsl_sf_mathieu_b_array (int order_min, int order_max,   Function
            double q, gsl_sf_mathieu_workspace *work, double
            result_array[])
  These routines compute a series of Mathieu characteristic values $a_n(q)$, $b_n(q)$
  for n from order_min to order_max inclusive, storing the results in the array
  result_array.

7.26.3 Angular Mathieu Functions

int gsl_sf_mathieu_ce (int n, double q, double x,   Function
            gsl_sf_result *result)
int gsl_sf_mathieu_se (int n, double q, double x,   Function
            gsl_sf_result *result)
  These routines compute the angular Mathieu functions $ce_n(q, x)$ and
  $se_n(q, x)$, respectively.

int gsl_sf_mathieu_ce_array (int nmin, int nmax, double q,   Function
            double x, gsl_sf_mathieu_workspace *work, double
            result_array[])
int gsl_sf_mathieu_se_array (int nmin, int nmax, double q,   Function
            double x, gsl_sf_mathieu_workspace *work, double
            result_array[])
  These routines compute a series of the angular Mathieu functions $ce_n(q, x)$
  and $se_n(q, x)$ of order n from nmin to nmax inclusive, storing the results in
  the array result_array.

7.26.4 Radial Mathieu Functions

int gsl_sf_mathieu_Mc (int j, int n, double q, double x,   Function
            gsl_sf_result *result)
int gsl_sf_mathieu_Ms (int j, int n, double q, double x,   Function
            gsl_sf_result *result)
  These routines compute the radial j-th kind Mathieu functions $Mc_n^{(j)}(q, x)$
  and $Ms_n^{(j)}(q, x)$ of order n.
  The allowed values of j are 1 and 2. The functions for $j = 3, 4$ can be
  computed as $M_n^{(3)} = M_n^{(1)} + iM_n^{(2)}$ and $M_n^{(4)} = M_n^{(1)} - iM_n^{(2)}$, where $M_n^{(j)} = $
  $Mc_n^{(j)}$ or $Ms_n^{(j)}$.

```
int gsl_sf_mathieu_Mc_array (int j, int nmin, int nmax,            Function
            double q, double x, gsl_sf_mathieu_workspace *work, double
            result_array[])
int gsl_sf_mathieu_Ms_array (int j, int nmin, int nmax,            Function
            double q, double x, gsl_sf_mathieu_workspace *work, double
            result_array[])
```
These routines compute a series of the radial Mathieu functions of kind j, with order from nmin to nmax inclusive, storing the results in the array result_array.

7.27 Power Function

The following functions are equivalent to the function gsl_pow_int (see Section 4.4 [Small integer powers], page 24) with an error estimate. These functions are declared in the header file 'gsl_sf_pow_int.h'.

```
double gsl_sf_pow_int (double x, int n)                             Function
int gsl_sf_pow_int_e (double x, int n, gsl_sf_result * result)     Function
```
These routines compute the power x^n for integer n. The power is computed using the minimum number of multiplications. For example, x^8 is computed as $((x^2)^2)^2$, requiring only 3 multiplications. For reasons of efficiency, these functions do not check for overflow or underflow conditions.

```
#include <gsl/gsl_sf_pow_int.h>
/* compute 3.0**12 */
double y = gsl_sf_pow_int(3.0, 12);
```

7.28 Psi (Digamma) Function

The polygamma functions of order n are defined by

$$\psi^{(n)}(x) = \left(\frac{d}{dx}\right)^n \psi(x) = \left(\frac{d}{dx}\right)^{n+1} \log(\Gamma(x))$$

where $\psi(x) = \Gamma'(x)/\Gamma(x)$ is known as the digamma function. These functions are declared in the header file 'gsl_sf_psi.h'.

7.28.1 Digamma Function

```
double gsl_sf_psi_int (int n)                                      Function
int gsl_sf_psi_int_e (int n, gsl_sf_result * result)              Function
```
These routines compute the digamma function $\psi(n)$ for positive integer n. The digamma function is also called the Psi function.

```
double gsl_sf_psi (double x)                                       Function
int gsl_sf_psi_e (double x, gsl_sf_result * result)              Function
```
These routines compute the digamma function $\psi(x)$ for general x, $x \neq 0$.

double gsl_sf_psi_1piy (double y)   Function
int gsl_sf_psi_1piy_e (double y, gsl_sf_result * *result*)   Function
    These routines compute the real part of the digamma function on the line $1 + iy$, $\mathrm{Re}[\psi(1 + iy)]$.

7.28.2 Trigamma Function

double gsl_sf_psi_1_int (int n)   Function
int gsl_sf_psi_1_int_e (int n, gsl_sf_result * *result*)   Function
    These routines compute the Trigamma function $\psi'(n)$ for positive integer n.

double gsl_sf_psi_1 (double x)   Function
int gsl_sf_psi_1_e (double x, gsl_sf_result * *result*)   Function
    These routines compute the Trigamma function $\psi'(x)$ for general x.

7.28.3 Polygamma Function

double gsl_sf_psi_n (int n, double x)   Function
int gsl_sf_psi_n_e (int n, double x, gsl_sf_result * *result*) Function
    These routines compute the polygamma function $\psi^{(n)}(x)$ for $n \geq 0$, $x > 0$.

7.29 Synchrotron Functions

    The functions described in this section are declared in the header file 'gsl_sf_synchrotron.h'.

double gsl_sf_synchrotron_1 (double x)   Function
int gsl_sf_synchrotron_1_e (double x, gsl_sf_result * *result*) Function
    These routines compute the first synchrotron function $x \int_x^\infty dt\, K_{5/3}(t)$ for $x \geq 0$.

double gsl_sf_synchrotron_2 (double x)   Function
int gsl_sf_synchrotron_2_e (double x, gsl_sf_result * *result*) Function
    These routines compute the second synchrotron function $x K_{2/3}(x)$ for $x \geq 0$.

7.30 Transport Functions

    The transport functions $J(n, x)$ are defined by the integral representations $J(n, x) := \int_0^x dt\, t^n e^t / (e^t - 1)^2$. They are declared in the header file 'gsl_sf_transport.h'.

double gsl_sf_transport_2 (double x)   Function
int gsl_sf_transport_2_e (double x, gsl_sf_result * *result*) Function
    These routines compute the transport function $J(2, x)$.

double gsl_sf_transport_3 (double x)   Function
int gsl_sf_transport_3_e (double x, gsl_sf_result * *result*) Function
    These routines compute the transport function $J(3, x)$.

double gsl_sf_transport_4 (double x)   Function
int gsl_sf_transport_4_e (double x, gsl_sf_result * result)   Function
    These routines compute the transport function $J(4, x)$.

double gsl_sf_transport_5 (double x)   Function
int gsl_sf_transport_5_e (double x, gsl_sf_result * result)   Function
    These routines compute the transport function $J(5, x)$.

7.31 Trigonometric Functions

The library includes its own trigonometric functions in order to provide consistency across platforms and reliable error estimates. These functions are declared in the header file 'gsl_sf_trig.h'.

7.31.1 Circular Trigonometric Functions

double gsl_sf_sin (double x)   Function
int gsl_sf_sin_e (double x, gsl_sf_result * result)   Function
    These routines compute the sine function $\sin(x)$.

double gsl_sf_cos (double x)   Function
int gsl_sf_cos_e (double x, gsl_sf_result * result)   Function
    These routines compute the cosine function $\cos(x)$.

double gsl_sf_hypot (double x, double y)   Function
int gsl_sf_hypot_e (double x, double y, gsl_sf_result * result) Function
    These routines compute the hypotenuse function $\sqrt{x^2 + y^2}$ avoiding overflow and underflow.

double gsl_sf_sinc (double x)   Function
int gsl_sf_sinc_e (double x, gsl_sf_result * result)   Function
    These routines compute $\mathrm{sinc}(x) = \sin(\pi x)/(\pi x)$ for any value of x.

7.31.2 Trigonometric Functions for Complex Arguments

int gsl_sf_complex_sin_e (double zr, double zi, gsl_sf_result *   Function
        szr, gsl_sf_result * szi)
    This function computes the complex sine, $\sin(z_r + iz_i)$ storing the real and imaginary parts in szr, szi.

int gsl_sf_complex_cos_e (double zr, double zi, gsl_sf_result *   Function
        czr, gsl_sf_result * czi)
    This function computes the complex cosine, $\cos(z_r + iz_i)$ storing the real and imaginary parts in szr, szi.

int gsl_sf_complex_logsin_e (double zr, double zi,   Function
        gsl_sf_result * lszr, gsl_sf_result * lszi)
    This function computes the logarithm of the complex sine, $\log(\sin(z_r + iz_i))$ storing the real and imaginary parts in szr, szi.

7.31.3 Hyperbolic Trigonometric Functions

double gsl_sf_lnsinh (double x)   Function
int gsl_sf_lnsinh_e (double x, gsl_sf_result * result)   Function
    These routines compute $\log(\sinh(x))$ for $x > 0$.

double gsl_sf_lncosh (double x)   Function
int gsl_sf_lncosh_e (double x, gsl_sf_result * result)   Function
    These routines compute $\log(\cosh(x))$ for any x.

7.31.4 Conversion Functions

int gsl_sf_polar_to_rect (double r, double theta,   Function
        gsl_sf_result * x, gsl_sf_result * y);
    This function converts the polar coordinates $(r, theta)$ to rectilinear coordi-
nates (x,y), $x = r\cos(\theta)$, $y = r\sin(\theta)$.

int gsl_sf_rect_to_polar (double x, double y, gsl_sf_result *   Function
        r, gsl_sf_result * theta)
    This function converts the rectilinear coordinates (x,y) to polar coordinates
$(r, theta)$, such that $x = r\cos(\theta)$, $y = r\sin(\theta)$. The argument theta lies in
the range $[-\pi, \pi]$.

7.31.5 Restriction Functions

double gsl_sf_angle_restrict_symm (double theta)   Function
int gsl_sf_angle_restrict_symm_e (double * theta)   Function
    These routines force the angle theta to lie in the range $(-\pi, \pi]$.

    Note that the mathematical value of π is slightly greater than M_PI, so the
machine numbers M_PI and -M_PI are included in the range.

double gsl_sf_angle_restrict_pos (double theta)   Function
int gsl_sf_angle_restrict_pos_e (double * theta)   Function
    These routines force the angle theta to lie in the range $[0, 2\pi)$.

    Note that the mathematical value of 2π is slightly greater than 2*M_PI, so
the machine number 2*M_PI is included in the range.

7.31.6 Trigonometric Functions With Error Estimates

int gsl_sf_sin_err_e (double x, double dx, gsl_sf_result *   Function
        result)
    This routine computes the sine of an angle x with an associated absolute
error dx, $\sin(x \pm dx)$. Note that this function is provided in the error-handling
form only since its purpose is to compute the propagated error.

int gsl_sf_cos_err_e (double x, double dx, gsl_sf_result *   Function
        result)
    This routine computes the cosine of an angle x with an associated absolute
error dx, $\cos(x \pm dx)$. Note that this function is provided in the error-
handling form only since its purpose is to compute the propagated error.

7.32 Zeta Functions

The Riemann zeta function is defined in Abramowitz & Stegun, Section 23.2. The functions described in this section are declared in the header file 'gsl_sf_zeta.h'.

7.32.1 Riemann Zeta Function

The Riemann zeta function is defined by the infinite sum $\zeta(s) = \sum_{k=1}^{\infty} k^{-s}$.

double gsl_sf_zeta_int (int n)   Function
int gsl_sf_zeta_int_e (int n, gsl_sf_result * result)   Function
    These routines compute the Riemann zeta function $\zeta(n)$ for integer n, $n \neq 1$.

double gsl_sf_zeta (double s)   Function
int gsl_sf_zeta_e (double s, gsl_sf_result * result)   Function
    These routines compute the Riemann zeta function $\zeta(s)$ for arbitrary s, $s \neq 1$.

7.32.2 Riemann Zeta Function Minus One

For large positive argument, the Riemann zeta function approaches one. In this region the fractional part is interesting, and therefore we need a function to evaluate it explicitly.

double gsl_sf_zetam1_int (int n)   Function
int gsl_sf_zetam1_int_e (int n, gsl_sf_result * result)   Function
    These routines compute $\zeta(n) - 1$ for integer n, $n \neq 1$.

double gsl_sf_zetam1 (double s)   Function
int gsl_sf_zetam1_e (double s, gsl_sf_result * result)   Function
    These routines compute $\zeta(s) - 1$ for arbitrary s, $s \neq 1$.

7.32.3 Hurwitz Zeta Function

The Hurwitz zeta function is defined by $\zeta(s,q) = \sum_{0}^{\infty}(k+q)^{-s}$.

double gsl_sf_hzeta (double s, double q)   Function
int gsl_sf_hzeta_e (double s, double q, gsl_sf_result * result)   Function
    These routines compute the Hurwitz zeta function $\zeta(s,q)$ for $s > 1$, $q > 0$.

7.32.4 Eta Function

The eta function is defined by $\eta(s) = (1 - 2^{1-s})\zeta(s)$.

double gsl_sf_eta_int (int n)   Function
int gsl_sf_eta_int_e (int n, gsl_sf_result * result)   Function
    These routines compute the eta function $\eta(n)$ for integer n.

double gsl_sf_eta (double s)   Function
int gsl_sf_eta_e (double s, gsl_sf_result * result)   Function
    These routines compute the eta function $\eta(s)$ for arbitrary s.

7.33 Examples

The following example demonstrates the use of the error handling form of the special functions, in this case to compute the Bessel function $J_0(5.0)$,

```
#include <stdio.h>
#include <gsl/gsl_errno.h>
#include <gsl/gsl_sf_bessel.h>

int
main (void)
{
  double x = 5.0;
  gsl_sf_result result;

  double expected = -0.17759677131433830434739701;

  int status = gsl_sf_bessel_J0_e (x, &result);

  printf ("status  = %s\n", gsl_strerror(status));
  printf ("J0(5.0) = %.18f\n"
          "        +/- % .18f\n",
          result.val, result.err);
  printf ("exact   = %.18f\n", expected);
  return status;
}
```

Here are the results of running the program,

```
$ ./a.out
status  = success
J0(5.0) = -0.177596771314338292
        +/-  0.000000000000000193
exact   = -0.177596771314338292
```

The next program computes the same quantity using the natural form of the function. In this case the error term *result.err* and return status are not accessible.

```
#include <stdio.h>
#include <gsl/gsl_sf_bessel.h>

int
main (void)
{
  double x = 5.0;
  double expected = -0.17759677131433830434739701;

  double y = gsl_sf_bessel_J0 (x);

  printf ("J0(5.0) = %.18f\n", y);
  printf ("exact   = %.18f\n", expected);
  return 0;
```

```
        }
```
The results of the function are the same,
```
      $ ./a.out
      J0(5.0) = -0.177596771314338292
      exact   = -0.177596771314338292
```

7.34 References and Further Reading

The library follows the conventions of *Abramowitz & Stegun* where possible,

Abramowitz & Stegun (eds.), *Handbook of Mathematical Functions*

The following papers contain information on the algorithms used to compute the special functions,

MISCFUN: A software package to compute uncommon special functions. *ACM Trans. Math. Soft.*, vol. 22, 1996, 288–301

G.N. Watson, A Treatise on the Theory of Bessel Functions, 2nd Edition (Cambridge University Press, 1944).

G. Nemeth, Mathematical Approximations of Special Functions, Nova Science Publishers, ISBN 1-56072-052-2

B.C. Carlson, Special Functions of Applied Mathematics (1977)

W.J. Thompson, Atlas for Computing Mathematical Functions, John Wiley & Sons, New York (1997).

Y.Y. Luke, Algorithms for the Computation of Mathematical Functions, Academic Press, New York (1977).

8 Vectors and Matrices

The functions described in this chapter provide a simple vector and matrix interface to ordinary C arrays. The memory management of these arrays is implemented using a single underlying type, known as a block. By writing your functions in terms of vectors and matrices you can pass a single structure containing both data and dimensions as an argument without needing additional function parameters. The structures are compatible with the vector and matrix formats used by BLAS routines.

8.1 Data types

All the functions are available for each of the standard data-types. The versions for double have the prefix gsl_block, gsl_vector and gsl_matrix. Similarly the versions for single-precision float arrays have the prefix gsl_block_float, gsl_vector_float and gsl_matrix_float. The full list of available types is given below,

gsl_block	double
gsl_block_float	float
gsl_block_long_double	long double
gsl_block_int	int
gsl_block_uint	unsigned int
gsl_block_long	long
gsl_block_ulong	unsigned long
gsl_block_short	short
gsl_block_ushort	unsigned short
gsl_block_char	char
gsl_block_uchar	unsigned char
gsl_block_complex	complex double
gsl_block_complex_float	complex float
gsl_block_complex_long_double	complex long double

Corresponding types exist for the gsl_vector and gsl_matrix functions.

8.2 Blocks

For consistency all memory is allocated through a gsl_block structure. The structure contains two components, the size of an area of memory and a pointer to the memory. The gsl_block structure looks like this,

```
typedef struct
{
  size_t size;
  double * data;
} gsl_block;
```

Vectors and matrices are made by *slicing* an underlying block. A slice is a set of elements formed from an initial offset and a combination of indices and step-sizes. In the case of a matrix the step-size for the column index represents the row-length. The step-size for a vector is known as the *stride*.

The functions for allocating and deallocating blocks are defined in the header file 'gsl_block.h'

8.2.1 Block allocation

The functions for allocating memory to a block follow the style of malloc and free. In addition they also perform their own error checking. If there is insufficient memory available to allocate a block then the functions call the GSL error handler (with an error number of GSL_ENOMEM) in addition to returning a null pointer. Thus if you use the library error handler to abort your program then it isn't necessary to check every alloc.

gsl_block * gsl_block_alloc (size_t n)   *Function*
This function allocates memory for a block of n double-precision elements, returning a pointer to the block struct. The block is not initialized and so the values of its elements are undefined. Use the function gsl_block_calloc if you want to ensure that all the elements are initialized to zero.

A null pointer is returned if insufficient memory is available to create the block.

gsl_block * gsl_block_calloc (size_t n)   *Function*
This function allocates memory for a block and initializes all the elements of the block to zero.

void gsl_block_free (gsl_block * b)   *Function*
This function frees the memory used by a block b previously allocated with gsl_block_alloc or gsl_block_calloc. The block b must be a valid block object (a null pointer is not allowed).

8.2.2 Reading and writing blocks

The library provides functions for reading and writing blocks to a file as binary data or formatted text.

int gsl_block_fwrite (FILE * *stream*, const gsl_block * b)   *Function*
This function writes the elements of the block b to the stream *stream* in binary format. The return value is 0 for success and GSL_EFAILED if there was a problem writing to the file. Since the data is written in the native binary format it may not be portable between different architectures.

int gsl_block_fread (FILE * *stream*, gsl_block * b)   *Function*
This function reads into the block b from the open stream *stream* in binary format. The block b must be preallocated with the correct length since the function uses the size of b to determine how many bytes to read. The return value is 0 for success and GSL_EFAILED if there was a problem reading from the file. The data is assumed to have been written in the native binary format on the same architecture.

int gsl_block_fprintf (FILE * *stream*, const gsl_block * *b*,   Function
        const char * *format*)

> This function writes the elements of the block *b* line-by-line to the stream
> *stream* using the format specifier *format*, which should be one of the %g, %e
> or %f formats for floating point numbers and %d for integers. The function
> returns 0 for success and GSL_EFAILED if there was a problem writing to the
> file.

int gsl_block_fscanf (FILE * *stream*, gsl_block * *b*)   Function

> This function reads formatted data from the stream *stream* into the block *b*.
> The block *b* must be preallocated with the correct length since the function
> uses the size of *b* to determine how many numbers to read. The function
> returns 0 for success and GSL_EFAILED if there was a problem reading from
> the file.

8.2.3 Example programs for blocks

The following program shows how to allocate a block,

```
#include <stdio.h>
#include <gsl/gsl_block.h>

int
main (void)
{
  gsl_block * b = gsl_block_alloc (100);

  printf ("length of block = %u\n", b->size);
  printf ("block data address = %#x\n", b->data);

  gsl_block_free (b);
  return 0;
}
```

Here is the output from the program,

```
length of block = 100
block data address = 0x804b0d8
```

8.3 Vectors

Vectors are defined by a gsl_vector structure which describes a slice of a
block. Different vectors can be created which point to the same block. A vector
slice is a set of equally-spaced elements of an area of memory.

The gsl_vector structure contains five components, the *size*, the *stride*, a
pointer to the memory where the elements are stored, *data*, a pointer to the
block owned by the vector, *block*, if any, and an ownership flag, *owner*. The
structure is very simple and looks like this,

```
typedef struct
{
  size_t size;
  size_t stride;
  double * data;
  gsl_block * block;
  int owner;
} gsl_vector;
```

The *size* is simply the number of vector elements. The range of valid indices runs from 0 to size-1. The *stride* is the step-size from one element to the next in physical memory, measured in units of the appropriate datatype. The pointer *data* gives the location of the first element of the vector in memory. The pointer *block* stores the location of the memory block in which the vector elements are located (if any). If the vector owns this block then the *owner* field is set to one and the block will be deallocated when the vector is freed. If the vector points to a block owned by another object then the *owner* field is zero and any underlying block will not be deallocated with the vector.

The functions for allocating and accessing vectors are defined in the header file 'gsl_vector.h'

8.3.1 Vector allocation

The functions for allocating memory to a vector follow the style of malloc and free. In addition they also perform their own error checking. If there is insufficient memory available to allocate a vector then the functions call the GSL error handler (with an error number of GSL_ENOMEM) in addition to returning a null pointer. Thus if you use the library error handler to abort your program then it isn't necessary to check every alloc.

gsl_vector * gsl_vector_alloc (size_t *n*)   Function
  This function creates a vector of length *n*, returning a pointer to a newly initialized vector struct. A new block is allocated for the elements of the vector, and stored in the *block* component of the vector struct. The block is "owned" by the vector, and will be deallocated when the vector is deallocated.

gsl_vector * gsl_vector_calloc (size_t *n*)   Function
  This function allocates memory for a vector of length *n* and initializes all the elements of the vector to zero.

void gsl_vector_free (gsl_vector * *v*)   Function
  This function frees a previously allocated vector *v*. If the vector was created using gsl_vector_alloc then the block underlying the vector will also be deallocated. If the vector has been created from another object then the memory is still owned by that object and will not be deallocated. The vector *v* must be a valid vector object (a null pointer is not allowed).

8.3.2 Accessing vector elements

Unlike FORTRAN compilers, C compilers do not usually provide support for range checking of vectors and matrices.[1] The functions gsl_vector_get and gsl_vector_set can perform portable range checking for you and report an error if you attempt to access elements outside the allowed range.

The functions for accessing the elements of a vector or matrix are defined in 'gsl_vector.h' and declared extern inline to eliminate function-call overhead. You must compile your program with the preprocessor macro HAVE_INLINE defined to use these functions.

If necessary you can turn off range checking completely without modifying any source files by recompiling your program with the preprocessor definition GSL_RANGE_CHECK_OFF. Provided your compiler supports inline functions the effect of turning off range checking is to replace calls to gsl_vector_get(v,i) by v->data[i*v->stride] and calls to gsl_vector_set(v,i,x) by v->data[i*v->stride]=x. Thus there should be no performance penalty for using the range checking functions when range checking is turned off.

If you use a C99 compiler which requires inline functions in header files to be declared inline instead of extern inline, define the macro GSL_C99_INLINE (see Section 2.5 [Inline functions], page 9). With GCC this is selected automatically when compiling in C99 mode (-std=c99).

If inline functions are not used, calls to the functions gsl_vector_get and gsl_vector_set will link to the compiled versions of these functions in the library itself. The range checking in these functions is controlled by the global integer variable gsl_check_range. It is enabled by default—to disable it, set gsl_check_range to zero. Due to function-call overhead, there is less benefit in disabling range checking here than for inline functions.

double gsl_vector_get (const gsl_vector * v, size_t i)   *Function*
    This function returns the i-th element of a vector v. If i lies outside the allowed range of 0 to $n-1$ then the error handler is invoked and 0 is returned. An inline version of this function is used when HAVE_INLINE is defined.

void gsl_vector_set (gsl_vector * v, size_t i, double x)   *Function*
    This function sets the value of the i-th element of a vector v to x. If i lies outside the allowed range of 0 to $n-1$ then the error handler is invoked. An inline version of this function is used when HAVE_INLINE is defined.

double * gsl_vector_ptr (gsl_vector * v, size_t i)   *Function*
const double * gsl_vector_const_ptr (const gsl_vector * v,   *Function*
       size_t i)
    These functions return a pointer to the i-th element of a vector v. If i lies outside the allowed range of 0 to $n-1$ then the error handler is invoked and a null pointer is returned. Inline versions of these functions are used when HAVE_INLINE is defined.

---

[1] Range checking is available in the GNU C Compiler bounds-checking extension, but it is not part of the default installation of GCC. Memory accesses can also be checked with Valgrind or the gcc -fmudflap memory protection option.

8.3.3 Initializing vector elements

void gsl_vector_set_all (gsl_vector * v, double x)   Function
    This function sets all the elements of the vector v to the value x.

void gsl_vector_set_zero (gsl_vector * v)   Function
    This function sets all the elements of the vector v to zero.

int gsl_vector_set_basis (gsl_vector * v, size_t i)   Function
    This function makes a basis vector by setting all the elements of the vector
    v to zero except for the i-th element which is set to one.

8.3.4 Reading and writing vectors

   The library provides functions for reading and writing vectors to a file as
binary data or formatted text.

int gsl_vector_fwrite (FILE * stream, const gsl_vector * v)   Function
    This function writes the elements of the vector v to the stream stream in
    binary format. The return value is 0 for success and GSL_EFAILED if there
    was a problem writing to the file. Since the data is written in the native
    binary format it may not be portable between different architectures.

int gsl_vector_fread (FILE * stream, gsl_vector * v)   Function
    This function reads into the vector v from the open stream stream in binary
    format. The vector v must be preallocated with the correct length since
    the function uses the size of v to determine how many bytes to read. The
    return value is 0 for success and GSL_EFAILED if there was a problem reading
    from the file. The data is assumed to have been written in the native binary
    format on the same architecture.

int gsl_vector_fprintf (FILE * stream, const gsl_vector * v,   Function
            const char * format)
    This function writes the elements of the vector v line-by-line to the stream
    stream using the format specifier format, which should be one of the %g, %e
    or %f formats for floating point numbers and %d for integers. The function
    returns 0 for success and GSL_EFAILED if there was a problem writing to the
    file.

int gsl_vector_fscanf (FILE * stream, gsl_vector * v)   Function
    This function reads formatted data from the stream stream into the vector v.
    The vector v must be preallocated with the correct length since the function
    uses the size of v to determine how many numbers to read. The function
    returns 0 for success and GSL_EFAILED if there was a problem reading from
    the file.

8.3.5 Vector views

In addition to creating vectors from slices of blocks it is also possible to slice vectors and create vector views. For example, a subvector of another vector can be described with a view, or two views can be made which provide access to the even and odd elements of a vector.

A vector view is a temporary object, stored on the stack, which can be used to operate on a subset of vector elements. Vector views can be defined for both constant and non-constant vectors, using separate types that preserve constness. A vector view has the type gsl_vector_view and a constant vector view has the type gsl_vector_const_view. In both cases the elements of the view can be accessed as a gsl_vector using the vector component of the view object. A pointer to a vector of type gsl_vector * or const gsl_vector * can be obtained by taking the address of this component with the & operator.

When using this pointer it is important to ensure that the view itself remains in scope—the simplest way to do so is by always writing the pointer as &view. vector, and never storing this value in another variable.

gsl_vector_view gsl_vector_subvector (gsl_vector * v, size_t   Function
        offset, size_t n)
gsl_vector_const_view gsl_vector_const_subvector (const   Function
        gsl_vector * v, size_t offset, size_t n)
These functions return a vector view of a subvector of another vector v. The start of the new vector is offset by offset elements from the start of the original vector. The new vector has n elements. Mathematically, the i-th element of the new vector v' is given by,

$$v'(i) = v->data[(offset + i)*v->stride]$$

where the index i runs from 0 to n−1.

The data pointer of the returned vector struct is set to null if the combined parameters (offset,n) overrun the end of the original vector.

The new vector is only a view of the block underlying the original vector, v. The block containing the elements of v is not owned by the new vector. When the view goes out of scope the original vector v and its block will continue to exist. The original memory can only be deallocated by freeing the original vector. Of course, the original vector should not be deallocated while the view is still in use.

The function gsl_vector_const_subvector is equivalent to gsl_vector_ subvector but can be used for vectors which are declared const.

gsl_vector_view gsl_vector_subvector_with_stride (gsl_vector   Function
        * v, size_t offset, size_t stride, size_t n)
gsl_vector_const_view gsl_vector_const_subvector_with_stride   Function
        (const gsl_vector * v, size_t offset, size_t stride, size_t n)
These functions return a vector view of a subvector of another vector v with an additional stride argument. The subvector is formed in the same way as for gsl_vector_subvector but the new vector has n elements with a step-size of stride from one element to the next in the original vector. Mathematically, the i-th element of the new vector v' is given by,

v'(i) = v->data[(offset + i*stride)*v->stride]

where the index *i* runs from 0 to n-1.

Note that subvector views give direct access to the underlying elements of the original vector. For example, the following code will zero the even elements of the vector v of length n, while leaving the odd elements untouched,

```
gsl_vector_view v_even
  = gsl_vector_subvector_with_stride (v, 0, 2, n/2);
gsl_vector_set_zero (&v_even.vector);
```

A vector view can be passed to any subroutine which takes a vector argument just as a directly allocated vector would be, using &*view*.vector. For example, the following code computes the norm of the odd elements of v using the BLAS routine DNRM2,

```
gsl_vector_view v_odd
  = gsl_vector_subvector_with_stride (v, 1, 2, n/2);
double r = gsl_blas_dnrm2 (&v_odd.vector);
```

The function gsl_vector_const_subvector_with_stride is equivalent to gsl_vector_subvector_with_stride but can be used for vectors which are declared const.

gsl_vector_view gsl_vector_complex_real (gsl_vector_complex   Function
    * *v*)

gsl_vector_const_view gsl_vector_complex_const_real (const   Function
    gsl_vector_complex * *v*)

These functions return a vector view of the real parts of the complex vector *v*.

The function gsl_vector_complex_const_real is equivalent to gsl_vector_complex_real but can be used for vectors which are declared const.

gsl_vector_view gsl_vector_complex_imag (gsl_vector_complex   Function
    * *v*)

gsl_vector_const_view gsl_vector_complex_const_imag (const   Function
    gsl_vector_complex * *v*)

These functions return a vector view of the imaginary parts of the complex vector *v*.

The function gsl_vector_complex_const_imag is equivalent to gsl_vector_complex_imag but can be used for vectors which are declared const.

gsl_vector_view gsl_vector_view_array (double * *base*, size_t   Function
    n)

gsl_vector_const_view gsl_vector_const_view_array (const   Function
    double * *base*, size_t *n*)

These functions return a vector view of an array. The start of the new vector is given by *base* and has *n* elements. Mathematically, the *i*-th element of the new vector v' is given by, v'(i) = base[i]

where the index *i* runs from 0 to n-1.

The array containing the elements of *v* is not owned by the new vector view. When the view goes out of scope the original array will continue to exist. The original memory can only be deallocated by freeing the original pointer *base*. Of course, the original array should not be deallocated while the view is still in use.

The function gsl_vector_const_view_array is equivalent to gsl_vector_view_array but can be used for arrays which are declared const.

gsl_vector_view gsl_vector_view_array_with_stride (double *   *Function*
         base, size_t *stride*, size_t *n*)
gsl_vector_const_view   *Function*
         gsl_vector_const_view_array_with_stride (const double *
         base, size_t *stride*, size_t *n*)

These functions return a vector view of an array *base* with an additional stride argument. The subvector is formed in the same way as for gsl_vector_view_array but the new vector has *n* elements with a step-size of *stride* from one element to the next in the original array. Mathematically, the *i*-th element of the new vector *v'* is given by,

    v'(i) = base[i*stride]

where the index *i* runs from 0 to n-1.

Note that the view gives direct access to the underlying elements of the original array. A vector view can be passed to any subroutine which takes a vector argument just as a directly allocated vector would be, using &*view*. vector.

The function gsl_vector_const_view_array_with_stride is equivalent to gsl_vector_view_array_with_stride but can be used for arrays which are declared const.

8.3.6 Copying vectors

Common operations on vectors such as addition and multiplication are available in the BLAS part of the library (see Chapter 12 [BLAS Support], page 137). However, it is useful to have a small number of utility functions which do not require the full BLAS code. The following functions fall into this category.

int gsl_vector_memcpy (gsl_vector * *dest*, const gsl_vector *   *Function*
         src)
This function copies the elements of the vector *src* into the vector *dest*. The two vectors must have the same length.

int gsl_vector_swap (gsl_vector * *v*, gsl_vector * *w*)   *Function*
This function exchanges the elements of the vectors *v* and *w* by copying. The two vectors must have the same length.

8.3.7 Exchanging elements

The following function can be used to exchange, or permute, the elements of a vector.

int gsl_vector_swap_elements (gsl_vector * v, size_t i, size_t   Function
       j)
    This function exchanges the i-th and j-th elements of the vector v in-place.

int gsl_vector_reverse (gsl_vector * v)   Function
    This function reverses the order of the elements of the vector v.

8.3.8 Vector operations

int gsl_vector_add (gsl_vector * a, const gsl_vector * b)   Function
    This function adds the elements of vector b to the elements of vector a,
    $a_i' = a_i + b_i$. The two vectors must have the same length.

int gsl_vector_sub (gsl_vector * a, const gsl_vector * b)   Function
    This function subtracts the elements of vector b from the elements of vector
    a, $a_i' = a_i - b_i$. The two vectors must have the same length.

int gsl_vector_mul (gsl_vector * a, const gsl_vector * b)   Function
    This function multiplies the elements of vector a by the elements of vector
    b, $a_i' = a_i * b_i$. The two vectors must have the same length.

int gsl_vector_div (gsl_vector * a, const gsl_vector * b)   Function
    This function divides the elements of vector a by the elements of vector b,
    $a_i' = a_i/b_i$. The two vectors must have the same length.

int gsl_vector_scale (gsl_vector * a, const double x)   Function
    This function multiplies the elements of vector a by the constant factor x,
    $a_i' = x a_i$.

int gsl_vector_add_constant (gsl_vector * a, const double x)   Function
    This function adds the constant value x to the elements of the vector a,
    $a_i' = a_i + x$.

8.3.9 Finding maximum and minimum elements of vectors

The following operations are only defined for real vectors.

double gsl_vector_max (const gsl_vector * v)   Function
    This function returns the maximum value in the vector v.

double gsl_vector_min (const gsl_vector * v)   Function
    This function returns the minimum value in the vector v.

void gsl_vector_minmax (const gsl_vector * v, double *   Function
       min_out, double * max_out)
    This function returns the minimum and maximum values in the vector v,
    storing them in min_out and max_out.

`size_t gsl_vector_max_index (const gsl_vector * v)`   Function
 This function returns the index of the maximum value in the vector v. When there are several equal maximum elements then the lowest index is returned.

`size_t gsl_vector_min_index (const gsl_vector * v)`   Function
 This function returns the index of the minimum value in the vector v. When there are several equal minimum elements then the lowest index is returned.

`void gsl_vector_minmax_index (const gsl_vector * v, size_t *`   Function
      `imin, size_t * imax)`
 This function returns the indices of the minimum and maximum values in the vector v, storing them in imin and imax. When there are several equal minimum or maximum elements then the lowest indices are returned.

8.3.10 Vector properties

The following functions are defined for real and complex vectors. For complex vectors both the real and imaginary parts must satisfy the conditions.

`int gsl_vector_isnull (const gsl_vector * v)`   Function
`int gsl_vector_ispos (const gsl_vector * v)`   Function
`int gsl_vector_isneg (const gsl_vector * v)`   Function
`int gsl_vector_isnonneg (const gsl_vector * v)`   Function
 These functions return 1 if all the elements of the vector v are zero, strictly positive, strictly negative, or non-negative respectively, and 0 otherwise.

8.3.11 Example programs for vectors

This program shows how to allocate, initialize and read from a vector using the functions gsl_vector_alloc, gsl_vector_set and gsl_vector_get.

```
#include <stdio.h>
#include <gsl/gsl_vector.h>

int
main (void)
{
  int i;
  gsl_vector * v = gsl_vector_alloc (3);

  for (i = 0; i < 3; i++)
    {
      gsl_vector_set (v, i, 1.23 + i);
    }

  for (i = 0; i < 100; i++) /* OUT OF RANGE ERROR */
    {
      printf ("v_%d = %g\n", i, gsl_vector_get (v, i));
    }

  gsl_vector_free (v);
```

```
      return 0;
  }
```

Here is the output from the program. The final loop attempts to read outside
the range of the vector v, and the error is trapped by the range-checking code
in gsl_vector_get.

```
$ ./a.out
v_0 = 1.23
v_1 = 2.23
v_2 = 3.23
gsl: vector_source.c:12: ERROR: index out of range
Default GSL error handler invoked.
Aborted (core dumped)
```

The next program shows how to write a vector to a file.

```
#include <stdio.h>
#include <gsl/gsl_vector.h>

int
main (void)
{
  int i;
  gsl_vector * v = gsl_vector_alloc (100);

  for (i = 0; i < 100; i++)
    {
      gsl_vector_set (v, i, 1.23 + i);
    }

  {
    FILE * f = fopen ("test.dat", "w");
    gsl_vector_fprintf (f, v, "%.5g");
    fclose (f);
  }

  gsl_vector_free (v);
  return 0;
}
```

After running this program the file 'test.dat' should contain the elements of
v, written using the format specifier %.5g. The vector could then be read back
in using the function gsl_vector_fscanf (f, v) as follows:

```
#include <stdio.h>
#include <gsl/gsl_vector.h>

int
main (void)
{
  int i;
  gsl_vector * v = gsl_vector_alloc (10);
```

```
{
   FILE * f = fopen ("test.dat", "r");
   gsl_vector_fscanf (f, v);
   fclose (f);
}

for (i = 0; i < 10; i++)
  {
    printf ("%g\n", gsl_vector_get(v, i));
  }

gsl_vector_free (v);
return 0;
}
```

8.4 Matrices

Matrices are defined by a gsl_matrix structure which describes a generalized slice of a block. Like a vector it represents a set of elements in an area of memory, but uses two indices instead of one.

The gsl_matrix structure contains six components, the two dimensions of the matrix, a physical dimension, a pointer to the memory where the elements of the matrix are stored, *data*, a pointer to the block owned by the matrix *block*, if any, and an ownership flag, *owner*. The physical dimension determines the memory layout and can differ from the matrix dimension to allow the use of submatrices. The gsl_matrix structure is very simple and looks like this,

```
typedef struct
{
  size_t size1;
  size_t size2;
  size_t tda;
  double * data;
  gsl_block * block;
  int owner;
} gsl_matrix;
```

Matrices are stored in row-major order, meaning that each row of elements forms a contiguous block in memory. This is the standard "C-language ordering" of two-dimensional arrays. Note that FORTRAN stores arrays in column-major order. The number of rows is *size1*. The range of valid row indices runs from 0 to size1-1. Similarly *size2* is the number of columns. The range of valid column indices runs from 0 to size2-1. The physical row dimension *tda*, or *trailing dimension*, specifies the size of a row of the matrix as laid out in memory.

For example, in the following matrix *size1* is 3, *size2* is 4, and *tda* is 8. The physical memory layout of the matrix begins in the top left hand-corner and proceeds from left to right along each row in turn.

```
00 01 02 03 XX XX XX XX
10 11 12 13 XX XX XX XX
20 21 22 23 XX XX XX XX
```

Each unused memory location is represented by "XX". The pointer *data* gives the location of the first element of the matrix in memory. The pointer *block* stores the location of the memory block in which the elements of the matrix are located (if any). If the matrix owns this block then the *owner* field is set to one and the block will be deallocated when the matrix is freed. If the matrix is only a slice of a block owned by another object then the *owner* field is zero and any underlying block will not be freed.

The functions for allocating and accessing matrices are defined in the header file 'gsl_matrix.h'

8.4.1 Matrix allocation

The functions for allocating memory to a matrix follow the style of malloc and free. They also perform their own error checking. If there is insufficient memory available to allocate a matrix then the functions call the GSL error handler (with an error number of GSL_ENOMEM) in addition to returning a null pointer. Thus if you use the library error handler to abort your program then it isn't necessary to check every alloc.

gsl_matrix * gsl_matrix_alloc (size_t *n1*, size_t *n2*)   Function
> This function creates a matrix of size *n1* rows by *n2* columns, returning a pointer to a newly initialized matrix struct. A new block is allocated for the elements of the matrix, and stored in the *block* component of the matrix struct. The block is "owned" by the matrix, and will be deallocated when the matrix is deallocated.

gsl_matrix * gsl_matrix_calloc (size_t *n1*, size_t *n2*)   Function
> This function allocates memory for a matrix of size *n1* rows by *n2* columns and initializes all the elements of the matrix to zero.

void gsl_matrix_free (gsl_matrix * *m*)   Function
> This function frees a previously allocated matrix *m*. If the matrix was created using gsl_matrix_alloc then the block underlying the matrix will also be deallocated. If the matrix has been created from another object then the memory is still owned by that object and will not be deallocated. The matrix *m* must be a valid matrix object (a null pointer is not allowed).

8.4.2 Accessing matrix elements

The functions for accessing the elements of a matrix use the same range checking system as vectors. You can turn off range checking by recompiling your program with the preprocessor definition GSL_RANGE_CHECK_OFF.

The elements of the matrix are stored in "C-order", where the second index moves continuously through memory. More precisely, the element accessed by the function gsl_matrix_get(m,i,j) and gsl_matrix_set(m,i,j,x) is

    m->data[i * m->tda + j]

where tda is the physical row-length of the matrix.

double gsl_matrix_get (const gsl_matrix * m, size_t i, size_t   Function
        j)
   This function returns the (i,j)-th element of a matrix m. If i or j lie outside
   the allowed range of 0 to $n1 - 1$ and 0 to $n2 - 1$ then the error handler is
   invoked and 0 is returned. An inline version of this function is used when
   HAVE_INLINE is defined.

void gsl_matrix_set (gsl_matrix * m, size_t i, size_t j, double   Function
        x)
   This function sets the value of the (i,j)-th element of a matrix m to x. If
   i or j lies outside the allowed range of 0 to $n1 - 1$ and 0 to $n2 - 1$ then
   the error handler is invoked. An inline version of this function is used when
   HAVE_INLINE is defined.

double * gsl_matrix_ptr (gsl_matrix * m, size_t i, size_t j)   Function
const double * gsl_matrix_const_ptr (const gsl_matrix * m,   Function
        size_t i, size_t j)
   These functions return a pointer to the (i,j)-th element of a matrix m. If
   i or j lie outside the allowed range of 0 to $n1 - 1$ and 0 to $n2 - 1$ then the
   error handler is invoked and a null pointer is returned. Inline versions of
   these functions are used when HAVE_INLINE is defined.

8.4.3 Initializing matrix elements

void gsl_matrix_set_all (gsl_matrix * m, double x)   Function
   This function sets all the elements of the matrix m to the value x.

void gsl_matrix_set_zero (gsl_matrix * m)   Function
   This function sets all the elements of the matrix m to zero.

void gsl_matrix_set_identity (gsl_matrix * m)   Function
   This function sets the elements of the matrix m to the corresponding elements of the identity matrix, $m(i,j) = \delta(i,j)$, i.e. a unit diagonal with
   all off-diagonal elements zero. This applies to both square and rectangular
   matrices.

8.4.4 Reading and writing matrices

The library provides functions for reading and writing matrices to a file as binary data or formatted text.

int gsl_matrix_fwrite (FILE * *stream*, const gsl_matrix * *m*)   Function
This function writes the elements of the matrix *m* to the stream *stream* in binary format. The return value is 0 for success and GSL_EFAILED if there was a problem writing to the file. Since the data is written in the native binary format it may not be portable between different architectures.

int gsl_matrix_fread (FILE * *stream*, gsl_matrix * *m*)   Function
This function reads into the matrix *m* from the open stream *stream* in binary format. The matrix *m* must be preallocated with the correct dimensions since the function uses the size of *m* to determine how many bytes to read. The return value is 0 for success and GSL_EFAILED if there was a problem reading from the file. The data is assumed to have been written in the native binary format on the same architecture.

int gsl_matrix_fprintf (FILE * *stream*, const gsl_matrix * *m*,   Function
          const char * *format*)
This function writes the elements of the matrix *m* line-by-line to the stream *stream* using the format specifier *format*, which should be one of the %g, %e or %f formats for floating point numbers and %d for integers. The function returns 0 for success and GSL_EFAILED if there was a problem writing to the file.

int gsl_matrix_fscanf (FILE * *stream*, gsl_matrix * *m*)   Function
This function reads formatted data from the stream *stream* into the matrix *m*. The matrix *m* must be preallocated with the correct dimensions since the function uses the size of *m* to determine how many numbers to read. The function returns 0 for success and GSL_EFAILED if there was a problem reading from the file.

8.4.5 Matrix views

A matrix view is a temporary object, stored on the stack, which can be used to operate on a subset of matrix elements. Matrix views can be defined for both constant and non-constant matrices using separate types that preserve constness. A matrix view has the type gsl_matrix_view and a constant matrix view has the type gsl_matrix_const_view. In both cases the elements of the view can by accessed using the matrix component of the view object. A pointer gsl_matrix * or const gsl_matrix * can be obtained by taking the address of the matrix component with the & operator. In addition to matrix views it is also possible to create vector views of a matrix, such as row or column views.

gsl_matrix_view gsl_matrix_submatrix (gsl_matrix * m, size_t   Function
        k1, size_t k2, size_t n1, size_t n2)

gsl_matrix_const_view gsl_matrix_const_submatrix (const   Function
        gsl_matrix * m, size_t k1, size_t k2, size_t n1, size_t n2)

These functions return a matrix view of a submatrix of the matrix m. The
upper-left element of the submatrix is the element (k1,k2) of the original
matrix. The submatrix has n1 rows and n2 columns. The physical number
of columns in memory given by tda is unchanged. Mathematically, the (i, j)-
th element of the new matrix is given by,

    m'(i,j) = m->data[(k1*m->tda + k2) + i*m->tda + j]

where the index i runs from 0 to n1-1 and the index j runs from 0 to n2-1.

The data pointer of the returned matrix struct is set to null if the combined
parameters (i,j,n1,n2,tda) overrun the ends of the original matrix.

The new matrix view is only a view of the block underlying the existing
matrix, m. The block containing the elements of m is not owned by the new
matrix view. When the view goes out of scope the original matrix m and its
block will continue to exist. The original memory can only be deallocated
by freeing the original matrix. Of course, the original matrix should not be
deallocated while the view is still in use.

The function gsl_matrix_const_submatrix is equivalent to gsl_matrix_
submatrix but can be used for matrices which are declared const.

gsl_matrix_view gsl_matrix_view_array (double * base, size_t   Function
        n1, size_t n2)

gsl_matrix_const_view gsl_matrix_const_view_array (const   Function
        double * base, size_t n1, size_t n2)

These functions return a matrix view of the array base. The matrix has n1
rows and n2 columns. The physical number of columns in memory is also
given by n2. Mathematically, the (i, j)-th element of the new matrix is given
by,

    m'(i,j) = base[i*n2 + j]

where the index i runs from 0 to n1-1 and the index j runs from 0 to n2-1.

The new matrix is only a view of the array base. When the view goes out
of scope the original array base will continue to exist. The original memory
can only be deallocated by freeing the original array. Of course, the original
array should not be deallocated while the view is still in use.

The function gsl_matrix_const_view_array is equivalent to gsl_matrix_
view_array but can be used for matrices which are declared const.

gsl_matrix_view gsl_matrix_view_array_with_tda (double *   Function
        base, size_t n1, size_t n2, size_t tda)

gsl_matrix_const_view gsl_matrix_const_view_array_with_tda   Function
        (const double * base, size_t n1, size_t n2, size_t tda)

These functions return a matrix view of the array base with a physical
number of columns tda which may differ from the corresponding dimension
of the matrix. The matrix has n1 rows and n2 columns, and the physical

number of columns in memory is given by *tda*. Mathematically, the (i, j)-th element of the new matrix is given by,

```
m'(i,j) = base[i*tda + j]
```

where the index i runs from 0 to n1-1 and the index j runs from 0 to n2-1.

The new matrix is only a view of the array *base*. When the view goes out of scope the original array *base* will continue to exist. The original memory can only be deallocated by freeing the original array. Of course, the original array should not be deallocated while the view is still in use.

The function `gsl_matrix_const_view_array_with_tda` is equivalent to `gsl_matrix_view_array_with_tda` but can be used for matrices which are declared const.

`gsl_matrix_view gsl_matrix_view_vector (gsl_vector * v,`   *Function*
        `size_t n1, size_t n2)`
`gsl_matrix_const_view gsl_matrix_const_view_vector (const`   *Function*
        `gsl_vector * v, size_t n1, size_t n2)`

These functions return a matrix view of the vector *v*. The matrix has *n1* rows and *n2* columns. The vector must have unit stride. The physical number of columns in memory is also given by *n2*. Mathematically, the (i, j)-th element of the new matrix is given by,

```
m'(i,j) = v->data[i*n2 + j]
```

where the index i runs from 0 to n1-1 and the index j runs from 0 to n2-1.

The new matrix is only a view of the vector *v*. When the view goes out of scope the original vector *v* will continue to exist. The original memory can only be deallocated by freeing the original vector. Of course, the original vector should not be deallocated while the view is still in use.

The function `gsl_matrix_const_view_vector` is equivalent to `gsl_matrix_view_vector` but can be used for matrices which are declared const.

`gsl_matrix_view gsl_matrix_view_vector_with_tda (gsl_vector`   *Function*
        `* v, size_t n1, size_t n2, size_t tda)`
`gsl_matrix_const_view gsl_matrix_const_view_vector_with_tda`   *Function*
        `(const gsl_vector * v, size_t n1, size_t n2, size_t tda)`

These functions return a matrix view of the vector *v* with a physical number of columns *tda* which may differ from the corresponding matrix dimension. The vector must have unit stride. The matrix has *n1* rows and *n2* columns, and the physical number of columns in memory is given by *tda*. Mathematically, the (i, j)-th element of the new matrix is given by,

```
m'(i,j) = v->data[i*tda + j]
```

where the index i runs from 0 to n1-1 and the index j runs from 0 to n2-1.

The new matrix is only a view of the vector *v*. When the view goes out of scope the original vector *v* will continue to exist. The original memory can only be deallocated by freeing the original vector. Of course, the original vector should not be deallocated while the view is still in use.

The function `gsl_matrix_const_view_vector_with_tda` is equivalent to `gsl_matrix_view_vector_with_tda` but can be used for matrices which are declared const.

8.4.6 Creating row and column views

In general there are two ways to access an object, by reference or by copying. The functions described in this section create vector views which allow access to a row or column of a matrix by reference. Modifying elements of the view is equivalent to modifying the matrix, since both the vector view and the matrix point to the same memory block.

gsl_vector_view gsl_matrix_row (gsl_matrix * m, size_t i)   Function
gsl_vector_const_view gsl_matrix_const_row (const gsl_matrix   Function
        * m, size_t i)

These functions return a vector view of the i-th row of the matrix m. The data pointer of the new vector is set to null if i is out of range.

The function gsl_vector_const_row is equivalent to gsl_matrix_row but can be used for matrices which are declared const.

gsl_vector_view gsl_matrix_column (gsl_matrix * m, size_t j)   Function
gsl_vector_const_view gsl_matrix_const_column (const   Function
        gsl_matrix * m, size_t j)

These functions return a vector view of the j-th column of the matrix m. The data pointer of the new vector is set to null if j is out of range.

The function gsl_vector_const_column is equivalent to gsl_matrix_ column but can be used for matrices which are declared const.

gsl_vector_view gsl_matrix_subrow (gsl_matrix * m, size_t i,   Function
        size_t $offset$, size_t n)
gsl_vector_const_view gsl_matrix_const_subrow (const   Function
        gsl_matrix * m, size_t i, size_t $offset$, size_t n)

These functions return a vector view of the i-th row of the matrix m beginning at $offset$ elements past the first column and containing n elements. The data pointer of the new vector is set to null if i, $offset$, or n are out of range.

The function gsl_vector_const_subrow is equivalent to gsl_matrix_ subrow but can be used for matrices which are declared const.

gsl_vector_view gsl_matrix_subcolumn (gsl_matrix * m, size_t   Function
        j, size_t $offset$, size_t n)
gsl_vector_const_view gsl_matrix_const_subcolumn (const   Function
        gsl_matrix * m, size_t j, size_t $offset$, size_t n)

These functions return a vector view of the j-th column of the matrix m beginning at $offset$ elements past the first row and containing n elements. The data pointer of the new vector is set to null if j, $offset$, or n are out of range.

The function gsl_vector_const_subcolumn is equivalent to gsl_matrix_ subcolumn but can be used for matrices which are declared const.

gsl_vector_view gsl_matrix_diagonal (gsl_matrix * m)   Function
gsl_vector_const_view gsl_matrix_const_diagonal (const   Function
          gsl_matrix * m)

These functions returns a vector view of the diagonal of the matrix m. The matrix m is not required to be square. For a rectangular matrix the length of the diagonal is the same as the smaller dimension of the matrix.

The function gsl_matrix_const_diagonal is equivalent to gsl_matrix_diagonal but can be used for matrices which are declared const.

gsl_vector_view gsl_matrix_subdiagonal (gsl_matrix * m,   Function
          size_t k)
gsl_vector_const_view gsl_matrix_const_subdiagonal (const   Function
          gsl_matrix * m, size_t k)

These functions return a vector view of the k-th subdiagonal of the matrix m. The matrix m is not required to be square. The diagonal of the matrix corresponds to $k = 0$.

The function gsl_matrix_const_subdiagonal is equivalent to gsl_matrix_subdiagonal but can be used for matrices which are declared const.

gsl_vector_view gsl_matrix_superdiagonal (gsl_matrix * m,   Function
          size_t k)
gsl_vector_const_view gsl_matrix_const_superdiagonal (const   Function
          gsl_matrix * m, size_t k)

These functions return a vector view of the k-th superdiagonal of the matrix m. The matrix m is not required to be square. The diagonal of the matrix corresponds to $k = 0$.

The function gsl_matrix_const_superdiagonal is equivalent to gsl_matrix_superdiagonal but can be used for matrices which are declared const.

8.4.7 Copying matrices

int gsl_matrix_memcpy (gsl_matrix * dest, const gsl_matrix *   Function
          src)
This function copies the elements of the matrix src into the matrix dest. The two matrices must have the same size.

int gsl_matrix_swap (gsl_matrix * m1, gsl_matrix * m2)   Function
This function exchanges the elements of the matrices m1 and m2 by copying. The two matrices must have the same size.

8.4.8 Copying rows and columns

The functions described in this section copy a row or column of a matrix into a vector. This allows the elements of the vector and the matrix to be modified independently. Note that if the matrix and the vector point to overlapping regions of memory then the result will be undefined. The same effect can be achieved with more generality using `gsl_vector_memcpy` with vector views of rows and columns.

int gsl_matrix_get_row (gsl_vector * v, const gsl_matrix * m,   *Function*
     size_t i)

This function copies the elements of the i-th row of the matrix m into the vector v. The length of the vector must be the same as the length of the row.

int gsl_matrix_get_col (gsl_vector * v, const gsl_matrix * m,   *Function*
     size_t j)

This function copies the elements of the j-th column of the matrix m into the vector v. The length of the vector must be the same as the length of the column.

int gsl_matrix_set_row (gsl_matrix * m, size_t i, const   *Function*
     gsl_vector * v)

This function copies the elements of the vector v into the i-th row of the matrix m. The length of the vector must be the same as the length of the row.

int gsl_matrix_set_col (gsl_matrix * m, size_t j, const   *Function*
     gsl_vector * v)

This function copies the elements of the vector v into the j-th column of the matrix m. The length of the vector must be the same as the length of the column.

8.4.9 Exchanging rows and columns

The following functions can be used to exchange the rows and columns of a matrix.

int gsl_matrix_swap_rows (gsl_matrix * m, size_t i, size_t j)   *Function*
This function exchanges the i-th and j-th rows of the matrix m in-place.

int gsl_matrix_swap_columns (gsl_matrix * m, size_t i, size_t   *Function*
     j)
This function exchanges the i-th and j-th columns of the matrix m in-place.

int gsl_matrix_swap_rowcol (gsl_matrix * m, size_t i, size_t   *Function*
     j)
This function exchanges the i-th row and j-th column of the matrix m in-place. The matrix must be square for this operation to be possible.

int gsl_matrix_transpose_memcpy (gsl_matrix * *dest*, const   Function
        gsl_matrix * *src*)
This function makes the matrix *dest* the transpose of the matrix *src* by
copying the elements of *src* into *dest*. This function works for all matrices provided that the dimensions of the matrix *dest* match the transposed
dimensions of the matrix *src*.

int gsl_matrix_transpose (gsl_matrix * *m*)   Function
This function replaces the matrix *m* by its transpose by copying the elements
of the matrix in-place. The matrix must be square for this operation to be
possible.

8.4.10 Matrix operations

The following operations are defined for real and complex matrices.

int gsl_matrix_add (gsl_matrix * a, const gsl_matrix * b)   Function
This function adds the elements of matrix b to the elements of matrix a,
$a'(i,j) = a(i,j) + b(i,j)$. The two matrices must have the same dimensions.

int gsl_matrix_sub (gsl_matrix * a, const gsl_matrix * b)   Function
This function subtracts the elements of matrix b from the elements of matrix a, $a'(i,j) = a(i,j) - b(i,j)$. The two matrices must have the same
dimensions.

int gsl_matrix_mul_elements (gsl_matrix * a, const gsl_matrix   Function
        * b)
This function multiplies the elements of matrix a by the elements of matrix
b, $a'(i,j) = a(i,j)*b(i,j)$. The two matrices must have the same dimensions.

int gsl_matrix_div_elements (gsl_matrix * a, const gsl_matrix   Function
        * b)
This function divides the elements of matrix a by the elements of matrix b,
$a'(i,j) = a(i,j)/b(i,j)$. The two matrices must have the same dimensions.

int gsl_matrix_scale (gsl_matrix * a, const double x)   Function
This function multiplies the elements of matrix a by the constant factor x,
$a'(i,j) = xa(i,j)$.

int gsl_matrix_add_constant (gsl_matrix * a, const double x)   Function
This function adds the constant value x to the elements of the matrix a,
$a'(i,j) = a(i,j) + x$.

8.4.11 Finding maximum and minimum elements of matrices

The following operations are only defined for real matrices.

double gsl_matrix_max (const gsl_matrix * m)   Function
  This function returns the maximum value in the matrix m.

double gsl_matrix_min (const gsl_matrix * m)   Function
  This function returns the minimum value in the matrix m.

void gsl_matrix_minmax (const gsl_matrix * m, double *   Function
        min_out, double * max_out)
  This function returns the minimum and maximum values in the matrix m,
  storing them in min_out and max_out.

void gsl_matrix_max_index (const gsl_matrix * m, size_t *   Function
        imax, size_t * jmax)
  This function returns the indices of the maximum value in the matrix m,
  storing them in imax and jmax. When there are several equal maximum
  elements then the first element found is returned, searching in row-major
  order.

void gsl_matrix_min_index (const gsl_matrix * m, size_t *   Function
        imin, size_t * jmin)
  This function returns the indices of the minimum value in the matrix m,
  storing them in imin and jmin. When there are several equal minimum
  elements then the first element found is returned, searching in row-major
  order.

void gsl_matrix_minmax_index (const gsl_matrix * m, size_t *   Function
        imin, size_t * jmin, size_t * imax, size_t * jmax)
  This function returns the indices of the minimum and maximum values in
  the matrix m, storing them in (imin,jmin) and (imax,jmax). When there
  are several equal minimum or maximum elements then the first elements
  found are returned, searching in row-major order.

8.4.12 Matrix properties

The following functions are defined for real and complex matrices. For complex matrices both the real and imaginary parts must satisfy the conditions.

int gsl_matrix_isnull (const gsl_matrix * m)   Function
int gsl_matrix_ispos (const gsl_matrix * m)   Function
int gsl_matrix_isneg (const gsl_matrix * m)   Function
int gsl_matrix_isnonneg (const gsl_matrix * m)   Function
  These functions return 1 if all the elements of the matrix m are zero, strictly
  positive, strictly negative, or non-negative respectively, and 0 otherwise. To
  test whether a matrix is positive-definite, use the Cholesky decomposition
  (see Section 13.5 [Cholesky Decomposition], page 160).

8.4.13 Example programs for matrices

The program below shows how to allocate, initialize and read from a matrix using the functions gsl_matrix_alloc, gsl_matrix_set and gsl_matrix_get.

```
#include <stdio.h>
#include <gsl/gsl_matrix.h>

int
main (void)
{
  int i, j;
  gsl_matrix * m = gsl_matrix_alloc (10, 3);

  for (i = 0; i < 10; i++)
    for (j = 0; j < 3; j++)
      gsl_matrix_set (m, i, j, 0.23 + 100*i + j);

  for (i = 0; i < 100; i++)   /* OUT OF RANGE ERROR */
    for (j = 0; j < 3; j++)
      printf ("m(%d,%d) = %g\n", i, j,
              gsl_matrix_get (m, i, j));

  gsl_matrix_free (m);

  return 0;
}
```

Here is the output from the program. The final loop attempts to read outside the range of the matrix m, and the error is trapped by the range-checking code in gsl_matrix_get.

```
$ ./a.out
m(0,0) = 0.23
m(0,1) = 1.23
m(0,2) = 2.23
m(1,0) = 100.23
m(1,1) = 101.23
m(1,2) = 102.23
...
m(9,2) = 902.23
gsl: matrix_source.c:13: ERROR: first index out of range
Default GSL error handler invoked.
Aborted (core dumped)
```

The next program shows how to write a matrix to a file.

```
#include <stdio.h>
#include <gsl/gsl_matrix.h>

int
main (void)
{
```

```
int i, j, k = 0;
gsl_matrix * m = gsl_matrix_alloc (100, 100);
gsl_matrix * a = gsl_matrix_alloc (100, 100);

for (i = 0; i < 100; i++)
  for (j = 0; j < 100; j++)
    gsl_matrix_set (m, i, j, 0.23 + i + j);

{
   FILE * f = fopen ("test.dat", "wb");
   gsl_matrix_fwrite (f, m);
   fclose (f);
}

{
   FILE * f = fopen ("test.dat", "rb");
   gsl_matrix_fread (f, a);
   fclose (f);
}

for (i = 0; i < 100; i++)
  for (j = 0; j < 100; j++)
    {
       double mij = gsl_matrix_get (m, i, j);
       double aij = gsl_matrix_get (a, i, j);
       if (mij != aij) k++;
    }

gsl_matrix_free (m);
gsl_matrix_free (a);

printf ("differences = %d (should be zero)\n", k);
return (k > 0);
}
```

After running this program the file 'test.dat' should contain the elements of m, written in binary format. The matrix which is read back in using the function gsl_matrix_fread should be exactly equal to the original matrix.

The following program demonstrates the use of vector views. The program computes the column norms of a matrix.

```
#include <math.h>
#include <stdio.h>
#include <gsl/gsl_matrix.h>
#include <gsl/gsl_blas.h>

int
main (void)
{
  size_t i,j;
```

```
gsl_matrix *m = gsl_matrix_alloc (10, 10);

for (i = 0; i < 10; i++)
  for (j = 0; j < 10; j++)
    gsl_matrix_set (m, i, j, sin (i) + cos (j));

for (j = 0; j < 10; j++)
  {
    gsl_vector_view column = gsl_matrix_column (m, j);
    double d;

    d = gsl_blas_dnrm2 (&column.vector);

    printf ("matrix column %d, norm = %g\n", j, d);
  }

gsl_matrix_free (m);

return 0;
}
```

Here is the output of the program,

```
$ ./a.out
matrix column 0, norm = 4.31461
matrix column 1, norm = 3.1205
matrix column 2, norm = 2.19316
matrix column 3, norm = 3.26114
matrix column 4, norm = 2.53416
matrix column 5, norm = 2.57281
matrix column 6, norm = 4.20469
matrix column 7, norm = 3.65202
matrix column 8, norm = 2.08524
matrix column 9, norm = 3.07313
```

The results can be confirmed using GNU OCTAVE,

```
$ octave
GNU Octave, version 2.0.16.92
octave> m = sin(0:9)' * ones(1,10)
              + ones(10,1) * cos(0:9);
octave> sqrt(sum(m.^2))
ans =
  4.3146  3.1205  2.1932  3.2611  2.5342  2.5728
  4.2047  3.6520  2.0852  3.0731
```

8.5 References and Further Reading

The block, vector and matrix objects in GSL follow the `valarray` model of C++. A description of this model can be found in the following reference,

B. Stroustrup, *The C++ Programming Language* (3rd Ed), Section 22.4 Vector Arithmetic. Addison-Wesley 1997, ISBN 0-201-88954-4.

9 Permutations

This chapter describes functions for creating and manipulating permutations. A permutation p is represented by an array of n integers in the range 0 to $n-1$, where each value p_i occurs once and only once. The application of a permutation p to a vector v yields a new vector v' where $v'_i = v_{p_i}$. For example, the array $(0, 1, 3, 2)$ represents a permutation which exchanges the last two elements of a four element vector. The corresponding identity permutation is $(0, 1, 2, 3)$.

Note that the permutations produced by the linear algebra routines correspond to the exchange of matrix columns, and so should be considered as applying to row-vectors in the form $v' = vP$ rather than column-vectors, when permuting the elements of a vector.

The functions described in this chapter are defined in the header file 'gsl_permutation.h'.

9.1 The Permutation struct

A permutation is defined by a structure containing two components, the size of the permutation and a pointer to the permutation array. The elements of the permutation array are all of type size_t. The gsl_permutation structure looks like this,

```
typedef struct
{
  size_t size;
  size_t * data;
} gsl_permutation;
```

9.2 Permutation allocation

gsl_permutation * gsl_permutation_alloc (size_t n)   Function
    This function allocates memory for a new permutation of size n. The permutation is not initialized and its elements are undefined. Use the function gsl_permutation_calloc if you want to create a permutation which is initialized to the identity. A null pointer is returned if insufficient memory is available to create the permutation.

gsl_permutation * gsl_permutation_calloc (size_t n)   Function
    This function allocates memory for a new permutation of size n and initializes it to the identity. A null pointer is returned if insufficient memory is available to create the permutation.

void gsl_permutation_init (gsl_permutation * p)   Function
    This function initializes the permutation p to the identity, i.e. $(0, 1, 2, \ldots, n-1)$.

void gsl_permutation_free (gsl_permutation * p)   Function
    This function frees all the memory used by the permutation p.

int gsl_permutation_memcpy (gsl_permutation * *dest*, const   *Function*
        gsl_permutation * *src*)
This function copies the elements of the permutation *src* into the permutation *dest*. The two permutations must have the same size.

9.3 Accessing permutation elements

The following functions can be used to access and manipulate permutations.

size_t gsl_permutation_get (const gsl_permutation * *p*, const   *Function*
        size_t *i*)
This function returns the value of the *i*-th element of the permutation *p*. If *i*
lies outside the allowed range of 0 to $n-1$ then the error handler is invoked
and 0 is returned. An inline version of this function is used when HAVE_
INLINE is defined.

int gsl_permutation_swap (gsl_permutation * *p*, const size_t *i*,   *Function*
        const size_t *j*)
This function exchanges the *i*-th and *j*-th elements of the permutation *p*.

9.4 Permutation properties

size_t gsl_permutation_size (const gsl_permutation * *p*)   *Function*
This function returns the size of the permutation *p*.

size_t * gsl_permutation_data (const gsl_permutation * *p*)   *Function*
This function returns a pointer to the array of elements in the permutation
p.

int gsl_permutation_valid (const gsl_permutation * *p*)   *Function*
This function checks that the permutation *p* is valid. The *n* elements should
contain each of the numbers 0 to $n-1$ once and only once.

9.5 Permutation functions

void gsl_permutation_reverse (gsl_permutation * *p*)   *Function*
This function reverses the elements of the permutation *p*.

int gsl_permutation_inverse (gsl_permutation * *inv*, const   *Function*
        gsl_permutation * *p*)
This function computes the inverse of the permutation *p*, storing the result
in *inv*.

int gsl_permutation_next (gsl_permutation * *p*)   *Function*
This function advances the permutation *p* to the next permutation in lexicographic order and returns GSL_SUCCESS. If no further permutations are
available it returns GSL_FAILURE and leaves *p* unmodified. Starting with
the identity permutation and repeatedly applying this function will iterate
through all possible permutations of a given order.

int gsl_permutation_prev (gsl_permutation * p)   Function
This function steps backwards from the permutation p to the previous permutation in lexicographic order, returning GSL_SUCCESS. If no previous permutation is available it returns GSL_FAILURE and leaves p unmodified.

9.6 Applying Permutations

int gsl_permute (const size_t * p, double * data, size_t *stride*,   Function
      size_t *n*)
This function applies the permutation p to the array *data* of size n with stride *stride*.

int gsl_permute_inverse (const size_t * p, double * data,   Function
      size_t *stride*, size_t *n*)
This function applies the inverse of the permutation p to the array *data* of size n with stride *stride*.

int gsl_permute_vector (const gsl_permutation * p, gsl_vector   Function
      * v)
This function applies the permutation p to the elements of the vector v, considered as a row-vector acted on by a permutation matrix from the right, $v' = vP$. The j-th column of the permutation matrix P is given by the p_j-th column of the identity matrix. The permutation p and the vector v must have the same length.

int gsl_permute_vector_inverse (const gsl_permutation * p,   Function
      gsl_vector * v)
This function applies the inverse of the permutation p to the elements of the vector v, considered as a row-vector acted on by an inverse permutation matrix from the right, $v' = vP^T$. Note that for permutation matrices the inverse is the same as the transpose. The j-th column of the permutation matrix P is given by the p_j-th column of the identity matrix. The permutation p and the vector v must have the same length.

int gsl_permutation_mul (gsl_permutation * p, const   Function
      gsl_permutation * pa, const gsl_permutation * pb)
This function combines the two permutations pa and pb into a single permutation p, where $p = pa.pb$. The permutation p is equivalent to applying pb first and then pa.

9.7 Reading and writing permutations

The library provides functions for reading and writing permutations to a file as binary data or formatted text.

int gsl_permutation_fwrite (FILE * *stream*, const   *Function*
        gsl_permutation * *p*)

This function writes the elements of the permutation *p* to the stream *stream* in binary format. The function returns GSL_EFAILED if there was a problem writing to the file. Since the data is written in the native binary format it may not be portable between different architectures.

int gsl_permutation_fread (FILE * *stream*, gsl_permutation * *p*)   *Function*

This function reads into the permutation *p* from the open stream *stream* in binary format. The permutation *p* must be preallocated with the correct length since the function uses the size of *p* to determine how many bytes to read. The function returns GSL_EFAILED if there was a problem reading from the file. The data is assumed to have been written in the native binary format on the same architecture.

int gsl_permutation_fprintf (FILE * *stream*, const   *Function*
        gsl_permutation * *p*, const char * *format*)

This function writes the elements of the permutation *p* line-by-line to the stream *stream* using the format specifier *format*, which should be suitable for a type of *size_t*. In ISO C99 the type modifier z represents size_t, so "%zu\n" is a suitable format.[1] The function returns GSL_EFAILED if there was a problem writing to the file.

int gsl_permutation_fscanf (FILE * *stream*, gsl_permutation *   *Function*
        p)

This function reads formatted data from the stream *stream* into the permutation *p*. The permutation *p* must be preallocated with the correct length since the function uses the size of *p* to determine how many numbers to read. The function returns GSL_EFAILED if there was a problem reading from the file.

---

[1] In versions of the GNU C library prior to the ISO C99 standard, the type modifier Z was used instead.

9.8 Permutations in cyclic form

A permutation can be represented in both *linear* and *cyclic* notations. The functions described in this section convert between the two forms. The linear notation is an index mapping, and has already been described above. The cyclic notation expresses a permutation as a series of circular rearrangements of groups of elements, or *cycles*.

For example, under the cycle (1 2 3), 1 is replaced by 2, 2 is replaced by 3 and 3 is replaced by 1 in a circular fashion. Cycles of different sets of elements can be combined independently, for example (1 2 3) (4 5) combines the cycle (1 2 3) with the cycle (4 5), which is an exchange of elements 4 and 5. A cycle of length one represents an element which is unchanged by the permutation and is referred to as a *singleton*.

It can be shown that every permutation can be decomposed into combinations of cycles. The decomposition is not unique, but can always be rearranged into a standard *canonical form* by a reordering of elements. The library uses the canonical form defined in Knuth's *Art of Computer Programming* (Vol 1, 3rd Ed, 1997) Section 1.3.3, p.178.

The procedure for obtaining the canonical form given by Knuth is,

1. Write all singleton cycles explicitly

2. Within each cycle, put the smallest number first

3. Order the cycles in decreasing order of the first number in the cycle.

For example, the linear representation (2 4 3 0 1) is represented as (1 4) (0 2 3) in canonical form. The permutation corresponds to an exchange of elements 1 and 4, and rotation of elements 0, 2 and 3.

The important property of the canonical form is that it can be reconstructed from the contents of each cycle without the brackets. In addition, by removing the brackets it can be considered as a linear representation of a different permutation. In the example given above the permutation (2 4 3 0 1) would become (1 4 0 2 3). This mapping has many applications in the theory of permutations.

int gsl_permutation_linear_to_canonical (gsl_permutation * q,   Function
        const gsl_permutation * p)
  This function computes the canonical form of the permutation p and stores it in the output argument q.

int gsl_permutation_canonical_to_linear (gsl_permutation * p,   Function
        const gsl_permutation * q)
  This function converts a permutation q in canonical form back into linear form storing it in the output argument p.

size_t gsl_permutation_inversions (const gsl_permutation * p)   Function
  This function counts the number of inversions in the permutation p. An inversion is any pair of elements that are not in order. For example, the permutation 2031 has three inversions, corresponding to the pairs (2,0) (2,1) and (3,1). The identity permutation has no inversions.

size_t gsl_permutation_linear_cycles (const gsl_permutation *   Function
         p)
    This function counts the number of cycles in the permutation p, given in
    linear form.

size_t gsl_permutation_canonical_cycles (const   Function
         gsl_permutation * q)
    This function counts the number of cycles in the permutation q, given in
    canonical form.

9.9 Examples

The example program below creates a random permutation (by shuffling the
elements of the identity) and finds its inverse.

```
#include <stdio.h>
#include <gsl/gsl_rng.h>
#include <gsl/gsl_randist.h>
#include <gsl/gsl_permutation.h>

int
main (void)
{
  const size_t N = 10;
  const gsl_rng_type * T;
  gsl_rng * r;

  gsl_permutation * p = gsl_permutation_alloc (N);
  gsl_permutation * q = gsl_permutation_alloc (N);

  gsl_rng_env_setup();
  T = gsl_rng_default;
  r = gsl_rng_alloc (T);

  printf ("initial permutation:");
  gsl_permutation_init (p);
  gsl_permutation_fprintf (stdout, p, " %u");
  printf ("\n");

  printf (" random permutation:");
  gsl_ran_shuffle (r, p->data, N, sizeof(size_t));
  gsl_permutation_fprintf (stdout, p, " %u");
  printf ("\n");

  printf ("inverse permutation:");
  gsl_permutation_inverse (q, p);
  gsl_permutation_fprintf (stdout, q, " %u");
  printf ("\n");
```

```
      gsl_permutation_free (p);
      gsl_permutation_free (q);
      gsl_rng_free (r);

      return 0;
    }
```

Here is the output from the program,

```
    $ ./a.out
    initial permutation: 0 1 2 3 4 5 6 7 8 9
     random permutation: 1 3 5 2 7 6 0 4 9 8
    inverse permutation: 6 0 3 1 7 2 5 4 9 8
```

The random permutation p[i] and its inverse q[i] are related through the identity p[q[i]] = i, which can be verified from the output.

The next example program steps forwards through all possible third order permutations, starting from the identity,

```
    #include <stdio.h>
    #include <gsl/gsl_permutation.h>

    int
    main (void)
    {
      gsl_permutation * p = gsl_permutation_alloc (3);

      gsl_permutation_init (p);

      do
        {
           gsl_permutation_fprintf (stdout, p, " %u");
           printf ("\n");
        }
      while (gsl_permutation_next(p) == GSL_SUCCESS);

      gsl_permutation_free (p);

      return 0;
    }
```

Here is the output from the program,

```
    $ ./a.out
     0 1 2
     0 2 1
     1 0 2
     1 2 0
     2 0 1
     2 1 0
```

The permutations are generated in lexicographic order. To reverse the sequence, begin with the final permutation (which is the reverse of the identity) and replace gsl_permutation_next with gsl_permutation_prev.

9.10 References and Further Reading

The subject of permutations is covered extensively in Knuth's *Sorting and Searching*,

> Donald E. Knuth, *The Art of Computer Programming: Sorting and Searching* (Vol 3, 3rd Ed, 1997), Addison-Wesley, ISBN 0201896850.

For the definition of the *canonical form* see,

> Donald E. Knuth, *The Art of Computer Programming: Fundamental Algorithms* (Vol 1, 3rd Ed, 1997), Addison-Wesley, ISBN 0201896850. Section 1.3.3, *An Unusual Correspondence*, p.178–179.

10 Combinations

This chapter describes functions for creating and manipulating combinations. A combination c is represented by an array of k integers in the range 0 to $n-1$, where each value c_i occurs at most once. The combination c corresponds to indices of k elements chosen from an n element vector. Combinations are useful for iterating over all k-element subsets of a set.

The functions described in this chapter are defined in the header file 'gsl_combination.h'.

10.1 The Combination struct

A combination is defined by a structure containing three components, the values of n and k, and a pointer to the combination array. The elements of the combination array are all of type size_t, and are stored in increasing order. The gsl_combination structure looks like this,

```
typedef struct
{
  size_t n;
  size_t k;
  size_t *data;
} gsl_combination;
```

10.2 Combination allocation

gsl_combination * gsl_combination_alloc (size_t n, size_t k)   Function
> This function allocates memory for a new combination with parameters n, k. The combination is not initialized and its elements are undefined. Use the function gsl_combination_calloc if you want to create a combination which is initialized to the lexicographically first combination. A null pointer is returned if insufficient memory is available to create the combination.

gsl_combination * gsl_combination_calloc (size_t n, size_t k)   Function
> This function allocates memory for a new combination with parameters n, k and initializes it to the lexicographically first combination. A null pointer is returned if insufficient memory is available to create the combination.

void gsl_combination_init_first (gsl_combination * c)   Function
> This function initializes the combination c to the lexicographically first combination, i.e. $(0, 1, 2, \ldots, k-1)$.

void gsl_combination_init_last (gsl_combination * c)   Function
> This function initializes the combination c to the lexicographically last combination, i.e. $(n-k, n-k+1, \ldots, n-1)$.

void gsl_combination_free (gsl_combination * c)   Function
> This function frees all the memory used by the combination c.

int gsl_combination_memcpy (gsl_combination * *dest*, const   Function
        gsl_combination * *src*)
  This function copies the elements of the combination *src* into the combination *dest*. The two combinations must have the same size.

10.3 Accessing combination elements

The following function can be used to access the elements of a combination.

size_t gsl_combination_get (const gsl_combination * *c*, const   Function
        size_t *i*)
  This function returns the value of the i-th element of the combination c. If i lies outside the allowed range of 0 to $k-1$ then the error handler is invoked and 0 is returned. An inline version of this function is used when HAVE_INLINE is defined.

10.4 Combination properties

size_t gsl_combination_n (const gsl_combination * *c*)   Function
  This function returns the range (n) of the combination c.

size_t gsl_combination_k (const gsl_combination * *c*)   Function
  This function returns the number of elements (k) in the combination c.

size_t * gsl_combination_data (const gsl_combination * *c*)   Function
  This function returns a pointer to the array of elements in the combination c.

int gsl_combination_valid (gsl_combination * *c*)   Function
  This function checks that the combination c is valid. The k elements should lie in the range 0 to $n-1$, with each value occurring once at most and in increasing order.

10.5 Combination functions

int gsl_combination_next (gsl_combination * *c*)   Function
  This function advances the combination c to the next combination in lexicographic order and returns GSL_SUCCESS. If no further combinations are available it returns GSL_FAILURE and leaves c unmodified. Starting with the first combination and repeatedly applying this function will iterate through all possible combinations of a given order.

int gsl_combination_prev (gsl_combination * *c*)   Function
  This function steps backwards from the combination c to the previous combination in lexicographic order, returning GSL_SUCCESS. If no previous combination is available it returns GSL_FAILURE and leaves c unmodified.

10.6 Reading and writing combinations

The library provides functions for reading and writing combinations to a file as binary data or formatted text.

int gsl_combination_fwrite (FILE * *stream*, const   *Function*
      gsl_combination * c)

    This function writes the elements of the combination c to the stream *stream* in binary format. The function returns GSL_EFAILED if there was a problem writing to the file. Since the data is written in the native binary format it may not be portable between different architectures.

int gsl_combination_fread (FILE * *stream*, gsl_combination * c)   *Function*

    This function reads elements from the open stream *stream* into the combination c in binary format. The combination c must be preallocated with correct values of n and k since the function uses the size of c to determine how many bytes to read. The function returns GSL_EFAILED if there was a problem reading from the file. The data is assumed to have been written in the native binary format on the same architecture.

int gsl_combination_fprintf (FILE * *stream*, const   *Function*
      gsl_combination * c, const char * *format*)

    This function writes the elements of the combination c line-by-line to the stream *stream* using the format specifier *format*, which should be suitable for a type of *size_t*. In ISO C99 the type modifier z represents size_t, so "%zu\n" is a suitable format.[1] The function returns GSL_EFAILED if there was a problem writing to the file.

int gsl_combination_fscanf (FILE * *stream*, gsl_combination *   *Function*
      c)

    This function reads formatted data from the stream *stream* into the combination c. The combination c must be preallocated with correct values of n and k since the function uses the size of c to determine how many numbers to read. The function returns GSL_EFAILED if there was a problem reading from the file.

---

[1] In versions of the GNU C library prior to the ISO C99 standard, the type modifier Z was used instead.

10.7 Examples

The example program below prints all subsets of the set $\{0, 1, 2, 3\}$ ordered by size. Subsets of the same size are ordered lexicographically.

```
#include <stdio.h>
#include <gsl/gsl_combination.h>

int
main (void)
{
  gsl_combination * c;
  size_t i;

  printf ("All subsets of {0,1,2,3} by size:\n") ;
  for (i = 0; i <= 4; i++)
    {
      c = gsl_combination_calloc (4, i);
      do
        {
          printf ("{");
          gsl_combination_fprintf (stdout, c, " %u");
          printf (" }\n");
        }
      while (gsl_combination_next (c) == GSL_SUCCESS);
      gsl_combination_free (c);
    }

  return 0;
}
```

Here is the output from the program,

```
$ ./a.out
All subsets of {0,1,2,3} by size:
{ }
{ 0 }
{ 1 }
{ 2 }
{ 3 }
{ 0 1 }
{ 0 2 }
{ 0 3 }
{ 1 2 }
{ 1 3 }
{ 2 3 }
{ 0 1 2 }
{ 0 1 3 }
{ 0 2 3 }
{ 1 2 3 }
{ 0 1 2 3 }
```

All 16 subsets are generated, and the subsets of each size are sorted lexicograph-ically.

10.8 References and Further Reading

Further information on combinations can be found in,

Donald L. Kreher, Douglas R. Stinson, *Combinatorial Algorithms: Generation, Enumeration and Search*, 1998, CRC Press LLC, ISBN 084933988X

11 Sorting

This chapter describes functions for sorting data, both directly and indirectly (using an index). All the functions use the *heapsort* algorithm. Heapsort is an $O(N \log N)$ algorithm which operates in-place and does not require any additional storage. It also provides consistent performance, the running time for its worst-case (ordered data) being not significantly longer than the average and best cases. Note that the heapsort algorithm does not preserve the relative ordering of equal elements—it is an *unstable* sort. However the resulting order of equal elements will be consistent across different platforms when using these functions.

11.1 Sorting objects

The following function provides a simple alternative to the standard library function qsort. It is intended for systems lacking qsort, not as a replacement for it. The function qsort should be used whenever possible, as it will be faster and can provide stable ordering of equal elements. Documentation for qsort is available in the *GNU C Library Reference Manual*.

The functions described in this section are defined in the header file 'gsl_heapsort.h'.

void gsl_heapsort (void * array, size_t *count*, size_t *size*,   *Function*
        gsl_comparison_fn_t *compare*)

This function sorts the *count* elements of the array *array*, each of size *size*, into ascending order using the comparison function *compare*. The type of the comparison function is defined by,

```
int (*gsl_comparison_fn_t) (const void * a,
                            const void * b)
```

A comparison function should return a negative integer if the first argument is less than the second argument, 0 if the two arguments are equal and a positive integer if the first argument is greater than the second argument.

For example, the following function can be used to sort doubles into ascending numerical order.

```
int
compare_doubles (const double * a,
                 const double * b)
{
    if (*a > *b)
       return 1;
    else if (*a < *b)
       return -1;
    else
       return 0;
}
```

The appropriate function call to perform the sort is,

```
gsl_heapsort (array, count, sizeof(double),
              compare_doubles);
```

Note that unlike qsort the heapsort algorithm cannot be made into a stable sort by pointer arithmetic. The trick of comparing pointers for equal elements in the comparison function does not work for the heapsort algorithm. The heapsort algorithm performs an internal rearrangement of the data which destroys its initial ordering.

int gsl_heapsort_index (size_t * p, const void * array, size_t   Function
        count, size_t size, gsl_comparison_fn_t compare)

This function indirectly sorts the count elements of the array array, each of size size, into ascending order using the comparison function compare. The resulting permutation is stored in p, an array of length n. The elements of p give the index of the array element which would have been stored in that position if the array had been sorted in place. The first element of p gives the index of the least element in array, and the last element of p gives the index of the greatest element in array. The array itself is not changed.

11.2 Sorting vectors

The following functions will sort the elements of an array or vector, either directly or indirectly. They are defined for all real and integer types using the normal suffix rules. For example, the float versions of the array functions are gsl_sort_float and gsl_sort_float_index. The corresponding vector functions are gsl_sort_vector_float and gsl_sort_vector_float_index. The prototypes are available in the header files 'gsl_sort_float.h' 'gsl_sort_vector_float.h'. The complete set of prototypes can be included using the header files 'gsl_sort.h' and 'gsl_sort_vector.h'.

There are no functions for sorting complex arrays or vectors, since the ordering of complex numbers is not uniquely defined. To sort a complex vector by magnitude compute a real vector containing the magnitudes of the complex elements, and sort this vector indirectly. The resulting index gives the appropriate ordering of the original complex vector.

void gsl_sort (double * data, size_t stride, size_t n)   Function

This function sorts the n elements of the array data with stride stride into ascending numerical order.

void gsl_sort_vector (gsl_vector * v)   Function

This function sorts the elements of the vector v into ascending numerical order.

void gsl_sort_index (size_t * p, const double * data, size_t   Function
        stride, size_t n)

This function indirectly sorts the n elements of the array data with stride stride into ascending order, storing the resulting permutation in p. The array p must be allocated with a sufficient length to store the n elements of the permutation. The elements of p give the index of the array element which would have been stored in that position if the array had been sorted in place. The array data is not changed.

int gsl_sort_vector_index (gsl_permutation * p, const   Function
    gsl_vector * v)

This function indirectly sorts the elements of the vector v into ascending order, storing the resulting permutation in p. The elements of p give the index of the vector element which would have been stored in that position if the vector had been sorted in place. The first element of p gives the index of the least element in v, and the last element of p gives the index of the greatest element in v. The vector v is not changed.

11.3 Selecting the k smallest or largest elements

The functions described in this section select the k smallest or largest elements of a data set of size N. The routines use an $O(kN)$ direct insertion algorithm which is suited to subsets that are small compared with the total size of the dataset. For example, the routines are useful for selecting the 10 largest values from one million data points, but not for selecting the largest 100,000 values. If the subset is a significant part of the total dataset it may be faster to sort all the elements of the dataset directly with an $O(N \log N)$ algorithm and obtain the smallest or largest values that way.

int gsl_sort_smallest (double * dest, size_t k, const double *   Function
    src, size_t stride, size_t n)

This function copies the k smallest elements of the array src, of size n and stride stride, in ascending numerical order into the array dest. The size k of the subset must be less than or equal to n. The data src is not modified by this operation.

int gsl_sort_largest (double * dest, size_t k, const double *   Function
    src, size_t stride, size_t n)

This function copies the k largest elements of the array src, of size n and stride stride, in descending numerical order into the array dest. k must be less than or equal to n. The data src is not modified by this operation.

int gsl_sort_vector_smallest (double * dest, size_t k, const   Function
    gsl_vector * v)

int gsl_sort_vector_largest (double * dest, size_t k, const   Function
    gsl_vector * v)

These functions copy the k smallest or largest elements of the vector v into the array dest. k must be less than or equal to the length of the vector v.

The following functions find the indices of the k smallest or largest elements of a dataset,

int gsl_sort_smallest_index (size_t * p, size_t k, const   Function
    double * src, size_t stride, size_t n)

This function stores the indices of the k smallest elements of the array src, of size n and stride stride, in the array p. The indices are chosen so that the corresponding data is in ascending numerical order. k must be less than or equal to n. The data src is not modified by this operation.

int gsl_sort_largest_index (size_t * p, size_t k, const double   Function
        * src, size_t stride, size_t n)

This function stores the indices of the k largest elements of the array src, of
size n and stride stride, in the array p. The indices are chosen so that the
corresponding data is in descending numerical order. k must be less than or
equal to n. The data src is not modified by this operation.

int gsl_sort_vector_smallest_index (size_t * p, size_t k,   Function
        const gsl_vector * v)
int gsl_sort_vector_largest_index (size_t * p, size_t k, const   Function
        gsl_vector * v)

These functions store the indices of the k smallest or largest elements of the
vector v in the array p. k must be less than or equal to the length of the
vector v.

11.4 Computing the rank

The rank of an element is its order in the sorted data. The rank is the inverse
of the index permutation, p. It can be computed using the following algorithm,

```
for (i = 0; i < p->size; i++)
{
    size_t pi = p->data[i];
    rank->data[pi] = i;
}
```

This can be computed directly from the function gsl_permutation_
inverse(rank,p).

The following function will print the rank of each element of the vector v,

```
void
print_rank (gsl_vector * v)
{
    size_t i;
    size_t n = v->size;
    gsl_permutation * perm = gsl_permutation_alloc(n);
    gsl_permutation * rank = gsl_permutation_alloc(n);

    gsl_sort_vector_index (perm, v);
    gsl_permutation_inverse (rank, perm);

    for (i = 0; i < n; i++)
     {
       double vi = gsl_vector_get(v, i);
       printf ("element = %d, value = %g, rank = %d\n",
               i, vi, rank->data[i]);
     }

    gsl_permutation_free (perm);
    gsl_permutation_free (rank);
}
```

11.5 Examples

The following example shows how to use the permutation p to print the elements of the vector v in ascending order,

```
gsl_sort_vector_index (p, v);

for (i = 0; i < v->size; i++)
{
    double vpi = gsl_vector_get (v, p->data[i]);
    printf ("order = %d, value = %g\n", i, vpi);
}
```

The next example uses the function `gsl_sort_smallest` to select the 5 smallest numbers from 100000 uniform random variates stored in an array,

```
#include <gsl/gsl_rng.h>
#include <gsl/gsl_sort_double.h>

int
main (void)
{
  const gsl_rng_type * T;
  gsl_rng * r;

  size_t i, k = 5, N = 100000;

  double * x = malloc (N * sizeof(double));
  double * small = malloc (k * sizeof(double));

  gsl_rng_env_setup();

  T = gsl_rng_default;
  r = gsl_rng_alloc (T);

  for (i = 0; i < N; i++)
    {
      x[i] = gsl_rng_uniform(r);
    }

  gsl_sort_smallest (small, k, x, 1, N);

  printf ("%d smallest values from %d\n", k, N);

  for (i = 0; i < k; i++)
    {
      printf ("%d: %.18f\n", i, small[i]);
    }

  free (x);
  free (small);
```

```
    gsl_rng_free (r);
    return 0;
}
```

The output lists the 5 smallest values, in ascending order,

```
$ ./a.out
5 smallest values from 100000
0: 0.000003489200025797
1: 0.000008199829608202
2: 0.000008953968062997
3: 0.000010712770745158
4: 0.000033531803637743
```

11.6 References and Further Reading

The subject of sorting is covered extensively in Knuth's *Sorting and Searching*,

> Donald E. Knuth, *The Art of Computer Programming: Sorting and Searching* (Vol 3, 3rd Ed, 1997), Addison-Wesley, ISBN 0201896850.

The Heapsort algorithm is described in the following book,

> Robert Sedgewick, *Algorithms in C*, Addison-Wesley, ISBN 0201514257.

12 BLAS Support

The Basic Linear Algebra Subprograms (BLAS) define a set of fundamental operations on vectors and matrices which can be used to create optimized higher-level linear algebra functionality.

The library provides a low-level layer which corresponds directly to the C-language BLAS standard, referred to here as "CBLAS", and a higher-level interface for operations on GSL vectors and matrices. Users who are interested in simple operations on GSL vector and matrix objects should use the high-level layer, which is declared in the file 'gsl_blas.h'. This should satisfy the needs of most users. Note that GSL matrices are implemented using dense-storage so the interface only includes the corresponding dense-storage BLAS functions. The full BLAS functionality for band-format and packed-format matrices is available through the low-level CBLAS interface.

The interface for the gsl_cblas layer is specified in the file 'gsl_cblas.h'. This interface corresponds to the BLAS Technical Forum's standard for the C interface to legacy BLAS implementations. Users who have access to other conforming CBLAS implementations can use these in place of the version provided by the library. Note that users who have only a Fortran BLAS library can use a CBLAS conformant wrapper to convert it into a CBLAS library. A reference CBLAS wrapper for legacy Fortran implementations exists as part of the CBLAS standard and can be obtained from Netlib. The complete set of CBLAS functions is listed in an appendix (see Appendix D [GSL CBLAS Library], page 503).

There are three levels of BLAS operations,

Level 1
> Vector operations, e.g. $y = \alpha x + y$

Level 2
> Matrix-vector operations, e.g. $y = \alpha A x + \beta y$

Level 3
> Matrix-matrix operations, e.g. $C = \alpha A B + C$

Each routine has a name which specifies the operation, the type of matrices involved and their precisions. Some of the most common operations and their names are given below,

DOT
> scalar product, $x^T y$

AXPY
> vector sum, $\alpha x + y$

MV
> matrix-vector product, Ax

SV
> matrix-vector solve, $inv(A)x$

MM

> matrix-matrix product, AB

SM

> matrix-matrix solve, $inv(A)B$

The types of matrices are,

GE

> general

GB

> general band

SY

> symmetric

SB

> symmetric band

SP

> symmetric packed

HE

> hermitian

HB

> hermitian band

HP

> hermitian packed

TR

> triangular

TB

> triangular band

TP

> triangular packed

Each operation is defined for four precisions,

S   single real

D   double real

C   single complex

Z   double complex

Thus, for example, the name SGEMM stands for "single-precision general matrix-matrix multiply" and ZGEMM stands for "double-precision complex matrix-matrix multiply".

Note that the vector and matrix arguments to BLAS functions must not be aliased, as the results are undefined when the underlying arrays overlap (see Section 2.11 [Aliasing of arrays], page 13).

12.1 GSL BLAS Interface

GSL provides dense vector and matrix objects, based on the relevant built-in types. The library provides an interface to the BLAS operations which apply to these objects. The interface to this functionality is given in the file 'gsl_blas.h'.

12.1.1 Level 1

int gsl_blas_sdsdot (float *alpha*, const gsl_vector_float * x,   Function
      const gsl_vector_float * y, float * *result*)
This function computes the sum $\alpha + x^T y$ for the vectors x and y, returning the result in *result*.

int gsl_blas_sdot (const gsl_vector_float * x, const   Function
      gsl_vector_float * y, float * *result*)
int gsl_blas_dsdot (const gsl_vector_float * x, const   Function
      gsl_vector_float * y, double * *result*)
int gsl_blas_ddot (const gsl_vector * x, const gsl_vector * y,   Function
      double * *result*)
These functions compute the scalar product $x^T y$ for the vectors x and y, returning the result in *result*.

int gsl_blas_cdotu (const gsl_vector_complex_float * x, const   Function
      gsl_vector_complex_float * y, gsl_complex_float * *dotu*)
int gsl_blas_zdotu (const gsl_vector_complex * x, const   Function
      gsl_vector_complex * y, gsl_complex * *dotu*)
These functions compute the complex scalar product $x^T y$ for the vectors x and y, returning the result in *result*

int gsl_blas_cdotc (const gsl_vector_complex_float * x, const   Function
      gsl_vector_complex_float * y, gsl_complex_float * *dotc*)
int gsl_blas_zdotc (const gsl_vector_complex * x, const   Function
      gsl_vector_complex * y, gsl_complex * *dotc*)
These functions compute the complex conjugate scalar product $x^H y$ for the vectors x and y, returning the result in *result*

float gsl_blas_snrm2 (const gsl_vector_float * x)   Function
double gsl_blas_dnrm2 (const gsl_vector * x)   Function
These functions compute the Euclidean norm $||x||_2 = \sqrt{\sum x_i^2}$ of the vector x.

float gsl_blas_scnrm2 (const gsl_vector_complex_float * x)   Function
double gsl_blas_dznrm2 (const gsl_vector_complex * x)   Function
These functions compute the Euclidean norm of the complex vector x,

$$||x||_2 = \sqrt{\sum (\mathrm{Re}(x_i)^2 + \mathrm{Im}(x_i)^2)}.$$

float gsl_blas_sasum (const gsl_vector_float * x)   Function
double gsl_blas_dasum (const gsl_vector * x)   Function
These functions compute the absolute sum $\sum |x_i|$ of the elements of the vector x.

float gsl_blas_scasum (const gsl_vector_complex_float * x)   Function
double gsl_blas_dzasum (const gsl_vector_complex * x)   Function
  These functions compute the sum of the magnitudes of the real and imaginary parts of the complex vector x, $\sum (|\text{Re}(x_i)| + |\text{Im}(x_i)|)$.

CBLAS_INDEX_t gsl_blas_isamax (const gsl_vector_float * x)   Function
CBLAS_INDEX_t gsl_blas_idamax (const gsl_vector * x)   Function
CBLAS_INDEX_t gsl_blas_icamax (const   Function
        gsl_vector_complex_float * x)
CBLAS_INDEX_t gsl_blas_izamax (const gsl_vector_complex * x)   Function
  These functions return the index of the largest element of the vector x. The largest element is determined by its absolute magnitude for real vectors and by the sum of the magnitudes of the real and imaginary parts $|\text{Re}(x_i)| + |\text{Im}(x_i)|$ for complex vectors. If the largest value occurs several times then the index of the first occurrence is returned.

int gsl_blas_sswap (gsl_vector_float * x, gsl_vector_float *   Function
        y)
int gsl_blas_dswap (gsl_vector * x, gsl_vector * y)   Function
int gsl_blas_cswap (gsl_vector_complex_float * x,   Function
        gsl_vector_complex_float * y)
int gsl_blas_zswap (gsl_vector_complex * x,   Function
        gsl_vector_complex * y)
  These functions exchange the elements of the vectors x and y.

int gsl_blas_scopy (const gsl_vector_float * x,   Function
        gsl_vector_float * y)
int gsl_blas_dcopy (const gsl_vector * x, gsl_vector * y)   Function
int gsl_blas_ccopy (const gsl_vector_complex_float * x,   Function
        gsl_vector_complex_float * y)
int gsl_blas_zcopy (const gsl_vector_complex * x,   Function
        gsl_vector_complex * y)
  These functions copy the elements of the vector x into the vector y.

int gsl_blas_saxpy (float alpha, const gsl_vector_float * x,   Function
        gsl_vector_float * y)
int gsl_blas_daxpy (double alpha, const gsl_vector * x,   Function
        gsl_vector * y)
int gsl_blas_caxpy (const gsl_complex_float alpha, const   Function
        gsl_vector_complex_float * x, gsl_vector_complex_float *
        y)
int gsl_blas_zaxpy (const gsl_complex alpha, const   Function
        gsl_vector_complex * x, gsl_vector_complex * y)
  These functions compute the sum $y = \alpha x + y$ for the vectors x and y.

void gsl_blas_sscal (float *alpha*, gsl_vector_float * *x*)   Function
void gsl_blas_dscal (double *alpha*, gsl_vector * *x*)   Function
void gsl_blas_cscal (const gsl_complex_float *alpha*,   Function
       gsl_vector_complex_float * *x*)
void gsl_blas_zscal (const gsl_complex *alpha*,   Function
       gsl_vector_complex * *x*)
void gsl_blas_csscal (float *alpha*, gsl_vector_complex_float *   Function
       x)
void gsl_blas_zdscal (double *alpha*, gsl_vector_complex * *x*)   Function
    These functions rescale the vector *x* by the multiplicative factor *alpha*.

int gsl_blas_srotg (float a[], float *b*[], float *c*[], float   Function
      s[])
int gsl_blas_drotg (double a[], double *b*[], double *c*[],   Function
      double *s*[])
    These functions compute a Givens rotation (c, s) which zeroes the vector
    (a, b),

$$\begin{pmatrix} c & s \\ -s & c \end{pmatrix} \begin{pmatrix} a \\ b \end{pmatrix} = \begin{pmatrix} r' \\ 0 \end{pmatrix}$$

    The variables *a* and *b* are overwritten by the routine.

int gsl_blas_srot (gsl_vector_float * *x*, gsl_vector_float * *y*,   Function
      float *c*, float *s*)
int gsl_blas_drot (gsl_vector * *x*, gsl_vector * *y*, const   Function
      double *c*, const double *s*)
    These functions apply a Givens rotation $(x', y') = (cx + sy, -sx + cy)$ to the
    vectors *x*, *y*.

int gsl_blas_srotmg (float *d1*[], float *d2*[], float *b1*[],   Function
      float *b2*, float *P*[])
int gsl_blas_drotmg (double *d1*[], double *d2*[], double *b1*[],   Function
      double *b2*, double *P*[])
    These functions compute a modified Givens transformation. The modified
    Givens transformation is defined in the original Level-1 BLAS specification,
    given in the references.

int gsl_blas_srotm (gsl_vector_float * *x*, gsl_vector_float *   Function
      y, const float *P*[])
int gsl_blas_drotm (gsl_vector * *x*, gsl_vector * *y*, const   Function
      double *P*[])
    These functions apply a modified Givens transformation.

12.1.2 Level 2

int gsl_blas_sgemv (CBLAS_TRANSPOSE_t *TransA*, float *alpha*,   Function
        const gsl_matrix_float * *A*, const gsl_vector_float * *x*,
        float *beta*, gsl_vector_float * *y*)

int gsl_blas_dgemv (CBLAS_TRANSPOSE_t *TransA*, double *alpha*,   Function
        const gsl_matrix * *A*, const gsl_vector * *x*, double *beta*,
        gsl_vector * *y*)

int gsl_blas_cgemv (CBLAS_TRANSPOSE_t *TransA*, const   Function
        gsl_complex_float *alpha*, const gsl_matrix_complex_float *
        A, const gsl_vector_complex_float * *x*, const
        gsl_complex_float *beta*, gsl_vector_complex_float * *y*)

int gsl_blas_zgemv (CBLAS_TRANSPOSE_t *TransA*, const   Function
        gsl_complex *alpha*, const gsl_matrix_complex * *A*, const
        gsl_vector_complex * *x*, const gsl_complex *beta*,
        gsl_vector_complex * *y*)

These functions compute the matrix-vector product and sum $y = \alpha op(A)x + \beta y$, where $op(A) = A$, A^T, A^H for *TransA* = CblasNoTrans, CblasTrans, CblasConjTrans.

int gsl_blas_strmv (CBLAS_UPLO_t *Uplo*, CBLAS_TRANSPOSE_t   Function
        TransA, CBLAS_DIAG_t *Diag*, const gsl_matrix_float * *A*,
        gsl_vector_float * *x*)

int gsl_blas_dtrmv (CBLAS_UPLO_t *Uplo*, CBLAS_TRANSPOSE_t   Function
        TransA, CBLAS_DIAG_t *Diag*, const gsl_matrix * *A*,
        gsl_vector * *x*)

int gsl_blas_ctrmv (CBLAS_UPLO_t *Uplo*, CBLAS_TRANSPOSE_t   Function
        TransA, CBLAS_DIAG_t *Diag*, const gsl_matrix_complex_float
        * *A*, gsl_vector_complex_float * *x*)

int gsl_blas_ztrmv (CBLAS_UPLO_t *Uplo*, CBLAS_TRANSPOSE_t   Function
        TransA, CBLAS_DIAG_t *Diag*, const gsl_matrix_complex * *A*,
        gsl_vector_complex * *x*)

These functions compute the matrix-vector product $x = op(A)x$ for the triangular matrix A, where $op(A) = A$, A^T, A^H for *TransA* = CblasNoTrans, CblasTrans, CblasConjTrans. When *Uplo* is CblasUpper then the upper triangle of A is used, and when *Uplo* is CblasLower then the lower triangle of A is used. If *Diag* is CblasNonUnit then the diagonal of the matrix is used, but if *Diag* is CblasUnit then the diagonal elements of the matrix A are taken as unity and are not referenced.

int gsl_blas_strsv (CBLAS_UPLO_t *Uplo*, CBLAS_TRANSPOSE_t   Function
     TransA, CBLAS_DIAG_t *Diag*, const gsl_matrix_float * *A*,
     gsl_vector_float * *x*)

int gsl_blas_dtrsv (CBLAS_UPLO_t *Uplo*, CBLAS_TRANSPOSE_t   Function
     TransA, CBLAS_DIAG_t *Diag*, const gsl_matrix * *A*,
     gsl_vector * *x*)

int gsl_blas_ctrsv (CBLAS_UPLO_t *Uplo*, CBLAS_TRANSPOSE_t   Function
     TransA, CBLAS_DIAG_t *Diag*, const gsl_matrix_complex_float
     * *A*, gsl_vector_complex_float * *x*)

int gsl_blas_ztrsv (CBLAS_UPLO_t *Uplo*, CBLAS_TRANSPOSE_t   Function
     TransA, CBLAS_DIAG_t *Diag*, const gsl_matrix_complex * *A*,
     gsl_vector_complex * *x*)

These functions compute $inv(op(A))x$ for x, where $op(A) = A$, A^T, A^H for *TransA* = CblasNoTrans, CblasTrans, CblasConjTrans. When *Uplo* is CblasUpper then the upper triangle of A is used, and when *Uplo* is CblasLower then the lower triangle of A is used. If *Diag* is CblasNonUnit then the diagonal of the matrix is used, but if *Diag* is CblasUnit then the diagonal elements of the matrix A are taken as unity and are not referenced.

int gsl_blas_ssymv (CBLAS_UPLO_t *Uplo*, float *alpha*, const   Function
     gsl_matrix_float * *A*, const gsl_vector_float * *x*, float
     beta, gsl_vector_float * *y*)

int gsl_blas_dsymv (CBLAS_UPLO_t *Uplo*, double *alpha*, const   Function
     gsl_matrix * *A*, const gsl_vector * *x*, double *beta*,
     gsl_vector * *y*)

These functions compute the matrix-vector product and sum $y = \alpha A x + \beta y$ for the symmetric matrix A. Since the matrix A is symmetric only its upper half or lower half need to be stored. When *Uplo* is CblasUpper then the upper triangle and diagonal of A are used, and when *Uplo* is CblasLower then the lower triangle and diagonal of A are used.

int gsl_blas_chemv (CBLAS_UPLO_t *Uplo*, const   Function
     gsl_complex_float *alpha*, const gsl_matrix_complex_float *
     A, const gsl_vector_complex_float * *x*, const
     gsl_complex_float *beta*, gsl_vector_complex_float * *y*)

int gsl_blas_zhemv (CBLAS_UPLO_t *Uplo*, const gsl_complex   Function
     alpha, const gsl_matrix_complex * *A*, const
     gsl_vector_complex * *x*, const gsl_complex *beta*,
     gsl_vector_complex * *y*)

These functions compute the matrix-vector product and sum $y = \alpha A x + \beta y$ for the hermitian matrix A. Since the matrix A is hermitian only its upper half or lower half need to be stored. When *Uplo* is CblasUpper then the upper triangle and diagonal of A are used, and when *Uplo* is CblasLower then the lower triangle and diagonal of A are used. The imaginary elements of the diagonal are automatically assumed to be zero and are not referenced.

```
int gsl_blas_sger (float alpha, const gsl_vector_float * x,          Function
          const gsl_vector_float * y, gsl_matrix_float * A)
int gsl_blas_dger (double alpha, const gsl_vector * x, const         Function
          gsl_vector * y, gsl_matrix * A)
int gsl_blas_cgeru (const gsl_complex_float alpha, const             Function
          gsl_vector_complex_float * x, const
          gsl_vector_complex_float * y, gsl_matrix_complex_float *
          A)
int gsl_blas_zgeru (const gsl_complex alpha, const                   Function
          gsl_vector_complex * x, const gsl_vector_complex * y,
          gsl_matrix_complex * A)
```
These functions compute the rank-1 update $A = \alpha x y^T + A$ of the matrix A.

```
int gsl_blas_cgerc (const gsl_complex_float alpha, const             Function
          gsl_vector_complex_float * x, const
          gsl_vector_complex_float * y, gsl_matrix_complex_float *
          A)
int gsl_blas_zgerc (const gsl_complex alpha, const                   Function
          gsl_vector_complex * x, const gsl_vector_complex * y,
          gsl_matrix_complex * A)
```
These functions compute the conjugate rank-1 update $A = \alpha x y^H + A$ of the matrix A.

```
int gsl_blas_ssyr (CBLAS_UPLO_t Uplo, float alpha, const            Function
          gsl_vector_float * x, gsl_matrix_float * A)
int gsl_blas_dsyr (CBLAS_UPLO_t Uplo, double alpha, const           Function
          gsl_vector * x, gsl_matrix * A)
```
These functions compute the symmetric rank-1 update $A = \alpha x x^T + A$ of the symmetric matrix A. Since the matrix A is symmetric only its upper half or lower half need to be stored. When Uplo is CblasUpper then the upper triangle and diagonal of A are used, and when Uplo is CblasLower then the lower triangle and diagonal of A are used.

```
int gsl_blas_cher (CBLAS_UPLO_t Uplo, float alpha, const           Function
          gsl_vector_complex_float * x, gsl_matrix_complex_float *
          A)
int gsl_blas_zher (CBLAS_UPLO_t Uplo, double alpha, const          Function
          gsl_vector_complex * x, gsl_matrix_complex * A)
```
These functions compute the hermitian rank-1 update $A = \alpha x x^H + A$ of the hermitian matrix A. Since the matrix A is hermitian only its upper half or lower half need to be stored. When Uplo is CblasUpper then the upper triangle and diagonal of A are used, and when Uplo is CblasLower then the lower triangle and diagonal of A are used. The imaginary elements of the diagonal are automatically set to zero.

int gsl_blas_ssyr2 (CBLAS_UPLO_t *Uplo*, float *alpha*, const   Function
      gsl_vector_float * *x*, const gsl_vector_float * *y*,
      gsl_matrix_float * *A*)

int gsl_blas_dsyr2 (CBLAS_UPLO_t *Uplo*, double *alpha*, const   Function
      gsl_vector * *x*, const gsl_vector * *y*, gsl_matrix * *A*)

These functions compute the symmetric rank-2 update $A = \alpha x y^T + \alpha y x^T + A$ of the symmetric matrix A. Since the matrix A is symmetric only its upper half or lower half need to be stored. When *Uplo* is CblasUpper then the upper triangle and diagonal of A are used, and when *Uplo* is CblasLower then the lower triangle and diagonal of A are used.

int gsl_blas_cher2 (CBLAS_UPLO_t *Uplo*, const   Function
      gsl_complex_float *alpha*, const gsl_vector_complex_float *
      x, const gsl_vector_complex_float * *y*,
      gsl_matrix_complex_float * *A*)

int gsl_blas_zher2 (CBLAS_UPLO_t *Uplo*, const gsl_complex   Function
      alpha, const gsl_vector_complex * *x*, const
      gsl_vector_complex * *y*, gsl_matrix_complex * *A*)

These functions compute the hermitian rank-2 update $A = \alpha x y^H + \alpha^* y x^H A$ of the hermitian matrix A. Since the matrix A is hermitian only its upper half or lower half need to be stored. When *Uplo* is CblasUpper then the upper triangle and diagonal of A are used, and when *Uplo* is CblasLower then the lower triangle and diagonal of A are used. The imaginary elements of the diagonal are automatically set to zero.

12.1.3 Level 3

int gsl_blas_sgemm (CBLAS_TRANSPOSE_t *TransA*,   Function
      CBLAS_TRANSPOSE_t *TransB*, float *alpha*, const
      gsl_matrix_float * *A*, const gsl_matrix_float * *B*, float
      beta, gsl_matrix_float * *C*)

int gsl_blas_dgemm (CBLAS_TRANSPOSE_t *TransA*,   Function
      CBLAS_TRANSPOSE_t *TransB*, double *alpha*, const gsl_matrix *
      A, const gsl_matrix * *B*, double *beta*, gsl_matrix * *C*)

int gsl_blas_cgemm (CBLAS_TRANSPOSE_t *TransA*,   Function
      CBLAS_TRANSPOSE_t *TransB*, const gsl_complex_float *alpha*,
      const gsl_matrix_complex_float * *A*, const
      gsl_matrix_complex_float * *B*, const gsl_complex_float
      beta, gsl_matrix_complex_float * *C*)

int gsl_blas_zgemm (CBLAS_TRANSPOSE_t *TransA*,   Function
      CBLAS_TRANSPOSE_t *TransB*, const gsl_complex *alpha*, const
      gsl_matrix_complex * *A*, const gsl_matrix_complex * *B*,
      const gsl_complex *beta*, gsl_matrix_complex * *C*)

These functions compute the general matrix-matrix product and sum $C = \alpha\, op(A)op(B) + \beta C$ where $op(A) = A$, A^T, A^H for *TransA* = CblasNoTrans, CblasTrans, CblasConjTrans and similarly for the parameter *TransB*.

int gsl_blas_ssymm (CBLAS_SIDE_t *Side*, CBLAS_UPLO_t *Uplo*,   Function
          float *alpha*, const gsl_matrix_float * *A*, const
          gsl_matrix_float * *B*, float *beta*, gsl_matrix_float * *C*)
int gsl_blas_dsymm (CBLAS_SIDE_t *Side*, CBLAS_UPLO_t *Uplo*,   Function
          double *alpha*, const gsl_matrix * *A*, const gsl_matrix * *B*,
          double *beta*, gsl_matrix * *C*)
int gsl_blas_csymm (CBLAS_SIDE_t *Side*, CBLAS_UPLO_t *Uplo*,   Function
          const gsl_complex_float *alpha*, const
          gsl_matrix_complex_float * *A*, const
          gsl_matrix_complex_float * *B*, const gsl_complex_float
          beta, gsl_matrix_complex_float * *C*)
int gsl_blas_zsymm (CBLAS_SIDE_t *Side*, CBLAS_UPLO_t *Uplo*,   Function
          const gsl_complex *alpha*, const gsl_matrix_complex * *A*,
          const gsl_matrix_complex * *B*, const gsl_complex *beta*,
          gsl_matrix_complex * *C*)

These functions compute the matrix-matrix product and sum $C = \alpha AB + \beta C$
for *Side* is CblasLeft and $C = \alpha BA + \beta C$ for *Side* is CblasRight, where the
matrix A is symmetric. When *Uplo* is CblasUpper then the upper triangle
and diagonal of A are used, and when *Uplo* is CblasLower then the lower
triangle and diagonal of A are used.

int gsl_blas_chemm (CBLAS_SIDE_t *Side*, CBLAS_UPLO_t *Uplo*,   Function
          const gsl_complex_float *alpha*, const
          gsl_matrix_complex_float * *A*, const
          gsl_matrix_complex_float * *B*, const gsl_complex_float
          beta, gsl_matrix_complex_float * *C*)
int gsl_blas_zhemm (CBLAS_SIDE_t *Side*, CBLAS_UPLO_t *Uplo*,   Function
          const gsl_complex *alpha*, const gsl_matrix_complex * *A*,
          const gsl_matrix_complex * *B*, const gsl_complex *beta*,
          gsl_matrix_complex * *C*)

These functions compute the matrix-matrix product and sum $C = \alpha AB + \beta C$
for *Side* is CblasLeft and $C = \alpha BA + \beta C$ for *Side* is CblasRight, where the
matrix A is hermitian. When *Uplo* is CblasUpper then the upper triangle
and diagonal of A are used, and when *Uplo* is CblasLower then the lower
triangle and diagonal of A are used. The imaginary elements of the diagonal
are automatically set to zero.

int gsl_blas_strmm (CBLAS_SIDE_t *Side*, CBLAS_UPLO_t *Uplo*,   Function
        CBLAS_TRANSPOSE_t *TransA*, CBLAS_DIAG_t *Diag*, float *alpha*,
        const gsl_matrix_float * *A*, gsl_matrix_float * *B*)

int gsl_blas_dtrmm (CBLAS_SIDE_t *Side*, CBLAS_UPLO_t *Uplo*,   Function
        CBLAS_TRANSPOSE_t *TransA*, CBLAS_DIAG_t *Diag*, double *alpha*,
        const gsl_matrix * *A*, gsl_matrix * *B*)

int gsl_blas_ctrmm (CBLAS_SIDE_t *Side*, CBLAS_UPLO_t *Uplo*,   Function
        CBLAS_TRANSPOSE_t *TransA*, CBLAS_DIAG_t *Diag*, const
        gsl_complex_float *alpha*, const gsl_matrix_complex_float *
        A, gsl_matrix_complex_float * *B*)

int gsl_blas_ztrmm (CBLAS_SIDE_t *Side*, CBLAS_UPLO_t *Uplo*,   Function
        CBLAS_TRANSPOSE_t *TransA*, CBLAS_DIAG_t *Diag*, const
        gsl_complex *alpha*, const gsl_matrix_complex * *A*,
        gsl_matrix_complex * *B*)

These functions compute the matrix-matrix product $B = \alpha op(A)B$ for *Side* is CblasLeft and $B = \alpha Bop(A)$ for *Side* is CblasRight. The matrix A is triangular and $op(A) = A$, A^T, A^H for *TransA* = CblasNoTrans, CblasTrans, CblasConjTrans. When *Uplo* is CblasUpper then the upper triangle of A is used, and when *Uplo* is CblasLower then the lower triangle of A is used. If *Diag* is CblasNonUnit then the diagonal of A is used, but if *Diag* is CblasUnit then the diagonal elements of the matrix A are taken as unity and are not referenced.

int gsl_blas_strsm (CBLAS_SIDE_t *Side*, CBLAS_UPLO_t *Uplo*,   Function
        CBLAS_TRANSPOSE_t *TransA*, CBLAS_DIAG_t *Diag*, float *alpha*,
        const gsl_matrix_float * *A*, gsl_matrix_float * *B*)

int gsl_blas_dtrsm (CBLAS_SIDE_t *Side*, CBLAS_UPLO_t *Uplo*,   Function
        CBLAS_TRANSPOSE_t *TransA*, CBLAS_DIAG_t *Diag*, double *alpha*,
        const gsl_matrix * *A*, gsl_matrix * *B*)

int gsl_blas_ctrsm (CBLAS_SIDE_t *Side*, CBLAS_UPLO_t *Uplo*,   Function
        CBLAS_TRANSPOSE_t *TransA*, CBLAS_DIAG_t *Diag*, const
        gsl_complex_float *alpha*, const gsl_matrix_complex_float *
        A, gsl_matrix_complex_float * *B*)

int gsl_blas_ztrsm (CBLAS_SIDE_t *Side*, CBLAS_UPLO_t *Uplo*,   Function
        CBLAS_TRANSPOSE_t *TransA*, CBLAS_DIAG_t *Diag*, const
        gsl_complex *alpha*, const gsl_matrix_complex * *A*,
        gsl_matrix_complex * *B*)

These functions compute the inverse-triangular-matrix matrix product $B = \alpha op(inv(A))B$ for *Side* is CblasLeft and $B = \alpha Bop(inv(A))$ for *Side* is CblasRight. The matrix A is triangular and $op(A) = A$, A^T, A^H for *TransA* = CblasNoTrans, CblasTrans, CblasConjTrans. When *Uplo* is CblasUpper then the upper triangle of A is used, and when *Uplo* is CblasLower then the lower triangle of A is used. If *Diag* is CblasNonUnit then the diagonal of A is used, but if *Diag* is CblasUnit then the diagonal elements of the matrix A are taken as unity and are not referenced.

int gsl_blas_ssyrk (CBLAS_UPLO_t *Uplo*, CBLAS_TRANSPOSE_t   Function
         Trans, float *alpha*, const gsl_matrix_float * *A*, float *beta*,
         gsl_matrix_float * *C*)

int gsl_blas_dsyrk (CBLAS_UPLO_t *Uplo*, CBLAS_TRANSPOSE_t   Function
         Trans, double *alpha*, const gsl_matrix * *A*, double *beta*,
         gsl_matrix * *C*)

int gsl_blas_csyrk (CBLAS_UPLO_t *Uplo*, CBLAS_TRANSPOSE_t   Function
         Trans, const gsl_complex_float *alpha*, const
         gsl_matrix_complex_float * *A*, const gsl_complex_float
         beta, gsl_matrix_complex_float * *C*)

int gsl_blas_zsyrk (CBLAS_UPLO_t *Uplo*, CBLAS_TRANSPOSE_t   Function
         Trans, const gsl_complex *alpha*, const gsl_matrix_complex *
         A, const gsl_complex *beta*, gsl_matrix_complex * *C*)

These functions compute a rank-k update of the symmetric matrix C, $C = \alpha A A^T + \beta C$ when *Trans* is CblasNoTrans and $C = \alpha A^T A + \beta C$ when *Trans* is CblasTrans. Since the matrix C is symmetric only its upper half or lower half need to be stored. When *Uplo* is CblasUpper then the upper triangle and diagonal of C are used, and when *Uplo* is CblasLower then the lower triangle and diagonal of C are used.

int gsl_blas_cherk (CBLAS_UPLO_t *Uplo*, CBLAS_TRANSPOSE_t   Function
         Trans, float *alpha*, const gsl_matrix_complex_float * *A*,
         float *beta*, gsl_matrix_complex_float * *C*)

int gsl_blas_zherk (CBLAS_UPLO_t *Uplo*, CBLAS_TRANSPOSE_t   Function
         Trans, double *alpha*, const gsl_matrix_complex * *A*, double
         beta, gsl_matrix_complex * *C*)

These functions compute a rank-k update of the hermitian matrix C, $C = \alpha A A^H + \beta C$ when *Trans* is CblasNoTrans and $C = \alpha A^H A + \beta C$ when *Trans* is CblasConjTrans. Since the matrix C is hermitian only its upper half or lower half need to be stored. When *Uplo* is CblasUpper then the upper triangle and diagonal of C are used, and when *Uplo* is CblasLower then the lower triangle and diagonal of C are used. The imaginary elements of the diagonal are automatically set to zero.

int gsl_blas_ssyr2k (CBLAS_UPLO_t *Uplo*, CBLAS_TRANSPOSE_t   Function
        Trans, float *alpha*, const gsl_matrix_float * *A*, const
        gsl_matrix_float * *B*, float *beta*, gsl_matrix_float * *C*)
int gsl_blas_dsyr2k (CBLAS_UPLO_t *Uplo*, CBLAS_TRANSPOSE_t   Function
        Trans, double *alpha*, const gsl_matrix * *A*, const gsl_matrix
        * *B*, double *beta*, gsl_matrix * *C*)
int gsl_blas_csyr2k (CBLAS_UPLO_t *Uplo*, CBLAS_TRANSPOSE_t   Function
        Trans, const gsl_complex_float *alpha*, const
        gsl_matrix_complex_float * *A*, const
        gsl_matrix_complex_float * *B*, const gsl_complex_float
        beta, gsl_matrix_complex_float * *C*)
int gsl_blas_zsyr2k (CBLAS_UPLO_t *Uplo*, CBLAS_TRANSPOSE_t   Function
        Trans, const gsl_complex *alpha*, const gsl_matrix_complex *
        A, const gsl_matrix_complex * *B*, const gsl_complex *beta*,
        gsl_matrix_complex * *C*)

These functions compute a rank-2k update of the symmetric matrix C, $C = \alpha A B^T + \alpha B A^T + \beta C$ when *Trans* is CblasNoTrans and $C = \alpha A^T B + \alpha B^T A + \beta C$ when *Trans* is CblasTrans. Since the matrix C is symmetric only its upper half or lower half need to be stored. When *Uplo* is CblasUpper then the upper triangle and diagonal of C are used, and when *Uplo* is CblasLower then the lower triangle and diagonal of C are used.

int gsl_blas_cher2k (CBLAS_UPLO_t *Uplo*, CBLAS_TRANSPOSE_t   Function
        Trans, const gsl_complex_float *alpha*, const
        gsl_matrix_complex_float * *A*, const
        gsl_matrix_complex_float * *B*, float *beta*,
        gsl_matrix_complex_float * *C*)
int gsl_blas_zher2k (CBLAS_UPLO_t *Uplo*, CBLAS_TRANSPOSE_t   Function
        Trans, const gsl_complex *alpha*, const gsl_matrix_complex *
        A, const gsl_matrix_complex * *B*, double *beta*,
        gsl_matrix_complex * *C*)

These functions compute a rank-2k update of the hermitian matrix C, $C = \alpha A B^H + \alpha^* B A^H + \beta C$ when *Trans* is CblasNoTrans and $C = \alpha A^H B + \alpha^* B^H A + \beta C$ when *Trans* is CblasConjTrans. Since the matrix C is hermitian only its upper half or lower half need to be stored. When *Uplo* is CblasUpper then the upper triangle and diagonal of C are used, and when *Uplo* is CblasLower then the lower triangle and diagonal of C are used. The imaginary elements of the diagonal are automatically set to zero.

12.2 Examples

The following program computes the product of two matrices using the Level-3 BLAS function DGEMM,

$$\begin{pmatrix} 0.11 & 0.12 & 0.13 \\ 0.21 & 0.22 & 0.23 \end{pmatrix} \begin{pmatrix} 1011 & 1012 \\ 1021 & 1022 \\ 1031 & 1031 \end{pmatrix} = \begin{pmatrix} 367.76 & 368.12 \\ 674.06 & 674.72 \end{pmatrix}$$

The matrices are stored in row major order, according to the C convention for arrays.

```
#include <stdio.h>
#include <gsl/gsl_blas.h>

int
main (void)
{
  double a[] = { 0.11, 0.12, 0.13,
                 0.21, 0.22, 0.23 };

  double b[] = { 1011, 1012,
                 1021, 1022,
                 1031, 1032 };

  double c[] = { 0.00, 0.00,
                 0.00, 0.00 };

  gsl_matrix_view A = gsl_matrix_view_array(a, 2, 3);
  gsl_matrix_view B = gsl_matrix_view_array(b, 3, 2);
  gsl_matrix_view C = gsl_matrix_view_array(c, 2, 2);

  /* Compute C = A B */

  gsl_blas_dgemm (CblasNoTrans, CblasNoTrans,
                  1.0, &A.matrix, &B.matrix,
                  0.0, &C.matrix);

  printf ("[ %g, %g\n", c[0], c[1]);
  printf ("  %g, %g ]\n", c[2], c[3]);

  return 0;
}
```

Here is the output from the program,

```
$ ./a.out
[ 367.76, 368.12
  674.06, 674.72 ]
```

12.3 References and Further Reading

Information on the BLAS standards, including both the legacy and updated interface standards, is available online from the BLAS Homepage and BLAS Technical Forum web-site.

BLAS Homepage
http://www.netlib.org/blas/

BLAS Technical Forum
http://www.netlib.org/blas/blast-forum/

The following papers contain the specifications for Level 1, Level 2 and Level 3 BLAS.

C. Lawson, R. Hanson, D. Kincaid, F. Krogh, "Basic Linear Algebra Subprograms for Fortran Usage", *ACM Transactions on Mathematical Software*, Vol. 5 (1979), Pages 308–325.

J.J. Dongarra, J. DuCroz, S. Hammarling, R. Hanson, "An Extended Set of Fortran Basic Linear Algebra Subprograms", *ACM Transactions on Mathematical Software*, Vol. 14, No. 1 (1988), Pages 1–32.

J.J. Dongarra, I. Duff, J. DuCroz, S. Hammarling, "A Set of Level 3 Basic Linear Algebra Subprograms", *ACM Transactions on Mathematical Software*, Vol. 16 (1990), Pages 1–28.

Postscript versions of the latter two papers are available from the website http://www.netlib.org/blas/. A CBLAS wrapper for Fortran BLAS libraries is available from the same location.

13 Linear Algebra

This chapter describes functions for solving linear systems. The library provides linear algebra operations which operate directly on the `gsl_vector` and `gsl_matrix` objects. These routines use the standard algorithms from Golub & Van Loan's *Matrix Computations* with Level-1 and Level-2 BLAS calls for efficiency.

The functions described in this chapter are declared in the header file 'gsl_linalg.h'.

13.1 LU Decomposition

A general square matrix A has an LU decomposition into upper and lower triangular matrices,

$$PA = LU$$

where P is a permutation matrix, L is unit lower triangular matrix and U is upper triangular matrix. For square matrices this decomposition can be used to convert the linear system $Ax = b$ into a pair of triangular systems ($Ly = Pb$, $Ux = y$), which can be solved by forward and back-substitution. Note that the LU decomposition is valid for singular matrices.

int gsl_linalg_LU_decomp (gsl_matrix * A, gsl_permutation *   Function
       p, int * *signum*)
int gsl_linalg_complex_LU_decomp (gsl_matrix_complex * A,   Function
       gsl_permutation * p, int * *signum*)

These functions factorize the square matrix A into the LU decomposition $PA = LU$. On output the diagonal and upper triangular part of the input matrix A contain the matrix U. The lower triangular part of the input matrix (excluding the diagonal) contains L. The diagonal elements of L are unity, and are not stored.

The permutation matrix P is encoded in the permutation p. The j-th column of the matrix P is given by the k-th column of the identity matrix, where $k = p_j$ the j-th element of the permutation vector. The sign of the permutation is given by *signum*. It has the value $(-1)^n$, where n is the number of interchanges in the permutation.

The algorithm used in the decomposition is Gaussian Elimination with partial pivoting (Golub & Van Loan, *Matrix Computations*, Algorithm 3.4.1).

int gsl_linalg_LU_solve (const gsl_matrix * LU, const   Function
       gsl_permutation * p, const gsl_vector * b, gsl_vector * x)
int gsl_linalg_complex_LU_solve (const gsl_matrix_complex *   Function
       LU, const gsl_permutation * p, const gsl_vector_complex *
       b, gsl_vector_complex * x)

These functions solve the square system $Ax = b$ using the LU decomposition of A into (LU, p) given by `gsl_linalg_LU_decomp` or `gsl_linalg_complex_LU_decomp` as input.

int gsl_linalg_LU_svx (const gsl_matrix * *LU*, const   Function
        gsl_permutation * p, gsl_vector * x)
int gsl_linalg_complex_LU_svx (const gsl_matrix_complex *   Function
        LU, const gsl_permutation * p, gsl_vector_complex * x)
These functions solve the square system $Ax = b$ in-place using the precomputed *LU* decomposition of A into (*LU*,p). On input x should contain the
right-hand side b, which is replaced by the solution on output.

int gsl_linalg_LU_refine (const gsl_matrix * *A*, const   Function
        gsl_matrix * *LU*, const gsl_permutation * p, const
        gsl_vector * b, gsl_vector * x, gsl_vector * *residual*)
int gsl_linalg_complex_LU_refine (const gsl_matrix_complex *   Function
        A, const gsl_matrix_complex * *LU*, const gsl_permutation *
        p, const gsl_vector_complex * b, gsl_vector_complex * x,
        gsl_vector_complex * *residual*)
These functions apply an iterative improvement to x, the solution of $Ax = b$,
from the precomputed *LU* decomposition of A into (*LU*,p). The initial
residual $r = Ax - b$ is also computed and stored in *residual*.

int gsl_linalg_LU_invert (const gsl_matrix * *LU*, const   Function
        gsl_permutation * p, gsl_matrix * *inverse*)
int gsl_linalg_complex_LU_invert (const gsl_matrix_complex *   Function
        LU, const gsl_permutation * p, gsl_matrix_complex *
        inverse)
These functions compute the inverse of a matrix A from its *LU* decomposition (*LU*,p), storing the result in the matrix *inverse*. The inverse is
computed by solving the system $Ax = b$ for each column of the identity
matrix. It is preferable to avoid direct use of the inverse whenever possible,
as the linear solver functions can obtain the same result more efficiently and
reliably (consult any introductory textbook on numerical linear algebra for
details).

double gsl_linalg_LU_det (gsl_matrix * *LU*, int *signum*)   Function
gsl_complex gsl_linalg_complex_LU_det (gsl_matrix_complex *   Function
        LU, int *signum*)
These functions compute the determinant of a matrix A from its *LU* decomposition, *LU*. The determinant is computed as the product of the diagonal
elements of U and the sign of the row permutation *signum*.

double gsl_linalg_LU_lndet (gsl_matrix * *LU*)   Function
double gsl_linalg_complex_LU_lndet (gsl_matrix_complex * *LU*)   Function
These functions compute the logarithm of the absolute value of the deter-
minant of a matrix A, $\ln|\det(A)|$, from its *LU* decomposition, *LU*. This
function may be useful if the direct computation of the determinant would
overflow or underflow.

int gsl_linalg_LU_sgndet (gsl_matrix * *LU*, int *signum*)   Function
gsl_complex gsl_linalg_complex_LU_sgndet (gsl_matrix_complex   Function
        * *LU*, int *signum*)
   These functions compute the sign or phase factor of the determinant of a
   matrix A, $\det(A)/|\det(A)|$, from its LU decomposition, LU.

13.2 QR Decomposition

   A general rectangular M-by-N matrix A has a QR decomposition into the
product of an orthogonal M-by-M square matrix Q (where $Q^TQ = I$) and an
M-by-N right-triangular matrix R,

$$A = QR$$

This decomposition can be used to convert the linear system $Ax = b$ into the
triangular system $Rx = Q^Tb$, which can be solved by back-substitution. Another use of the QR decomposition is to compute an orthonormal basis for a set
of vectors. The first N columns of Q form an orthonormal basis for the range
of A, $ran(A)$, when A has full column rank.

int gsl_linalg_QR_decomp (gsl_matrix * *A*, gsl_vector * *tau*)   Function
   This function factorizes the M-by-N matrix A into the QR decomposition
   $A = QR$. On output the diagonal and upper triangular part of the input
   matrix contain the matrix R. The vector *tau* and the columns of the lower
   triangular part of the matrix A contain the Householder coefficients and
   Householder vectors which encode the orthogonal matrix Q. The vector *tau*
   must be of length $k = \min(M, N)$. The matrix Q is related to these components by, $Q = Q_k...Q_2Q_1$ where $Q_i = I - \tau_iv_iv_i^T$ and v_i is the Householder
   vector $v_i = (0, ..., 1, A(i + 1, i), A(i + 2, i), ..., A(m, i))$. This is the same
   storage scheme as used by LAPACK.

   The algorithm used to perform the decomposition is Householder QR (Golub
   & Van Loan, *Matrix Computations*, Algorithm 5.2.1).

int gsl_linalg_QR_solve (const gsl_matrix * *QR*, const   Function
        gsl_vector * *tau*, const gsl_vector * *b*, gsl_vector * *x*)
   This function solves the square system $Ax = b$ using the QR decomposition
   of A held in (QR, tau) which must have been computed previously with
   gsl_linalg_QR_decomp. The least-squares solution for rectangular systems
   can be found using gsl_linalg_QR_lssolve.

int gsl_linalg_QR_svx (const gsl_matrix * *QR*, const   Function
        gsl_vector * *tau*, gsl_vector * *x*)
   This function solves the square system $Ax = b$ in-place using the QR decomposition of A held in (QR,tau) which must have been computed previously
   by gsl_linalg_QR_decomp. On input x should contain the right-hand side
   b, which is replaced by the solution on output.

int gsl_linalg_QR_lssolve (const gsl_matrix * QR, const   *Function*
          gsl_vector * *tau*, const gsl_vector * b, gsl_vector * x,
          gsl_vector * *residual*)

This function finds the least squares solution to the overdetermined system
$Ax = b$ where the matrix A has more rows than columns. The least squares
solution minimizes the Euclidean norm of the residual, $\|Ax - b\|$. The routine
requires as input the QR decomposition of A into (QR, tau) given by gsl_
linalg_QR_decomp. The solution is returned in x. The residual is computed
as a by-product and stored in *residual*.

int gsl_linalg_QR_QTvec (const gsl_matrix * QR, const   *Function*
          gsl_vector * *tau*, gsl_vector * v)

This function applies the matrix Q^T encoded in the decomposition (QR, tau)
to the vector v, storing the result $Q^T v$ in v. The matrix multiplication is
carried out directly using the encoding of the Householder vectors without
needing to form the full matrix Q^T.

int gsl_linalg_QR_Qvec (const gsl_matrix * QR, const   *Function*
          gsl_vector * *tau*, gsl_vector * v)

This function applies the matrix Q encoded in the decomposition (QR, tau)
to the vector v, storing the result Qv in v. The matrix multiplication is
carried out directly using the encoding of the Householder vectors without
needing to form the full matrix Q.

int gsl_linalg_QR_QTmat (const gsl_matrix * QR, const   *Function*
          gsl_vector * *tau*, gsl_matrix * A)

This function applies the matrix Q^T encoded in the decomposition (QR, tau)
to the matrix A, storing the result $Q^T A$ in A. The matrix multiplication is
carried out directly using the encoding of the Householder vectors without
needing to form the full matrix Q^T.

int gsl_linalg_QR_Rsolve (const gsl_matrix * QR, const   *Function*
          gsl_vector * b, gsl_vector * x)

This function solves the triangular system $Rx = b$ for x. It may be useful
if the product $b' = Q^T b$ has already been computed using gsl_linalg_QR_
QTvec.

int gsl_linalg_QR_Rsvx (const gsl_matrix * QR, gsl_vector *   *Function*
          x)

This function solves the triangular system $Rx = b$ for x in-place. On input
x should contain the right-hand side b and is replaced by the solution on
output. This function may be useful if the product $b' = Q^T b$ has already
been computed using gsl_linalg_QR_QTvec.

int gsl_linalg_QR_unpack (const gsl_matrix * QR, const   *Function*
          gsl_vector * *tau*, gsl_matrix * Q, gsl_matrix * R)

This function unpacks the encoded QR decomposition (QR, tau) into the
matrices Q and R, where Q is M-by-M and R is M-by-N.

int gsl_linalg_QR_QRsolve (gsl_matrix * Q, gsl_matrix * R,   Function
      const gsl_vector * b, gsl_vector * x)

This function solves the system $Rx = Q^T b$ for x. It can be used when the QR decomposition of a matrix is available in unpacked form as (Q, R).

int gsl_linalg_QR_update (gsl_matrix * Q, gsl_matrix * R,   Function
      gsl_vector * w, const gsl_vector * v)

This function performs a rank-1 update wv^T of the QR decomposition (Q, R). The update is given by $Q'R' = Q(R + wv^T)$ where the output matrices Q' and R' are also orthogonal and right triangular. Note that w is destroyed by the update.

int gsl_linalg_R_solve (const gsl_matrix * R, const   Function
      gsl_vector * b, gsl_vector * x)

This function solves the triangular system $Rx = b$ for the N-by-N matrix R.

int gsl_linalg_R_svx (const gsl_matrix * R, gsl_vector * x)   Function

This function solves the triangular system $Rx = b$ in-place. On input x should contain the right-hand side b, which is replaced by the solution on output.

13.3 QR Decomposition with Column Pivoting

The QR decomposition can be extended to the rank deficient case by introducing a column permutation P,

$$AP = QR$$

The first r columns of Q form an orthonormal basis for the range of A for a matrix with column rank r. This decomposition can also be used to convert the linear system $Ax = b$ into the triangular system $Ry = Q^T b, x = Py$, which can be solved by back-substitution and permutation. We denote the QR decomposition with column pivoting by QRP^T since $A = QRP^T$.

int gsl_linalg_QRPT_decomp (gsl_matrix * A, gsl_vector * tau,   Function
      gsl_permutation * p, int * signum, gsl_vector * norm)

This function factorizes the M-by-N matrix A into the QRP^T decomposition $A = QRP^T$. On output the diagonal and upper triangular part of the input matrix contain the matrix R. The permutation matrix P is stored in the permutation p. The sign of the permutation is given by signum. It has the value $(-1)^n$, where n is the number of interchanges in the permutation. The vector tau and the columns of the lower triangular part of the matrix A contain the Householder coefficients and vectors which encode the orthogonal matrix Q. The vector tau must be of length $k = \min(M, N)$. The matrix Q is related to these components by, $Q = Q_k...Q_2 Q_1$ where $Q_i = I - \tau_i v_i v_i^T$ and v_i is the Householder vector $v_i = (0, ..., 1, A(i + 1, i), A(i + 2, i), ..., A(m, i))$. This is the same storage scheme as used by LAPACK. The vector norm is a workspace of length N used for column pivoting.

The algorithm used to perform the decomposition is Householder QR with column pivoting (Golub & Van Loan, *Matrix Computations*, Algorithm 5.4.1).

int gsl_linalg_QRPT_decomp2 (const gsl_matrix * A, gsl_matrix   *Function*
      * q, gsl_matrix * r, gsl_vector * tau, gsl_permutation * p,
      int * $signum$, gsl_vector * $norm$)
This function factorizes the matrix A into the decomposition $A = QRP^T$ without modifying A itself and storing the output in the separate matrices q and r.

int gsl_linalg_QRPT_solve (const gsl_matrix * QR, const   *Function*
           gsl_vector * tau, const gsl_permutation * p, const
           gsl_vector * b, gsl_vector * x)
This function solves the square system $Ax = b$ using the QRP^T decomposition of A held in (QR, tau, p) which must have been computed previously by gsl_linalg_QRPT_decomp.

int gsl_linalg_QRPT_svx (const gsl_matrix * QR, const   *Function*
           gsl_vector * tau, const gsl_permutation * p, gsl_vector * x)
This function solves the square system $Ax = b$ in-place using the QRP^T decomposition of A held in (QR,tau,p). On input x should contain the right-hand side b, which is replaced by the solution on output.

int gsl_linalg_QRPT_QRsolve (const gsl_matrix * Q, const   *Function*
           gsl_matrix * R, const gsl_permutation * p, const gsl_vector
           * b, gsl_vector * x)
This function solves the square system $RP^T x = Q^T b$ for x. It can be used when the QR decomposition of a matrix is available in unpacked form as (Q, R).

int gsl_linalg_QRPT_update (gsl_matrix * Q, gsl_matrix * R,   *Function*
           const gsl_permutation * p, gsl_vector * w, const gsl_vector
           * v)
This function performs a rank-1 update wv^T of the QRP^T decomposition (Q, R, p). The update is given by $Q'R' = Q(R + wv^T P)$ where the output matrices Q' and R' are also orthogonal and right triangular. Note that w is destroyed by the update. The permutation p is not changed.

int gsl_linalg_QRPT_Rsolve (const gsl_matrix * QR, const   *Function*
           gsl_permutation * p, const gsl_vector * b, gsl_vector * x)
This function solves the triangular system $RP^T x = b$ for the N-by-N matrix R contained in QR.

int gsl_linalg_QRPT_Rsvx (const gsl_matrix * QR, const   *Function*
           gsl_permutation * p, gsl_vector * x)
This function solves the triangular system $RP^T x = b$ in-place for the N-by-N matrix R contained in QR. On input x should contain the right-hand side b, which is replaced by the solution on output.

13.4 Singular Value Decomposition

A general rectangular M-by-N matrix A has a singular value decomposition (SVD) into the product of an M-by-N orthogonal matrix U, an N-by-N diagonal matrix of singular values S and the transpose of an N-by-N orthogonal square matrix V,

$$A = USV^T$$

The singular values $\sigma_i = S_{ii}$ are all non-negative and are generally chosen to form a non-increasing sequence $\sigma_1 \geq \sigma_2 \geq ... \geq \sigma_N \geq 0$.

The singular value decomposition of a matrix has many practical uses. The condition number of the matrix is given by the ratio of the largest singular value to the smallest singular value. The presence of a zero singular value indicates that the matrix is singular. The number of non-zero singular values indicates the rank of the matrix. In practice singular value decomposition of a rank-deficient matrix will not produce exact zeroes for singular values, due to finite numerical precision. Small singular values should be edited by choosing a suitable tolerance.

For a rank-deficient matrix, the null space of A is given by the columns of V corresponding to the zero singular values. Similarly, the range of A is given by columns of U corresponding to the non-zero singular values.

Note that the routines here compute the "thin" version of the SVD with U as M-by-N orthogonal matrix. This allows in-place computation and is the most commonly-used form in practice. Mathematically, the "full" SVD is defined with U as an M-by-M orthogonal matrix and S as an M-by-N diagonal matrix (with additional rows of zeros).

int gsl_linalg_SV_decomp (gsl_matrix * A, gsl_matrix * V,   Function
        gsl_vector * S, gsl_vector * work)

    This function factorizes the M-by-N matrix A into the singular value decomposition $A = USV^T$ for $M \geq N$. On output the matrix A is replaced by U. The diagonal elements of the singular value matrix S are stored in the vector S. The singular values are non-negative and form a non-increasing sequence from S_1 to S_N. The matrix V contains the elements of V in un-transposed form. To form the product USV^T it is necessary to take the transpose of V. A workspace of length N is required in *work*.

    This routine uses the Golub-Reinsch SVD algorithm.

int gsl_linalg_SV_decomp_mod (gsl_matrix * A, gsl_matrix * X,   Function
        gsl_matrix * V, gsl_vector * S, gsl_vector * work)

    This function computes the SVD using the modified Golub-Reinsch algorithm, which is faster for $M \gg N$. It requires the vector *work* of length N and the N-by-N matrix X as additional working space.

int gsl_linalg_SV_decomp_jacobi (gsl_matrix * A, gsl_matrix *   Function
          V, gsl_vector * S)
    This function computes the SVD of the M-by-N matrix A using one-sided
    Jacobi orthogonalization for $M \geq N$. The Jacobi method can compute
    singular values to higher relative accuracy than Golub-Reinsch algorithms
    (see references for details).

int gsl_linalg_SV_solve (gsl_matrix * U, gsl_matrix * V,   Function
          gsl_vector * S, const gsl_vector * b, gsl_vector * x)
    This function solves the system $Ax = b$ using the singular value decompo-
    sition (U, S, V) held in A which must have been computed previously by
    gsl_linalg_SV_decomp.

    Only non-zero singular values are used in computing the solution. The parts
    of the solution corresponding to singular values of zero are ignored. Other
    singular values can be edited out by setting them to zero before calling this
    function.

    In the over-determined case where A has more rows than columns the system
    is solved in the least squares sense, returning the solution x which minimizes
    $||Ax - b||_2$.

13.5 Cholesky Decomposition

A symmetric, positive definite square matrix A has a Cholesky decomposition
into a product of a lower triangular matrix L and its transpose L^T,

$$A = LL^T$$

This is sometimes referred to as taking the square-root of a matrix. The
Cholesky decomposition can only be carried out when all the eigenvalues of
the matrix are positive. This decomposition can be used to convert the linear
system $Ax = b$ into a pair of triangular systems $(Ly = b, L^T x = y)$, which can
be solved by forward and back-substitution.

int gsl_linalg_cholesky_decomp (gsl_matrix * A)   Function
int gsl_linalg_complex_cholesky_decomp (gsl_matrix_complex *   Function
          A)
    These functions factorize the symmetric, positive-definite square matrix A
    into the Cholesky decomposition $A = LL^T$ (or $A = LL^\dagger$ for the complex
    case). On input, the values from the diagonal and lower-triangular part of
    the matrix A are used (the upper triangular part is ignored). On output the
    diagonal and lower triangular part of the input matrix A contain the matrix
    L, while the upper triangular part of the input matrix is overwritten with
    L^T (the diagonal terms being identical for both L and L^T). If the matrix
    is not positive-definite then the decomposition will fail, returning the error
    code GSL_EDOM.

    When testing whether a matrix is positive-definite, disable the error handler
    first to avoid triggering an error.

int gsl_linalg_cholesky_solve (const gsl_matrix * *cholesky*,   Function
        const gsl_vector * *b*, gsl_vector * *x*)

int gsl_linalg_complex_cholesky_solve (const   Function
        gsl_matrix_complex * *cholesky*, const gsl_vector_complex *
        b, gsl_vector_complex * *x*)

These functions solve the system $Ax = b$ using the Cholesky decomposition of A held in the matrix *cholesky* which must have been previously computed by gsl_linalg_cholesky_decomp or gsl_linalg_complex_cholesky_decomp.

int gsl_linalg_cholesky_svx (const gsl_matrix * *cholesky*,   Function
        gsl_vector * *x*)

int gsl_linalg_complex_cholesky_svx (const   Function
        gsl_matrix_complex * *cholesky*, gsl_vector_complex * *x*)

These functions solve the system $Ax = b$ in-place using the Cholesky decomposition of A held in the matrix *cholesky* which must have been previously computed by by gsl_linalg_cholesky_decomp or gsl_linalg_complex_cholesky_decomp. On input x should contain the right-hand side b, which is replaced by the solution on output.

int gsl_linalg_cholesky_invert (gsl_matrix * *cholesky*)   Function

This function computes the inverse of the matrix *cholesky* which must have been previously computed by gsl_linalg_cholesky_decomp. The inverse of the original matrix A is stored in *cholesky* on output.

13.6 Tridiagonal Decomposition of Real Symmetric Matrices

A symmetric matrix A can be factorized by similarity transformations into the form,

$$A = QTQ^T$$

where Q is an orthogonal matrix and T is a symmetric tridiagonal matrix.

int gsl_linalg_symmtd_decomp (gsl_matrix * *A*, gsl_vector *   Function
        tau)

This function factorizes the symmetric square matrix A into the symmetric tridiagonal decomposition QTQ^T. On output the diagonal and subdiagonal part of the input matrix A contain the tridiagonal matrix T. The remaining lower triangular part of the input matrix contains the Householder vectors which, together with the Householder coefficients *tau*, encode the orthogonal matrix Q. This storage scheme is the same as used by LAPACK. The upper triangular part of A is not referenced.

int gsl_linalg_symmtd_unpack (const gsl_matrix * A, const   Function
        gsl_vector * tau, gsl_matrix * Q, gsl_vector * diag,
        gsl_vector * subdiag)
    This function unpacks the encoded symmetric tridiagonal decomposition (A,
    tau) obtained from gsl_linalg_symmtd_decomp into the orthogonal matrix
    Q, the vector of diagonal elements diag and the vector of subdiagonal ele-
    ments subdiag.

int gsl_linalg_symmtd_unpack_T (const gsl_matrix * A,   Function
        gsl_vector * diag, gsl_vector * subdiag)
    This function unpacks the diagonal and subdiagonal of the encoded symmet-
    ric tridiagonal decomposition (A, tau) obtained from gsl_linalg_symmtd_
    decomp into the vectors diag and subdiag.

13.7 Tridiagonal Decomposition of Hermitian Matrices

A hermitian matrix A can be factorized by similarity transformations into
the form,

$$A = UTU^T$$

where U is a unitary matrix and T is a real symmetric tridiagonal matrix.

int gsl_linalg_hermtd_decomp (gsl_matrix_complex * A,   Function
        gsl_vector_complex * tau)
    This function factorizes the hermitian matrix A into the symmetric tridiag-
    onal decomposition UTU^T. On output the real parts of the diagonal and
    subdiagonal part of the input matrix A contain the tridiagonal matrix T.
    The remaining lower triangular part of the input matrix contains the House-
    holder vectors which, together with the Householder coefficients tau, encode
    the orthogonal matrix Q. This storage scheme is the same as used by LA-
    PACK. The upper triangular part of A and imaginary parts of the diagonal
    are not referenced.

int gsl_linalg_hermtd_unpack (const gsl_matrix_complex * A,   Function
        const gsl_vector_complex * tau, gsl_matrix_complex * Q,
        gsl_vector * diag, gsl_vector * subdiag)
    This function unpacks the encoded tridiagonal decomposition (A, tau) ob-
    tained from gsl_linalg_hermtd_decomp into the unitary matrix U, the real
    vector of diagonal elements diag and the real vector of subdiagonal elements
    subdiag.

int gsl_linalg_hermtd_unpack_T (const gsl_matrix_complex *   Function
        A, gsl_vector * diag, gsl_vector * subdiag)
    This function unpacks the diagonal and subdiagonal of the encoded tridiag-
    onal decomposition (A, tau) obtained from the gsl_linalg_hermtd_decomp
    into the real vectors diag and subdiag.

13.8 Hessenberg Decomposition of Real Matrices

A general real matrix A can be decomposed by orthogonal similarity transformations into the form

$$A = UHU^T$$

where U is orthogonal and H is an upper Hessenberg matrix, meaning that it has zeros below the first subdiagonal. The Hessenberg reduction is the first step in the Schur decomposition for the nonsymmetric eigenvalue problem, but has applications in other areas as well.

int gsl_linalg_hessenberg_decomp (gsl_matrix * A, gsl_vector   Function
        * tau)

> This function computes the Hessenberg decomposition of the matrix A by applying the similarity transformation $H = U^T A U$. On output, H is stored in the upper portion of A. The information required to construct the matrix U is stored in the lower triangular portion of A. U is a product of $N - 2$ Householder matrices. The Householder vectors are stored in the lower portion of A (below the subdiagonal) and the Householder coefficients are stored in the vector tau. tau must be of length N.

int gsl_linalg_hessenberg_unpack (gsl_matrix * H, gsl_vector   Function
        * tau, gsl_matrix * U)

> This function constructs the orthogonal matrix U from the information stored in the Hessenberg matrix H along with the vector tau. H and tau are outputs from gsl_linalg_hessenberg_decomp.

int gsl_linalg_hessenberg_unpack_accum (gsl_matrix * H,   Function
        gsl_vector * tau, gsl_matrix * V)

> This function is similar to gsl_linalg_hessenberg_unpack, except it accumulates the matrix U into V, so that $V' = VU$. The matrix V must be initialized prior to calling this function. Setting V to the identity matrix provides the same result as gsl_linalg_hessenberg_unpack. If H is order N, then V must have N columns but may have any number of rows.

int gsl_linalg_hessenberg_set_zero (gsl_matrix * H)   Function

> This function sets the lower triangular portion of H, below the subdiagonal, to zero. It is useful for clearing out the Householder vectors after calling gsl_linalg_hessenberg_decomp.

13.9 Hessenberg-Triangular Decomposition of Real Matrices

A general real matrix pair (A, B) can be decomposed by orthogonal similarity transformations into the form

$$A = UHV^T$$

$$B = URV^T$$

where U and V are orthogonal, H is an upper Hessenberg matrix, and R is upper triangular. The Hessenberg-Triangular reduction is the first step in the generalized Schur decomposition for the generalized eigenvalue problem.

int gsl_linalg_hesstri_decomp (gsl_matrix * A, gsl_matrix *   Function
        B, gsl_matrix * U, gsl_matrix * V, gsl_vector * work)

This function computes the Hessenberg-Triangular decomposition of the matrix pair (A, B). On output, H is stored in A, and R is stored in B. If U and V are provided (they may be null), the similarity transformations are stored in them. Additional workspace of length N is needed in work.

13.10 Bidiagonalization

A general matrix A can be factorized by similarity transformations into the form,

$$A = UBV^T$$

where U and V are orthogonal matrices and B is a N-by-N bidiagonal matrix with non-zero entries only on the diagonal and superdiagonal. The size of U is M-by-N and the size of V is N-by-N.

int gsl_linalg_bidiag_decomp (gsl_matrix * A, gsl_vector *   Function
        tau_U, gsl_vector * tau_V)

This function factorizes the M-by-N matrix A into bidiagonal form UBV^T. The diagonal and superdiagonal of the matrix B are stored in the diagonal and superdiagonal of A. The orthogonal matrices U and V are stored as compressed Householder vectors in the remaining elements of A. The Householder coefficients are stored in the vectors tau_U and tau_V. The length of tau_U must equal the number of elements in the diagonal of A and the length of tau_V should be one element shorter.

int gsl_linalg_bidiag_unpack (const gsl_matrix * A, const   Function
        gsl_vector * tau_U, gsl_matrix * U, const gsl_vector *
        tau_V, gsl_matrix * V, gsl_vector * diag, gsl_vector *
        superdiag)

This function unpacks the bidiagonal decomposition of A produced by gsl_linalg_bidiag_decomp, (A, tau_U, tau_V) into the separate orthogonal matrices U, V and the diagonal vector diag and superdiagonal superdiag. Note that U is stored as a compact M-by-N orthogonal matrix satisfying $U^T U = I$ for efficiency.

int gsl_linalg_bidiag_unpack2 (gsl_matrix * A, gsl_vector *   Function
        tau_U, gsl_vector * tau_V, gsl_matrix * V)
This function unpacks the bidiagonal decomposition of A produced by gsl_
linalg_bidiag_decomp, (A, tau_U, tau_V) into the separate orthogonal matrices U, V and the diagonal vector diag and superdiagonal superdiag. The
matrix U is stored in-place in A.

int gsl_linalg_bidiag_unpack_B (const gsl_matrix * A,   Function
        gsl_vector * diag, gsl_vector * superdiag)
This function unpacks the diagonal and superdiagonal of the bidiagonal decomposition of A from gsl_linalg_bidiag_decomp, into the diagonal vector
diag and superdiagonal vector superdiag.

13.11 Householder Transformations

A Householder transformation is a rank-1 modification of the identity matrix
which can be used to zero out selected elements of a vector. A Householder
matrix P takes the form,

$$P = I - \tau v v^T$$

where v is a vector (called the *Householder vector*) and $\tau = 2/(v^T v)$. The
functions described in this section use the rank-1 structure of the Householder
matrix to create and apply Householder transformations efficiently.

double gsl_linalg_householder_transform (gsl_vector * v)   Function
gsl_complex gsl_linalg_complex_householder_transform   Function
        (gsl_vector_complex * v)
This function prepares a Householder transformation $P = I - \tau v v^T$ which
can be used to zero all the elements of the input vector except the first.
On output the transformation is stored in the vector v and the scalar τ is
returned.

int gsl_linalg_householder_hm (double tau, const gsl_vector *   Function
        v, gsl_matrix * A)
int gsl_linalg_complex_householder_hm (gsl_complex tau, const   Function
        gsl_vector_complex * v, gsl_matrix_complex * A)
This function applies the Householder matrix P defined by the scalar *tau*
and the vector *v* to the left-hand side of the matrix A. On output the result
PA is stored in A.

int gsl_linalg_householder_mh (double tau, const gsl_vector *   Function
        v, gsl_matrix * A)
int gsl_linalg_complex_householder_mh (gsl_complex tau, const   Function
        gsl_vector_complex * v, gsl_matrix_complex * A)
This function applies the Householder matrix P defined by the scalar *tau*
and the vector *v* to the right-hand side of the matrix A. On output the
result AP is stored in A.

int gsl_linalg_householder_hv (double tau, const gsl_vector *   Function
        v, gsl_vector * w)
int gsl_linalg_complex_householder_hv (gsl_complex tau, const   Function
        gsl_vector_complex * v, gsl_vector_complex * w)
This function applies the Householder transformation P defined by the scalar
tau and the vector v to the vector w. On output the result Pw is stored in
w.

13.12 Householder solver for linear systems

int gsl_linalg_HH_solve (gsl_matrix * A, const gsl_vector *   Function
        b, gsl_vector * x)
This function solves the system $Ax = b$ directly using Householder transformations. On output the solution is stored in x and b is not modified. The
matrix A is destroyed by the Householder transformations.

int gsl_linalg_HH_svx (gsl_matrix * A, gsl_vector * x)   Function
This function solves the system $Ax = b$ in-place using Householder transformations. On input x should contain the right-hand side b, which is replaced
by the solution on output. The matrix A is destroyed by the Householder
transformations.

13.13 Tridiagonal Systems

The functions described in this section efficiently solve symmetric, nonsymmetric and cyclic tridiagonal systems with minimal storage. Note that the
current implementations of these functions use a variant of Cholesky decomposition, so the tridiagonal matrix must be positive definite. For non-positive
definite matrices, the functions return the error code GSL_ESING.

int gsl_linalg_solve_tridiag (const gsl_vector * diag, const   Function
        gsl_vector * e, const gsl_vector * f, const gsl_vector * b,
        gsl_vector * x)
This function solves the general N-by-N system $Ax = b$ where A is tridiagonal ($N \geq 2$). The super-diagonal and sub-diagonal vectors e and f must
be one element shorter than the diagonal vector $diag$. The form of A for the
4-by-4 case is shown below,

$$A = \begin{pmatrix} d_0 & e_0 & 0 & 0 \\ f_0 & d_1 & e_1 & 0 \\ 0 & f_1 & d_2 & e_2 \\ 0 & 0 & f_2 & d_3 \end{pmatrix}$$

int gsl_linalg_solve_symm_tridiag (const gsl_vector * diag,   Function
        const gsl_vector * e, const gsl_vector * b, gsl_vector * x)
This function solves the general N-by-N system $Ax = b$ where A is symmetric tridiagonal ($N \geq 2$). The off-diagonal vector e must be one element

shorter than the diagonal vector *diag*. The form of A for the 4-by-4 case is shown below,

$$A = \begin{pmatrix} d_0 & e_0 & 0 & 0 \\ e_0 & d_1 & e_1 & 0 \\ 0 & e_1 & d_2 & e_2 \\ 0 & 0 & e_2 & d_3 \end{pmatrix}$$

int gsl_linalg_solve_cyc_tridiag (const gsl_vector * *diag*,   Function
        const gsl_vector * e, const gsl_vector * f, const gsl_vector
        * b, gsl_vector * x)

This function solves the general N-by-N system $Ax = b$ where A is cyclic tridiagonal ($N \geq 3$). The cyclic super-diagonal and sub-diagonal vectors e and f must have the same number of elements as the diagonal vector *diag*. The form of A for the 4-by-4 case is shown below,

$$A = \begin{pmatrix} d_0 & e_0 & 0 & f_3 \\ f_0 & d_1 & e_1 & 0 \\ 0 & f_1 & d_2 & e_2 \\ e_3 & 0 & f_2 & d_3 \end{pmatrix}$$

int gsl_linalg_solve_symm_cyc_tridiag (const gsl_vector *   Function
        diag, const gsl_vector * e, const gsl_vector * b, gsl_vector
        * x)

This function solves the general N-by-N system $Ax = b$ where A is symmetric cyclic tridiagonal ($N \geq 3$). The cyclic off-diagonal vector e must have the same number of elements as the diagonal vector *diag*. The form of A for the 4-by-4 case is shown below,

$$A = \begin{pmatrix} d_0 & e_0 & 0 & e_3 \\ e_0 & d_1 & e_1 & 0 \\ 0 & e_1 & d_2 & e_2 \\ e_3 & 0 & e_2 & d_3 \end{pmatrix}$$

13.14 Balancing

The process of balancing a matrix applies similarity transformations to make the rows and columns have comparable norms. This is useful, for example, to reduce roundoff errors in the solution of eigenvalue problems. Balancing a matrix A consists of replacing A with a similar matrix

$$A' = D^{-1}AD$$

where D is a diagonal matrix whose entries are powers of the floating point radix.

int gsl_linalg_balance_matrix (gsl_matrix * A, gsl_vector *   Function
        D)

This function replaces the matrix A with its balanced counterpart and stores the diagonal elements of the similarity transformation into the vector D.

13.15 Examples

The following program solves the linear system $Ax = b$. The system to be solved is,

$$\begin{pmatrix} 0.18 & 0.60 & 0.57 & 0.96 \\ 0.41 & 0.24 & 0.99 & 0.58 \\ 0.14 & 0.30 & 0.97 & 0.66 \\ 0.51 & 0.13 & 0.19 & 0.85 \end{pmatrix} \begin{pmatrix} x_0 \\ x_1 \\ x_2 \\ x_3 \end{pmatrix} = \begin{pmatrix} 1.0 \\ 2.0 \\ 3.0 \\ 4.0 \end{pmatrix}$$

and the solution is found using LU decomposition of the matrix A.

```
#include <stdio.h>
#include <gsl/gsl_linalg.h>

int
main (void)
{
  double a_data[] = { 0.18, 0.60, 0.57, 0.96,
                      0.41, 0.24, 0.99, 0.58,
                      0.14, 0.30, 0.97, 0.66,
                      0.51, 0.13, 0.19, 0.85 };

  double b_data[] = { 1.0, 2.0, 3.0, 4.0 };

  gsl_matrix_view m
    = gsl_matrix_view_array (a_data, 4, 4);

  gsl_vector_view b
    = gsl_vector_view_array (b_data, 4);

  gsl_vector *x = gsl_vector_alloc (4);

  int s;

  gsl_permutation * p = gsl_permutation_alloc (4);

  gsl_linalg_LU_decomp (&m.matrix, p, &s);

  gsl_linalg_LU_solve (&m.matrix, p, &b.vector, x);

  printf ("x = \n");
  gsl_vector_fprintf (stdout, x, "%g");

  gsl_permutation_free (p);
  gsl_vector_free (x);
  return 0;
}
```

Here is the output from the program,

```
x = -4.05205
-12.6056
1.66091
8.69377
```

This can be verified by multiplying the solution x by the original matrix A using GNU OCTAVE,

```
octave> A = [ 0.18, 0.60, 0.57, 0.96;
              0.41, 0.24, 0.99, 0.58;
              0.14, 0.30, 0.97, 0.66;
              0.51, 0.13, 0.19, 0.85 ];

octave> x = [ -4.05205; -12.6056; 1.66091; 8.69377];

octave> A * x
ans =
   1.0000
   2.0000
   3.0000
   4.0000
```

This reproduces the original right-hand side vector, b, in accordance with the equation $Ax = b$.

13.16 References and Further Reading

Further information on the algorithms described in this section can be found in the following book,

G. H. Golub, C. F. Van Loan, *Matrix Computations* (3rd Ed, 1996), Johns Hopkins University Press, ISBN 0-8018-5414-8.

The LAPACK library is described in the following manual,

LAPACK Users' Guide (Third Edition, 1999), Published by SIAM, ISBN 0-89871-447-8.

http://www.netlib.org/lapack

The LAPACK source code can be found at the website above, along with an online copy of the users guide.

The Modified Golub-Reinsch algorithm is described in the following paper,

T.F. Chan, "An Improved Algorithm for Computing the Singular Value Decomposition", *ACM Transactions on Mathematical Software*, 8 (1982), pp 72–83.

The Jacobi algorithm for singular value decomposition is described in the following papers,

J.C. Nash, "A one-sided transformation method for the singular value decomposition and algebraic eigenproblem", *Computer Journal*, Volume 18, Number 1 (1975), p 74–76

J.C. Nash and S. Shlien "Simple algorithms for the partial singular value decomposition", *Computer Journal*, Volume 30 (1987), p 268–275.

James Demmel, Kresimir Veselic, "Jacobi's Method is more accurate than QR", *Lapack Working Note 15* (LAWN-15), October 1989. Available from netlib, `http://www.netlib.org/lapack/` in the `lawns` or `lawnspdf` directories.

14 Eigensystems

This chapter describes functions for computing eigenvalues and eigenvectors of matrices. There are routines for real symmetric, real nonsymmetric, complex hermitian, real generalized symmetric-definite, complex generalized hermitian-definite, and real generalized nonsymmetric eigensystems. Eigenvalues can be computed with or without eigenvectors. The hermitian and real symmetric matrix algorithms are symmetric bidiagonalization followed by QR reduction. The nonsymmetric algorithm is the Francis QR double-shift. The generalized nonsymmetric algorithm is the QZ method due to Moler and Stewart.

The functions described in this chapter are declared in the header file 'gsl_eigen.h'.

14.1 Real Symmetric Matrices

For real symmetric matrices, the library uses the symmetric bidiagonalization and QR reduction method. This is described in Golub & van Loan, section 8.3. The computed eigenvalues are accurate to an absolute accuracy of $\epsilon||A||_2$, where ϵ is the machine precision.

gsl_eigen_symm_workspace * gsl_eigen_symm_alloc (const   *Function*
     size_t n)
     This function allocates a workspace for computing eigenvalues of n-by-n real symmetric matrices. The size of the workspace is $O(2n)$.

void gsl_eigen_symm_free (gsl_eigen_symm_workspace * w)   *Function*
     This function frees the memory associated with the workspace w.

int gsl_eigen_symm (gsl_matrix * A, gsl_vector * eval,   *Function*
     gsl_eigen_symm_workspace * w)
     This function computes the eigenvalues of the real symmetric matrix A. Additional workspace of the appropriate size must be provided in w. The diagonal and lower triangular part of A are destroyed during the computation, but the strict upper triangular part is not referenced. The eigenvalues are stored in the vector eval and are unordered.

gsl_eigen_symmv_workspace * gsl_eigen_symmv_alloc (const   *Function*
     size_t n)
     This function allocates a workspace for computing eigenvalues and eigenvectors of n-by-n real symmetric matrices. The size of the workspace is $O(4n)$.

void gsl_eigen_symmv_free (gsl_eigen_symmv_workspace * w)   *Function*
     This function frees the memory associated with the workspace w.

int gsl_eigen_symmv (gsl_matrix * A, gsl_vector * *eval*,   Function
         gsl_matrix * *evec*, gsl_eigen_symmv_workspace * *w*)
   This function computes the eigenvalues and eigenvectors of the real symmetric matrix A. Additional workspace of the appropriate size must be provided
   in *w*. The diagonal and lower triangular part of A are destroyed during the
   computation, but the strict upper triangular part is not referenced. The
   eigenvalues are stored in the vector *eval* and are unordered. The corresponding eigenvectors are stored in the columns of the matrix *evec*. For example,
   the eigenvector in the first column corresponds to the first eigenvalue. The
   eigenvectors are guaranteed to be mutually orthogonal and normalised to
   unit magnitude.

14.2 Complex Hermitian Matrices

   For hermitian matrices, the library uses the complex form of the symmetric
bidiagonalization and QR reduction method.

gsl_eigen_herm_workspace * gsl_eigen_herm_alloc (const   Function
         size_t *n*)
   This function allocates a workspace for computing eigenvalues of n-by-n
   complex hermitian matrices. The size of the workspace is $O(3n)$.

void gsl_eigen_herm_free (gsl_eigen_herm_workspace * *w*)   Function
   This function frees the memory associated with the workspace *w*.

int gsl_eigen_herm (gsl_matrix_complex * A, gsl_vector * *eval*,   Function
         gsl_eigen_herm_workspace * *w*)
   This function computes the eigenvalues of the complex hermitian matrix A.
   Additional workspace of the appropriate size must be provided in *w*. The
   diagonal and lower triangular part of A are destroyed during the computation, but the strict upper triangular part is not referenced. The imaginary
   parts of the diagonal are assumed to be zero and are not referenced. The
   eigenvalues are stored in the vector *eval* and are unordered.

gsl_eigen_hermv_workspace * gsl_eigen_hermv_alloc (const   Function
         size_t *n*)
   This function allocates a workspace for computing eigenvalues and eigenvectors of n-by-n complex hermitian matrices. The size of the workspace is
   $O(5n)$.

void gsl_eigen_hermv_free (gsl_eigen_hermv_workspace * *w*)   Function
   This function frees the memory associated with the workspace *w*.

int gsl_eigen_hermv (gsl_matrix_complex * A, gsl_vector *   Function
         eval, gsl_matrix_complex * *evec*, gsl_eigen_hermv_workspace
         * *w*)
   This function computes the eigenvalues and eigenvectors of the complex hermitian matrix A. Additional workspace of the appropriate size must be provided in *w*. The diagonal and lower triangular part of A are destroyed during
   the computation, but the strict upper triangular part is not referenced. The

imaginary parts of the diagonal are assumed to be zero and are not referenced. The eigenvalues are stored in the vector *eval* and are unordered. The
corresponding complex eigenvectors are stored in the columns of the matrix
evec. For example, the eigenvector in the first column corresponds to the
first eigenvalue. The eigenvectors are guaranteed to be mutually orthogonal
and normalised to unit magnitude.

14.3 Real Nonsymmetric Matrices

The solution of the real nonsymmetric eigensystem problem for a matrix A
involves computing the Schur decomposition

$$A = ZTZ^T$$

where Z is an orthogonal matrix of Schur vectors and T, the Schur form, is quasi
upper triangular with diagonal 1-by-1 blocks which are real eigenvalues of A,
and diagonal 2-by-2 blocks whose eigenvalues are complex conjugate eigenvalues
of A. The algorithm used is the double-shift Francis method.

gsl_eigen_nonsymm_workspace * gsl_eigen_nonsymm_alloc (const   Function
       size_t n)
    This function allocates a workspace for computing eigenvalues of n-by-n real
    nonsymmetric matrices. The size of the workspace is $O(2n)$.

void gsl_eigen_nonsymm_free (gsl_eigen_nonsymm_workspace *   Function
       w)
    This function frees the memory associated with the workspace w.

void gsl_eigen_nonsymm_params (const int *compute_t*, const int   Function
       balance, gsl_eigen_nonsymm_workspace * w)
    This function sets some parameters which determine how the eigenvalue
    problem is solved in subsequent calls to gsl_eigen_nonsymm.

    If *compute_t* is set to 1, the full Schur form T will be computed by gsl_
    eigen_nonsymm. If it is set to 0, T will not be computed (this is the default
    setting). Computing the full Schur form T requires approximately 1.5–2
    times the number of flops.

    If *balance* is set to 1, a balancing transformation is applied to the matrix
    prior to computing eigenvalues. This transformation is designed to make
    the rows and columns of the matrix have comparable norms, and can re-
    sult in more accurate eigenvalues for matrices whose entries vary widely in
    magnitude. See Section 13.14 [Balancing], page 167 for more information.
    Note that the balancing transformation does not preserve the orthogonality
    of the Schur vectors, so if you wish to compute the Schur vectors with gsl_
    eigen_nonsymm_Z you will obtain the Schur vectors of the balanced matrix
    instead of the original matrix. The relationship will be

$$T = Q^t D^{-1} A D Q$$

where Q is the matrix of Schur vectors for the balanced matrix, and D is the balancing transformation. Then `gsl_eigen_nonsymm_Z` will compute a matrix Z which satisfies

$$T = Z^{-1} A Z$$

with $Z = DQ$. Note that Z will not be orthogonal. For this reason, balancing is not performed by default.

int **gsl_eigen_nonsymm** (gsl_matrix * A, gsl_vector_complex *   Function
        eval, gsl_eigen_nonsymm_workspace * *w*)
This function computes the eigenvalues of the real nonsymmetric matrix A and stores them in the vector *eval*. If T is desired, it is stored in the upper portion of A on output. Otherwise, on output, the diagonal of A will contain the 1-by-1 real eigenvalues and 2-by-2 complex conjugate eigenvalue systems, and the rest of A is destroyed. In rare cases, this function may fail to find all eigenvalues. If this happens, an error code is returned and the number of converged eigenvalues is stored in `w->n_evals`. The converged eigenvalues are stored in the beginning of *eval*.

int **gsl_eigen_nonsymm_Z** (gsl_matrix * A, gsl_vector_complex *   Function
        eval, gsl_matrix * Z, gsl_eigen_nonsymm_workspace * *w*)
This function is identical to `gsl_eigen_nonsymm` except that it also computes the Schur vectors and stores them into Z.

gsl_eigen_nonsymmv_workspace * **gsl_eigen_nonsymmv_alloc**   Function
        (const size_t *n*)
This function allocates a workspace for computing eigenvalues and eigenvectors of n-by-n real nonsymmetric matrices. The size of the workspace is $O(5n)$.

void **gsl_eigen_nonsymmv_free** (gsl_eigen_nonsymmv_workspace *   Function
        w)
This function frees the memory associated with the workspace *w*.

int **gsl_eigen_nonsymmv** (gsl_matrix * A, gsl_vector_complex *   Function
        eval, gsl_matrix_complex * *evec*,
        gsl_eigen_nonsymmv_workspace * *w*)
This function computes eigenvalues and right eigenvectors of the n-by-n real nonsymmetric matrix A. It first calls `gsl_eigen_nonsymm` to compute the eigenvalues, Schur form T, and Schur vectors. Then it finds eigenvectors of T and backtransforms them using the Schur vectors. The Schur vectors are destroyed in the process, but can be saved by using `gsl_eigen_nonsymmv_Z`. The computed eigenvectors are normalized to have unit magnitude. On output, the upper portion of A contains the Schur form T. If `gsl_eigen_nonsymm` fails, no eigenvectors are computed, and an error code is returned.

int **gsl_eigen_nonsymmv_Z** (gsl_matrix * A, gsl_vector_complex   Function
        * *eval*, gsl_matrix_complex * *evec*, gsl_matrix * Z,
        gsl_eigen_nonsymmv_workspace * *w*)
This function is identical to `gsl_eigen_nonsymmv` except that it also saves the Schur vectors into Z.

14.4 Real Generalized Symmetric-Definite Eigensystems

The real generalized symmetric-definite eigenvalue problem is to find eigenvalues λ and eigenvectors x such that

$$Ax = \lambda Bx$$

where A and B are symmetric matrices, and B is positive-definite. This problem reduces to the standard symmetric eigenvalue problem by applying the Cholesky decomposition to B:

$$Ax = \lambda Bx$$

$$Ax = \lambda LL^t x$$

$$\left(L^{-1}AL^{-t}\right) L^t x = \lambda L^t x$$

Therefore, the problem becomes $Cy = \lambda y$ where $C = L^{-1}AL^{-t}$ is symmetric, and $y = L^t x$. The standard symmetric eigensolver can be applied to the matrix C. The resulting eigenvectors are backtransformed to find the vectors of the original problem. The eigenvalues and eigenvectors of the generalized symmetric-definite eigenproblem are always real.

gsl_eigen_gensymm_workspace * gsl_eigen_gensymm_alloc (const   Function
    size_t n)
    This function allocates a workspace for computing eigenvalues of n-by-n real generalized symmetric-definite eigensystems. The size of the workspace is $O(2n)$.

void gsl_eigen_gensymm_free (gsl_eigen_gensymm_workspace *   Function
    w)
    This function frees the memory associated with the workspace w.

int gsl_eigen_gensymm (gsl_matrix * A, gsl_matrix * B,   Function
    gsl_vector * eval, gsl_eigen_gensymm_workspace * w)
    This function computes the eigenvalues of the real generalized symmetric-definite matrix pair (A, B), and stores them in eval, using the method outlined above. On output, B contains its Cholesky decomposition and A is destroyed.

gsl_eigen_gensymmv_workspace * gsl_eigen_gensymmv_alloc   Function
    (const size_t n)
    This function allocates a workspace for computing eigenvalues and eigenvectors of n-by-n real generalized symmetric-definite eigensystems. The size of the workspace is $O(4n)$.

void gsl_eigen_gensymmv_free (gsl_eigen_gensymmv_workspace *   Function
    w)
    This function frees the memory associated with the workspace w.

int gsl_eigen_gensymmv (gsl_matrix * A, gsl_matrix * B,   Function
        gsl_vector * eval, gsl_matrix * evec,
        gsl_eigen_gensymmv_workspace * w)
    This function computes the eigenvalues and eigenvectors of the real gener-
    alized symmetric-definite matrix pair (A, B), and stores them in eval and
    evec respectively. The computed eigenvectors are normalized to have unit
    magnitude. On output, B contains its Cholesky decomposition and A is
    destroyed.

14.5 Complex Generalized Hermitian-Definite Eigensystems

The complex generalized hermitian-definite eigenvalue problem is to find
eigenvalues λ and eigenvectors x such that

$$Ax = \lambda Bx$$

where A and B are hermitian matrices, and B is positive-definite. Similarly
to the real case, this can be reduced to $Cy = \lambda y$ where $C = L^{-1}AL^{-\dagger}$ is
hermitian, and $y = L^{\dagger}x$. The standard hermitian eigensolver can be applied to
the matrix C. The resulting eigenvectors are backtransformed to find the vectors
of the original problem. The eigenvalues of the generalized hermitian-definite
eigenproblem are always real.

gsl_eigen_genherm_workspace * gsl_eigen_genherm_alloc (const   Function
        size_t n)
    This function allocates a workspace for computing eigenvalues of n-by-n com-
    plex generalized hermitian-definite eigensystems. The size of the workspace
    is $O(3n)$.

void gsl_eigen_genherm_free (gsl_eigen_genherm_workspace *   Function
        w)
    This function frees the memory associated with the workspace w.

int gsl_eigen_genherm (gsl_matrix_complex * A,   Function
        gsl_matrix_complex * B, gsl_vector * eval,
        gsl_eigen_genherm_workspace * w)
    This function computes the eigenvalues of the complex generalized
    hermitian-definite matrix pair (A, B), and stores them in eval, using the
    method outlined above. On output, B contains its Cholesky decomposition
    and A is destroyed.

gsl_eigen_genhermv_workspace * gsl_eigen_genhermv_alloc   Function
        (const size_t n)
    This function allocates a workspace for computing eigenvalues and eigen-
    vectors of n-by-n complex generalized hermitian-definite eigensystems. The
    size of the workspace is $O(5n)$.

void gsl_eigen_genhermv_free (gsl_eigen_genhermv_workspace *   Function
        w)
    This function frees the memory associated with the workspace w.

```
int gsl_eigen_genhermv (gsl_matrix_complex * A,                    Function
        gsl_matrix_complex * B, gsl_vector * eval,
        gsl_matrix_complex * evec, gsl_eigen_genhermv_workspace *
        w)
```
This function computes the eigenvalues and eigenvectors of the complex generalized hermitian-definite matrix pair (A, B), and stores them in $eval$ and $evec$ respectively. The computed eigenvectors are normalized to have unit magnitude. On output, B contains its Cholesky decomposition and A is destroyed.

14.6 Real Generalized Nonsymmetric Eigensystems

Given two square matrices (A, B), the generalized nonsymmetric eigenvalue problem is to find eigenvalues λ and eigenvectors x such that

$$Ax = \lambda Bx$$

We may also define the problem as finding eigenvalues μ and eigenvectors y such that

$$\mu Ay = By$$

Note that these two problems are equivalent (with $\lambda = 1/\mu$) if neither λ nor μ is zero. If say, λ is zero, then it is still a well defined eigenproblem, but its alternate problem involving μ is not. Therefore, to allow for zero (and infinite) eigenvalues, the problem which is actually solved is

$$\beta Ax = \alpha Bx$$

The eigensolver routines below will return two values α and β and leave it to the user to perform the divisions $\lambda = \alpha/\beta$ and $\mu = \beta/\alpha$.

If the determinant of the matrix pencil $A - \lambda B$ is zero for all λ, the problem is said to be singular; otherwise it is called regular. Singularity normally leads to some $\alpha = \beta = 0$ which means the eigenproblem is ill-conditioned and generally does not have well defined eigenvalue solutions. The routines below are intended for regular matrix pencils and could yield unpredictable results when applied to singular pencils.

The solution of the real generalized nonsymmetric eigensystem problem for a matrix pair (A, B) involves computing the generalized Schur decomposition

$$A = QSZ^T$$

$$B = QTZ^T$$

where Q and Z are orthogonal matrices of left and right Schur vectors respectively, and (S, T) is the generalized Schur form whose diagonal elements give the α and β values. The algorithm used is the QZ method due to Moler and Stewart (see references).

gsl_eigen_gen_workspace * gsl_eigen_gen_alloc (const size_t   *Function*
          n)

This function allocates a workspace for computing eigenvalues of n-by-n real generalized nonsymmetric eigensystems. The size of the workspace is $O(n)$.

void gsl_eigen_gen_free (gsl_eigen_gen_workspace * *w*)   *Function*

This function frees the memory associated with the workspace *w*.

void gsl_eigen_gen_params (const int *compute_s*, const int   *Function*
          compute_t, const int *balance*, gsl_eigen_gen_workspace * *w*)

This function sets some parameters which determine how the eigenvalue problem is solved in subsequent calls to gsl_eigen_gen.

If *compute_s* is set to 1, the full Schur form S will be computed by gsl_eigen_gen. If it is set to 0, S will not be computed (this is the default setting). S is a quasi upper triangular matrix with 1-by-1 and 2-by-2 blocks on its diagonal. 1-by-1 blocks correspond to real eigenvalues, and 2-by-2 blocks correspond to complex eigenvalues.

If *compute_t* is set to 1, the full Schur form T will be computed by gsl_eigen_gen. If it is set to 0, T will not be computed (this is the default setting). T is an upper triangular matrix with non-negative elements on its diagonal. Any 2-by-2 blocks in S will correspond to a 2-by-2 diagonal block in T.

The *balance* parameter is currently ignored, since generalized balancing is not yet implemented.

int gsl_eigen_gen (gsl_matrix * *A*, gsl_matrix * *B*,   *Function*
          gsl_vector_complex * *alpha*, gsl_vector * *beta*,
          gsl_eigen_gen_workspace * *w*)

This function computes the eigenvalues of the real generalized nonsymmetric matrix pair (A, B), and stores them as pairs in (*alpha*, *beta*), where *alpha* is complex and *beta* is real. If β_i is non-zero, then $\lambda = \alpha_i/\beta_i$ is an eigenvalue. Likewise, if α_i is non-zero, then $\mu = \beta_i/\alpha_i$ is an eigenvalue of the alternate problem $\mu Ay = By$. The elements of *beta* are normalized to be non-negative.

If S is desired, it is stored in A on output. If T is desired, it is stored in B on output. The ordering of eigenvalues in (*alpha*, *beta*) follows the ordering of the diagonal blocks in the Schur forms S and T. In rare cases, this function may fail to find all eigenvalues. If this occurs, an error code is returned.

int gsl_eigen_gen_QZ (gsl_matrix * *A*, gsl_matrix * *B*,   *Function*
          gsl_vector_complex * *alpha*, gsl_vector * *beta*, gsl_matrix *
          Q, gsl_matrix * *Z*, gsl_eigen_gen_workspace * *w*)

This function is identical to gsl_eigen_gen except that it also computes the left and right Schur vectors and stores them into Q and Z respectively.

gsl_eigen_genv_workspace * gsl_eigen_genv_alloc (const   *Function*
          size_t *n*)

This function allocates a workspace for computing eigenvalues and eigenvectors of n-by-n real generalized nonsymmetric eigensystems. The size of the workspace is $O(7n)$.

void gsl_eigen_genv_free (gsl_eigen_genv_workspace * w)   Function
   This function frees the memory associated with the workspace w.

int gsl_eigen_genv (gsl_matrix * A, gsl_matrix * B,   Function
        gsl_vector_complex * alpha, gsl_vector * beta,
        gsl_matrix_complex * evec, gsl_eigen_genv_workspace * w)
   This function computes eigenvalues and right eigenvectors of the n-by-n real
   generalized nonsymmetric matrix pair (A, B). The eigenvalues are stored in
   (alpha, beta) and the eigenvectors are stored in evec. It first calls gsl_
   eigen_gen to compute the eigenvalues, Schur forms, and Schur vectors.
   Then it finds eigenvectors of the Schur forms and backtransforms them using
   the Schur vectors. The Schur vectors are destroyed in the process, but can
   be saved by using gsl_eigen_genv_QZ. The computed eigenvectors are normalized to have unit magnitude. On output, (A, B) contains the generalized
   Schur form (S, T). If gsl_eigen_gen fails, no eigenvectors are computed,
   and an error code is returned.

int gsl_eigen_genv_QZ (gsl_matrix * A, gsl_matrix * B,   Function
        gsl_vector_complex * alpha, gsl_vector * beta,
        gsl_matrix_complex * evec, gsl_matrix * Q, gsl_matrix * Z,
        gsl_eigen_genv_workspace * w)
   This function is identical to gsl_eigen_genv except that it also computes
   the left and right Schur vectors and stores them into Q and Z respectively.

14.7 Sorting Eigenvalues and Eigenvectors

int gsl_eigen_symmv_sort (gsl_vector * eval, gsl_matrix * evec,   Function
        gsl_eigen_sort_t sort_type)
   This function simultaneously sorts the eigenvalues stored in the vector eval
   and the corresponding real eigenvectors stored in the columns of the matrix evec into ascending or descending order according to the value of the
   parameter sort_type,

   GSL_EIGEN_SORT_VAL_ASC
      ascending order in numerical value

   GSL_EIGEN_SORT_VAL_DESC
      descending order in numerical value

   GSL_EIGEN_SORT_ABS_ASC
      ascending order in magnitude

   GSL_EIGEN_SORT_ABS_DESC
      descending order in magnitude

int gsl_eigen_hermv_sort (gsl_vector * eval,   Function
        gsl_matrix_complex * evec, gsl_eigen_sort_t sort_type)
   This function simultaneously sorts the eigenvalues stored in the vector eval
   and the corresponding complex eigenvectors stored in the columns of the
   matrix evec into ascending or descending order according to the value of the
   parameter sort_type as shown above.

int gsl_eigen_nonsymmv_sort (gsl_vector_complex * *eval*,   Function
          gsl_matrix_complex * *evec*, gsl_eigen_sort_t *sort_type*)
     This function simultaneously sorts the eigenvalues stored in the vector *eval*
     and the corresponding complex eigenvectors stored in the columns of the
     matrix *evec* into ascending or descending order according to the value of
     the parameter *sort_type* as shown above. Only GSL_EIGEN_SORT_ABS_ASC
     and GSL_EIGEN_SORT_ABS_DESC are supported due to the eigenvalues being
     complex.

int gsl_eigen_gensymmv_sort (gsl_vector * *eval*, gsl_matrix *   Function
          evec, gsl_eigen_sort_t *sort_type*)
     This function simultaneously sorts the eigenvalues stored in the vector *eval*
     and the corresponding real eigenvectors stored in the columns of the ma-
     trix *evec* into ascending or descending order according to the value of the
     parameter *sort_type* as shown above.

int gsl_eigen_genhermv_sort (gsl_vector * *eval*,   Function
          gsl_matrix_complex * *evec*, gsl_eigen_sort_t *sort_type*)
     This function simultaneously sorts the eigenvalues stored in the vector *eval*
     and the corresponding complex eigenvectors stored in the columns of the
     matrix *evec* into ascending or descending order according to the value of the
     parameter *sort_type* as shown above.

int gsl_eigen_genv_sort (gsl_vector_complex * *alpha*,   Function
          gsl_vector * *beta*, gsl_matrix_complex * *evec*,
          gsl_eigen_sort_t *sort_type*)
     This function simultaneously sorts the eigenvalues stored in the vectors (*al-
     pha*, *beta*) and the corresponding complex eigenvectors stored in the columns
     of the matrix *evec* into ascending or descending order according to the value
     of the parameter *sort_type* as shown above. Only GSL_EIGEN_SORT_ABS_ASC
     and GSL_EIGEN_SORT_ABS_DESC are supported due to the eigenvalues being
     complex.

14.8 Examples

The following program computes the eigenvalues and eigenvectors of the 4-th
order Hilbert matrix, $H(i,j) = 1/(i+j+1)$.

```
#include <stdio.h>
#include <gsl/gsl_math.h>
#include <gsl/gsl_eigen.h>

int
main (void)
{
  double data[] = { 1.0  , 1/2.0, 1/3.0, 1/4.0,
                    1/2.0, 1/3.0, 1/4.0, 1/5.0,
                    1/3.0, 1/4.0, 1/5.0, 1/6.0,
                    1/4.0, 1/5.0, 1/6.0, 1/7.0 };
```

```
    gsl_matrix_view m
      = gsl_matrix_view_array (data, 4, 4);

    gsl_vector *eval = gsl_vector_alloc (4);
    gsl_matrix *evec = gsl_matrix_alloc (4, 4);

    gsl_eigen_symmv_workspace * w =
      gsl_eigen_symmv_alloc (4);

    gsl_eigen_symmv (&m.matrix, eval, evec, w);

    gsl_eigen_symmv_free (w);

    gsl_eigen_symmv_sort (eval, evec,
                          GSL_EIGEN_SORT_ABS_ASC);

    {
      int i;

      for (i = 0; i < 4; i++)
        {
          double eval_i
             = gsl_vector_get (eval, i);
          gsl_vector_view evec_i
             = gsl_matrix_column (evec, i);

          printf ("eigenvalue = %g\n", eval_i);
          printf ("eigenvector = \n");
          gsl_vector_fprintf (stdout,
                              &evec_i.vector, "%g");
        }
    }

    gsl_vector_free (eval);
    gsl_matrix_free (evec);

    return 0;
  }
```

Here is the beginning of the output from the program,

```
$ ./a.out
eigenvalue = 9.67023e-05
eigenvector =
-0.0291933
0.328712
-0.791411
0.514553
...
```

This can be compared with the corresponding output from GNU OCTAVE,

```
octave> [v,d] = eig(hilb(4));
octave> diag(d)
ans =

   9.6702e-05
   6.7383e-03
   1.6914e-01
   1.5002e+00

octave> v
v =

   0.029193   0.179186  -0.582076   0.792608
  -0.328712  -0.741918   0.370502   0.451923
   0.791411   0.100228   0.509579   0.322416
  -0.514553   0.638283   0.514048   0.252161
```

Note that the eigenvectors can differ by a change of sign, since the sign of an eigenvector is arbitrary.

The following program illustrates the use of the nonsymmetric eigensolver, by computing the eigenvalues and eigenvectors of the Vandermonde matrix $V(x; i, j) = x_i^{n-j}$ with $x = (-1, -2, 3, 4)$.

```
#include <stdio.h>
#include <gsl/gsl_math.h>
#include <gsl/gsl_eigen.h>

int
main (void)
{
  double data[] = { -1.0,  1.0, -1.0,  1.0,
                    -8.0,  4.0, -2.0,  1.0,
                    27.0,  9.0,  3.0,  1.0,
                    64.0, 16.0,  4.0,  1.0 };

  gsl_matrix_view m
    = gsl_matrix_view_array (data, 4, 4);

  gsl_vector_complex *eval = gsl_vector_complex_alloc (4);
  gsl_matrix_complex *evec = gsl_matrix_complex_alloc (4, 4);

  gsl_eigen_nonsymmv_workspace * w =
    gsl_eigen_nonsymmv_alloc (4);

  gsl_eigen_nonsymmv (&m.matrix, eval, evec, w);

  gsl_eigen_nonsymmv_free (w);

  gsl_eigen_nonsymmv_sort (eval, evec,
```

```
                                           GSL_EIGEN_SORT_ABS_DESC);

            {
              int i, j;

              for (i = 0; i < 4; i++)
                {
                  gsl_complex eval_i
                    = gsl_vector_complex_get (eval, i);
                  gsl_vector_complex_view evec_i
                    = gsl_matrix_complex_column (evec, i);

                  printf ("eigenvalue = %g + %gi\n",
                          GSL_REAL(eval_i), GSL_IMAG(eval_i));
                  printf ("eigenvector = \n");
                  for (j = 0; j < 4; ++j)
                    {
                      gsl_complex z =
                        gsl_vector_complex_get(&evec_i.vector, j);
                      printf("%g + %gi\n", GSL_REAL(z), GSL_IMAG(z));
                    }
                }
            }

            gsl_vector_complex_free(eval);
            gsl_matrix_complex_free(evec);

            return 0;
          }
```

Here is the beginning of the output from the program,

```
    $ ./a.out
    eigenvalue = -6.41391 + 0i
    eigenvector =
    -0.0998822 + 0i
    -0.111251 + 0i
    0.292501 + 0i
    0.944505 + 0i
    eigenvalue = 5.54555 + 3.08545i
    eigenvector =
    -0.043487 + -0.0076308i
    0.0642377 + -0.142127i
    -0.515253 + 0.0405118i
    -0.840592 + -0.00148565i
    ...
```

This can be compared with the corresponding output from GNU OCTAVE,

```
octave> [v,d] = eig(vander([-1 -2 3 4]));
octave> diag(d)
ans =

  -6.4139 + 0.0000i
   5.5456 + 3.0854i
   5.5456 - 3.0854i
   2.3228 + 0.0000i

octave> v
v =

  Columns 1 through 3:

  -0.09988 + 0.00000i   -0.04350 - 0.00755i   -0.04350 + 0.00755i
  -0.11125 + 0.00000i    0.06399 - 0.14224i    0.06399 + 0.14224i
   0.29250 + 0.00000i   -0.51518 + 0.04142i   -0.51518 - 0.04142i
   0.94451 + 0.00000i   -0.84059 + 0.00000i   -0.84059 - 0.00000i

  Column 4:

  -0.14493 + 0.00000i
   0.35660 + 0.00000i
   0.91937 + 0.00000i
   0.08118 + 0.00000i
```

Note that the eigenvectors corresponding to the eigenvalue $5.54555 + 3.08545i$ differ by the multiplicative constant $0.9999984 + 0.0017674i$ which is an arbitrary phase factor of magnitude 1.

14.9 References and Further Reading

Further information on the algorithms described in this section can be found in the following book,

> G. H. Golub, C. F. Van Loan, *Matrix Computations* (3rd Ed, 1996), Johns Hopkins University Press, ISBN 0-8018-5414-8.

Further information on the generalized eigensystems QZ algorithm can be found in this paper,

> C. Moler, G. Stewart, "An Algorithm for Generalized Matrix Eigenvalue Problems", SIAM J. Numer. Anal., Vol 10, No 2, 1973.

Eigensystem routines for very large matrices can be found in the Fortran library LAPACK. The LAPACK library is described in,

> *LAPACK Users' Guide* (Third Edition, 1999), Published by SIAM, ISBN 0-89871-447-8.

> http://www.netlib.org/lapack

The LAPACK source code can be found at the website above along with an online copy of the users guide.

15 Fast Fourier Transforms (FFTs)

This chapter describes functions for performing Fast Fourier Transforms (FFTs). The library includes radix-2 routines (for lengths which are a power of two) and mixed-radix routines (which work for any length). For efficiency there are separate versions of the routines for real data and for complex data. The mixed-radix routines are a reimplementation of the FFTPACK library of Paul Swarztrauber. Fortran code for FFTPACK is available on Netlib (FFTPACK also includes some routines for sine and cosine transforms but these are currently not available in GSL). For details and derivations of the underlying algorithms consult the document *GSL FFT Algorithms* (see Section 15.8 [FFT References and Further Reading], page 202)

15.1 Mathematical Definitions

Fast Fourier Transforms are efficient algorithms for calculating the discrete fourier transform (DFT),

$$x_j = \sum_{k=0}^{N-1} z_k \exp(-2\pi i j k/N)$$

The DFT usually arises as an approximation to the continuous fourier transform when functions are sampled at discrete intervals in space or time. The naive evaluation of the discrete fourier transform is a matrix-vector multiplication $W\vec{z}$. A general matrix-vector multiplication takes $O(N^2)$ operations for N data-points. Fast fourier transform algorithms use a divide-and-conquer strategy to factorize the matrix W into smaller sub-matrices, corresponding to the integer factors of the length N. If N can be factorized into a product of integers $f_1 f_2 \ldots f_n$ then the DFT can be computed in $O(N \sum f_i)$ operations. For a radix-2 FFT this gives an operation count of $O(N \log_2 N)$.

All the FFT functions offer three types of transform: forwards, inverse and backwards, based on the same mathematical definitions. The definition of the *forward fourier transform*, $x = \mathrm{FFT}(z)$, is,

$$x_j = \sum_{k=0}^{N-1} z_k \exp(-2\pi i j k/N)$$

and the definition of the *inverse fourier transform*, $x = \mathrm{IFFT}(z)$, is,

$$z_j = \frac{1}{N} \sum_{k=0}^{N-1} x_k \exp(2\pi i j k/N).$$

The factor of $1/N$ makes this a true inverse. For example, a call to gsl_fft_complex_forward followed by a call to gsl_fft_complex_inverse should return the original data (within numerical errors).

In general there are two possible choices for the sign of the exponential in the transform/ inverse-transform pair. GSL follows the same convention as FFT-PACK, using a negative exponential for the forward transform. The advantage of this convention is that the inverse transform recreates the original function with simple fourier synthesis. Numerical Recipes uses the opposite convention, a positive exponential in the forward transform.

The *backwards FFT* is simply our terminology for an unscaled version of the inverse FFT,

$$z_j^{backwards} = \sum_{k=0}^{N-1} x_k \exp(2\pi ijk/N).$$

When the overall scale of the result is unimportant it is often convenient to use the backwards FFT instead of the inverse to save unnecessary divisions.

15.2 Overview of complex data FFTs

The inputs and outputs for the complex FFT routines are *packed* arrays of floating point numbers. In a packed array the real and imaginary parts of each complex number are placed in alternate neighboring elements. For example, the following definition of a packed array of length 6,

```
double x[3*2];
gsl_complex_packed_array data = x;
```

can be used to hold an array of three complex numbers, z[3], in the following way,

```
data[0] = Re(z[0])
data[1] = Im(z[0])
data[2] = Re(z[1])
data[3] = Im(z[1])
data[4] = Re(z[2])
data[5] = Im(z[2])
```

The array indices for the data have the same ordering as those in the definition of the DFT—i.e. there are no index transformations or permutations of the data.

A *stride* parameter allows the user to perform transforms on the elements z[stride*i] instead of z[i]. A stride greater than 1 can be used to take an in-place FFT of the column of a matrix. A stride of 1 accesses the array without any additional spacing between elements.

To perform an FFT on a vector argument, such as gsl_vector_complex * v, use the following definitions (or their equivalents) when calling the functions described in this chapter:

```
gsl_complex_packed_array data = v->data;
size_t stride = v->stride;
size_t n = v->size;
```

For physical applications it is important to remember that the index appearing in the DFT does not correspond directly to a physical frequency. If the time-step of the DFT is Δ then the frequency-domain includes both positive and negative frequencies, ranging from $-1/(2\Delta)$ through 0 to $+1/(2\Delta)$. The positive frequencies are stored from the beginning of the array up to the middle, and the negative frequencies are stored backwards from the end of the array.

Here is a table which shows the layout of the array *data*, and the correspondence between the time-domain data z, and the frequency-domain data x.

```
index     z                 x = FFT(z)

0         z(t = 0)          x(f = 0)
1         z(t = 1)          x(f = 1/(N Delta))
2         z(t = 2)          x(f = 2/(N Delta))
.         ........          .................
N/2       z(t = N/2)        x(f = +1/(2 Delta),
                                 -1/(2 Delta))
.         ........          .................
N-3       z(t = N-3)        x(f = -3/(N Delta))
N-2       z(t = N-2)        x(f = -2/(N Delta))
N-1       z(t = N-1)        x(f = -1/(N Delta))
```

When N is even the location $N/2$ contains the most positive and negative frequencies $(+1/(2\Delta), -1/(2\Delta))$ which are equivalent. If N is odd then general structure of the table above still applies, but $N/2$ does not appear.

15.3 Radix-2 FFT routines for complex data

The radix-2 algorithms described in this section are simple and compact, although not necessarily the most efficient. They use the Cooley-Tukey algorithm to compute in-place complex FFTs for lengths which are a power of 2—no additional storage is required. The corresponding self-sorting mixed-radix routines offer better performance at the expense of requiring additional working space.

All the functions described in this section are declared in the header file 'gsl_fft_complex.h'.

int gsl_fft_complex_radix2_forward (gsl_complex_packed_array   *Function*
        data, size_t *stride*, size_t *n*)

int gsl_fft_complex_radix2_transform   *Function*
        (gsl_complex_packed_array *data*, size_t *stride*, size_t *n*,
        gsl_fft_direction *sign*)

int gsl_fft_complex_radix2_backward   *Function*
        (gsl_complex_packed_array *data*, size_t *stride*, size_t *n*)

int gsl_fft_complex_radix2_inverse (gsl_complex_packed_array   *Function*
        data, size_t *stride*, size_t *n*)

These functions compute forward, backward and inverse FFTs of length n with stride *stride*, on the packed complex array *data* using an in-place radix-2 decimation-in-time algorithm. The length of the transform n is restricted to powers of two. For the transform version of the function the *sign* argument can be either forward (-1) or backward $(+1)$.

The functions return a value of GSL_SUCCESS if no errors were detected, or GSL_EDOM if the length of the data n is not a power of two.

int gsl_fft_complex_radix2_dif_forward   Function
        (gsl_complex_packed_array *data*, size_t *stride*, size_t *n*)
int gsl_fft_complex_radix2_dif_transform   Function
        (gsl_complex_packed_array *data*, size_t *stride*, size_t *n*,
        gsl_fft_direction *sign*)
int gsl_fft_complex_radix2_dif_backward   Function
        (gsl_complex_packed_array *data*, size_t *stride*, size_t *n*)
int gsl_fft_complex_radix2_dif_inverse   Function
        (gsl_complex_packed_array *data*, size_t *stride*, size_t *n*)
    These are decimation-in-frequency versions of the radix-2 FFT functions.

Here is an example program which computes the FFT of a short pulse in a
sample of length 128. To make the resulting fourier transform real the pulse is
defined for equal positive and negative times ($-10 \ldots 10$), where the negative
times wrap around the end of the array.

```
#include <stdio.h>
#include <math.h>
#include <gsl/gsl_errno.h>
#include <gsl/gsl_fft_complex.h>

#define REAL(z,i) ((z)[2*(i)])
#define IMAG(z,i) ((z)[2*(i)+1])

int
main (void)
{
  int i; double data[2*128];

  for (i = 0; i < 128; i++)
    {
        REAL(data,i) = 0.0; IMAG(data,i) = 0.0;
    }

  REAL(data,0) = 1.0;

  for (i = 1; i <= 10; i++)
    {
        REAL(data,i) = REAL(data,128-i) = 1.0;
    }

  for (i = 0; i < 128; i++)
    {
      printf ("%d %e %e\n", i,
              REAL(data,i), IMAG(data,i));
    }
  printf ("\n");

  gsl_fft_complex_radix2_forward (data, 1, 128);
```

```
for (i = 0; i < 128; i++)
  {
    printf ("%d %e %e\n", i,
            REAL(data,i)/sqrt(128),
            IMAG(data,i)/sqrt(128));
  }

return 0;
}
```

Note that we have assumed that the program is using the default error handler (which calls abort for any errors). If you are not using a safe error handler you would need to check the return status of `gsl_fft_complex_radix2_forward`.

The transformed data is rescaled by $1/\sqrt{N}$ so that it fits on the same plot as the input. Only the real part is shown, by the choice of the input data the imaginary part is zero. Allowing for the wrap-around of negative times at $t = 128$, and working in units of k/N, the DFT approximates the continuum fourier transform, giving a modulated sine function.

$$\int_{-a}^{+a} e^{-2\pi ikx} dx = \frac{\sin(2\pi ka)}{\pi k}$$

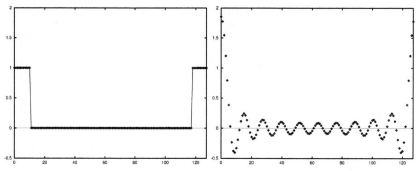

A pulse and its discrete fourier transform, from the example program.

15.4 Mixed-radix FFT routines for complex data

This section describes mixed-radix FFT algorithms for complex data. The mixed-radix functions work for FFTs of any length. They are a reimplementation of Paul Swarztrauber's Fortran FFTPACK library. The theory is explained in the review article *Self-sorting Mixed-radix FFTs* by Clive Temperton. The routines here use the same indexing scheme and basic algorithms as FFTPACK.

The mixed-radix algorithm is based on sub-transform modules—highly optimized small length FFTs which are combined to create larger FFTs. There are efficient modules for factors of 2, 3, 4, 5, 6 and 7. The modules for the composite factors of 4 and 6 are faster than combining the modules for $2*2$ and $2*3$.

For factors which are not implemented as modules there is a fall-back to a general length-n module which uses Singleton's method for efficiently computing a DFT. This module is $O(n^2)$, and slower than a dedicated module would be but works for any length n. Of course, lengths which use the general length-n module will still be factorized as much as possible. For example, a length of 143 will be factorized into $11 * 13$. Large prime factors are the worst case scenario, e.g. as found in $n = 2 * 3 * 99991$, and should be avoided because their $O(n^2)$ scaling will dominate the run-time (consult the document *GSL FFT Algorithms* included in the GSL distribution if you encounter this problem).

The mixed-radix initialization function gsl_fft_complex_wavetable_alloc returns the list of factors chosen by the library for a given length N. It can be used to check how well the length has been factorized, and estimate the run-time. To a first approximation the run-time scales as $N \sum f_i$, where the f_i are the factors of N. For programs under user control you may wish to issue a warning that the transform will be slow when the length is poorly factorized. If you frequently encounter data lengths which cannot be factorized using the existing small-prime modules consult *GSL FFT Algorithms* for details on adding support for other factors.

All the functions described in this section are declared in the header file 'gsl_fft_complex.h'.

gsl_fft_complex_wavetable * gsl_fft_complex_wavetable_alloc   Function
            (size_t n)

   This function prepares a trigonometric lookup table for a complex FFT of length n. The function returns a pointer to the newly allocated gsl_fft_complex_wavetable if no errors were detected, and a null pointer in the case of error. The length n is factorized into a product of subtransforms, and the factors and their trigonometric coefficients are stored in the wavetable. The trigonometric coefficients are computed using direct calls to sin and cos, for accuracy. Recursion relations could be used to compute the lookup table faster, but if an application performs many FFTs of the same length then this computation is a one-off overhead which does not affect the final throughput.

   The wavetable structure can be used repeatedly for any transform of the same length. The table is not modified by calls to any of the other FFT functions. The same wavetable can be used for both forward and backward (or inverse) transforms of a given length.

void gsl_fft_complex_wavetable_free   *Function*
        (gsl_fft_complex_wavetable * *wavetable*)
This function frees the memory associated with the wavetable *wavetable*. The wavetable can be freed if no further FFTs of the same length will be needed.

These functions operate on a gsl_fft_complex_wavetable structure which contains internal parameters for the FFT. It is not necessary to set any of the components directly but it can sometimes be useful to examine them. For example, the chosen factorization of the FFT length is given and can be used to provide an estimate of the run-time or numerical error. The wavetable structure is declared in the header file 'gsl_fft_complex.h'.

gsl_fft_complex_wavetable   Data Type
This is a structure that holds the factorization and trigonometric lookup tables for the mixed radix fft algorithm. It has the following components:

size_t n
    This is the number of complex data points

size_t nf
    This is the number of factors that the length n was decomposed into.

size_t factor[64]
    This is the array of factors. Only the first nf elements are used.

gsl_complex * trig
    This is a pointer to a preallocated trigonometric lookup table of n complex elements.

gsl_complex * twiddle[64]
    This is an array of pointers into trig, giving the twiddle factors for each pass.

The mixed radix algorithms require additional working space to hold the intermediate steps of the transform.

gsl_fft_complex_workspace * gsl_fft_complex_workspace_alloc   *Function*
        (size_t *n*)
This function allocates a workspace for a complex transform of length *n*.

void gsl_fft_complex_workspace_free   *Function*
        (gsl_fft_complex_workspace * *workspace*)
This function frees the memory associated with the workspace *workspace*. The workspace can be freed if no further FFTs of the same length will be needed.

The following functions compute the transform,

int gsl_fft_complex_forward (gsl_complex_packed_array *data*,   Function
          size_t *stride*, size_t *n*, const gsl_fft_complex_wavetable *
          wavetable, gsl_fft_complex_workspace * *work*)
int gsl_fft_complex_transform (gsl_complex_packed_array   Function
          data, size_t *stride*, size_t *n*, const
          gsl_fft_complex_wavetable * *wavetable*,
          gsl_fft_complex_workspace * *work*, gsl_fft_direction *sign*)
int gsl_fft_complex_backward (gsl_complex_packed_array *data*,   Function
          size_t *stride*, size_t *n*, const gsl_fft_complex_wavetable *
          wavetable, gsl_fft_complex_workspace * *work*)
int gsl_fft_complex_inverse (gsl_complex_packed_array *data*,   Function
          size_t *stride*, size_t *n*, const gsl_fft_complex_wavetable *
          wavetable, gsl_fft_complex_workspace * *work*)

These functions compute forward, backward and inverse FFTs of length n
with stride *stride*, on the packed complex array *data*, using a mixed radix
decimation-in-frequency algorithm. There is no restriction on the length n.
Efficient modules are provided for subtransforms of length 2, 3, 4, 5, 6 and 7.
Any remaining factors are computed with a slow, $O(n^2)$, general-n module.
The caller must supply a *wavetable* containing the trigonometric lookup
tables and a workspace *work*. For the transform version of the function the
sign argument can be either forward (-1) or backward ($+1$).

The functions return a value of 0 if no errors were detected. The following
gsl_errno conditions are defined for these functions:

GSL_EDOM
          The length of the data n is not a positive integer (i.e. n is zero).

GSL_EINVAL
          The length of the data n and the length used to compute the given
          wavetable do not match.

Here is an example program which computes the FFT of a short pulse in a
sample of length 630 ($= 2 * 3 * 3 * 5 * 7$) using the mixed-radix algorithm.

```
#include <stdio.h>
#include <math.h>
#include <gsl/gsl_errno.h>
#include <gsl/gsl_fft_complex.h>

#define REAL(z,i) ((z)[2*(i)])
#define IMAG(z,i) ((z)[2*(i)+1])

int
main (void)
{
  int i;
  const int n = 630;
  double data[2*n];

  gsl_fft_complex_wavetable * wavetable;
```

```
      gsl_fft_complex_workspace * workspace;

      for (i = 0; i < n; i++)
        {
          REAL(data,i) = 0.0;
          IMAG(data,i) = 0.0;
        }

      data[0] = 1.0;

      for (i = 1; i <= 10; i++)
        {
          REAL(data,i) = REAL(data,n-i) = 1.0;
        }

      for (i = 0; i < n; i++)
        {
          printf ("%d: %e %e\n", i, REAL(data,i),
                                    IMAG(data,i));
        }
      printf ("\n");

      wavetable = gsl_fft_complex_wavetable_alloc (n);
      workspace = gsl_fft_complex_workspace_alloc (n);

      for (i = 0; i < wavetable->nf; i++)
        {
          printf ("# factor %d: %d\n", i,
                  wavetable->factor[i]);
        }

      gsl_fft_complex_forward (data, 1, n,
                              wavetable, workspace);

      for (i = 0; i < n; i++)
        {
          printf ("%d: %e %e\n", i, REAL(data,i),
                                    IMAG(data,i));
        }

      gsl_fft_complex_wavetable_free (wavetable);
      gsl_fft_complex_workspace_free (workspace);
      return 0;
    }
```

Note that we have assumed that the program is using the default gsl error handler (which calls abort for any errors). If you are not using a safe error handler you would need to check the return status of all the gsl routines.

15.5 Overview of real data FFTs

The functions for real data are similar to those for complex data. However, there is an important difference between forward and inverse transforms. The fourier transform of a real sequence is not real. It is a complex sequence with a special symmetry:

$$z_k = z^*_{N-k}$$

A sequence with this symmetry is called *conjugate-complex* or *half-complex*. This different structure requires different storage layouts for the forward transform (from real to half-complex) and inverse transform (from half-complex back to real). As a consequence the routines are divided into two sets: functions in `gsl_fft_real` which operate on real sequences and functions in `gsl_fft_halfcomplex` which operate on half-complex sequences.

Functions in `gsl_fft_real` compute the frequency coefficients of a real sequence. The half-complex coefficients c of a real sequence x are given by fourier analysis,

$$c_k = \sum_{j=0}^{N-1} x_j \exp(-2\pi i j k/N)$$

Functions in `gsl_fft_halfcomplex` compute inverse or backwards transforms. They reconstruct real sequences by fourier synthesis from their half-complex frequency coefficients, c,

$$x_j = \frac{1}{N} \sum_{k=0}^{N-1} c_k \exp(2\pi i j k/N)$$

The symmetry of the half-complex sequence implies that only half of the complex numbers in the output need to be stored. The remaining half can be reconstructed using the half-complex symmetry condition. This works for all lengths, even and odd—when the length is even the middle value where $k = N/2$ is also real. Thus only N real numbers are required to store the half-complex sequence, and the transform of a real sequence can be stored in the same size array as the original data.

The precise storage arrangements depend on the algorithm, and are different for radix-2 and mixed-radix routines. The radix-2 function operates in-place, which constrains the locations where each element can be stored. The restriction forces real and imaginary parts to be stored far apart. The mixed-radix algorithm does not have this restriction, and it stores the real and imaginary parts of a given term in neighboring locations (which is desirable for better locality of memory accesses).

15.6 Radix-2 FFT routines for real data

This section describes radix-2 FFT algorithms for real data. They use the Cooley-Tukey algorithm to compute in-place FFTs for lengths which are a power of 2.

The radix-2 FFT functions for real data are declared in the header files 'gsl_fft_real.h'

int gsl_fft_real_radix2_transform (double *data*[], size_t   Function
         stride, size_t *n*)

This function computes an in-place radix-2 FFT of length n and stride *stride* on the real array *data*. The output is a half-complex sequence, which is stored in-place. The arrangement of the half-complex terms uses the following scheme: for $k < N/2$ the real part of the k-th term is stored in location k, and the corresponding imaginary part is stored in location $N - k$. Terms with $k > N/2$ can be reconstructed using the symmetry $z_k = z^*_{N-k}$. The terms for $k = 0$ and $k = N/2$ are both purely real, and count as a special case. Their real parts are stored in locations 0 and $N/2$ respectively, while their imaginary parts which are zero are not stored.

The following table shows the correspondence between the output *data* and the equivalent results obtained by considering the input data as a complex sequence with zero imaginary part (assuming *stride=1*),

```
    complex[0].real    =    data[0]
    complex[0].imag    =    0
    complex[1].real    =    data[1]
    complex[1].imag    =    data[N-1]
    ...............         ...............
    complex[k].real    =    data[k]
    complex[k].imag    =    data[N-k]
    ...............         ...............
    complex[N/2].real  =    data[N/2]
    complex[N/2].imag  =    0
    ...............         ...............
    complex[k'].real   =    data[k]        k' = N - k
    complex[k'].imag   =    -data[N-k]
    ...............         ...............
    complex[N-1].real  =    data[1]
    complex[N-1].imag  =    -data[N-1]
```

Note that the output data can be converted into the full complex sequence using the function gsl_fft_halfcomplex_radix2_unpack described below.

The radix-2 FFT functions for halfcomplex data are declared in the header file 'gsl_fft_halfcomplex.h'.

int gsl_fft_halfcomplex_radix2_inverse (double *data*[], size_t   Function
          stride, size_t *n*)
int gsl_fft_halfcomplex_radix2_backward (double *data*[],   Function
          size_t *stride*, size_t *n*)

These functions compute the inverse or backwards in-place radix-2 FFT of
length *n* and stride *stride* on the half-complex sequence *data* stored according
the output scheme used by gsl_fft_real_radix2. The result is a real array
stored in natural order.

int gsl_fft_halfcomplex_radix2_unpack (const double   Function
          halfcomplex_coefficient[], gsl_complex_packed_array
          complex_coefficient, size_t *stride*, size_t *n*)

This function converts *halfcomplex_coefficient*, an array of half-complex coefficients as returned by gsl_fft_real_radix2_transform, into an ordinary
complex array, *complex_coefficient*. It fills in the complex array using the
symmetry $z_k = z^*_{N-k}$ to reconstruct the redundant elements. The algorithm
for the conversion is,

```
complex_coefficient[0].real
  = halfcomplex_coefficient[0];
complex_coefficient[0].imag
  = 0.0;

for (i = 1; i < n - i; i++)
  {
    double hc_real
      = halfcomplex_coefficient[i*stride];
    double hc_imag
      = halfcomplex_coefficient[(n-i)*stride];
    complex_coefficient[i*stride].real = hc_real;
    complex_coefficient[i*stride].imag = hc_imag;
    complex_coefficient[(n - i)*stride].real = hc_real;
    complex_coefficient[(n - i)*stride].imag = -hc_imag;
  }

if (i == n - i)
  {
    complex_coefficient[i*stride].real
      = halfcomplex_coefficient[(n - 1)*stride];
    complex_coefficient[i*stride].imag
      = 0.0;
  }
```

15.7 Mixed-radix FFT routines for real data

This section describes mixed-radix FFT algorithms for real data. The mixed-radix functions work for FFTs of any length. They are a reimplementation of the real-FFT routines in the Fortran FFTPACK library by Paul Swarztrauber. The theory behind the algorithm is explained in the article *Fast Mixed-Radix Real Fourier Transforms* by Clive Temperton. The routines here use the same indexing scheme and basic algorithms as FFTPACK.

The functions use the FFTPACK storage convention for half-complex sequences. In this convention the half-complex transform of a real sequence is stored with frequencies in increasing order, starting at zero, with the real and imaginary parts of each frequency in neighboring locations. When a value is known to be real the imaginary part is not stored. The imaginary part of the zero-frequency component is never stored. It is known to be zero (since the zero frequency component is simply the sum of the input data (all real)). For a sequence of even length the imaginary part of the frequency $n/2$ is not stored either, since the symmetry $z_k = z_{N-k}^*$ implies that this is purely real too.

The storage scheme is best shown by some examples. The table below shows the output for an odd-length sequence, $n = 5$. The two columns give the correspondence between the 5 values in the half-complex sequence returned by gsl_fft_real_transform, *halfcomplex[]* and the values *complex[]* that would be returned if the same real input sequence were passed to gsl_fft_complex_backward as a complex sequence (with imaginary parts set to 0),

```
complex[0].real  =  halfcomplex[0]
complex[0].imag  =  0
complex[1].real  =  halfcomplex[1]
complex[1].imag  =  halfcomplex[2]
complex[2].real  =  halfcomplex[3]
complex[2].imag  =  halfcomplex[4]
complex[3].real  =  halfcomplex[3]
complex[3].imag  = -halfcomplex[4]
complex[4].real  =  halfcomplex[1]
complex[4].imag  = -halfcomplex[2]
```

The upper elements of the *complex* array, complex[3] and complex[4] are filled in using the symmetry condition. The imaginary part of the zero-frequency term complex[0].imag is known to be zero by the symmetry.

The next table shows the output for an even-length sequence, $n = 6$ In the even case there are two values which are purely real,

```
complex[0].real  =  halfcomplex[0]
complex[0].imag  =  0
complex[1].real  =  halfcomplex[1]
complex[1].imag  =  halfcomplex[2]
complex[2].real  =  halfcomplex[3]
complex[2].imag  =  halfcomplex[4]
complex[3].real  =  halfcomplex[5]
complex[3].imag  =  0
complex[4].real  =  halfcomplex[3]
complex[4].imag  = -halfcomplex[4]
```

```
complex[5].real  =  halfcomplex[1]
complex[5].imag  =  -halfcomplex[2]
```
The upper elements of the *complex* array, complex[4] and complex[5] are filled in using the symmetry condition. Both complex[0].imag and complex[3].imag are known to be zero.

All these functions are declared in the header files 'gsl_fft_real.h' and 'gsl_fft_halfcomplex.h'.

gsl_fft_real_wavetable * gsl_fft_real_wavetable_alloc   Function
      (size_t *n*)
gsl_fft_halfcomplex_wavetable *   Function
      gsl_fft_halfcomplex_wavetable_alloc (size_t *n*)
    These functions prepare trigonometric lookup tables for an FFT of size *n*
    real elements. The functions return a pointer to the newly allocated struct
    if no errors were detected, and a null pointer in the case of error. The length
    n is factorized into a product of subtransforms, and the factors and their
    trigonometric coefficients are stored in the wavetable. The trigonometric
    coefficients are computed using direct calls to sin and cos, for accuracy.
    Recursion relations could be used to compute the lookup table faster, but if
    an application performs many FFTs of the same length then computing the
    wavetable is a one-off overhead which does not affect the final throughput.

    The wavetable structure can be used repeatedly for any transform of the
    same length. The table is not modified by calls to any of the other FFT
    functions. The appropriate type of wavetable must be used for forward real
    or inverse half-complex transforms.

void gsl_fft_real_wavetable_free (gsl_fft_real_wavetable *   Function
      wavetable)
void gsl_fft_halfcomplex_wavetable_free   Function
      (gsl_fft_halfcomplex_wavetable * *wavetable*)
    These functions free the memory associated with the wavetable *wavetable*.
    The wavetable can be freed if no further FFTs of the same length will be
    needed.

The mixed radix algorithms require additional working space to hold the intermediate steps of the transform,

gsl_fft_real_workspace * gsl_fft_real_workspace_alloc   Function
      (size_t *n*)
    This function allocates a workspace for a real transform of length *n*. The
    same workspace can be used for both forward real and inverse halfcomplex
    transforms.

void gsl_fft_real_workspace_free (gsl_fft_real_workspace *   Function
      workspace)
    This function frees the memory associated with the workspace *workspace*.
    The workspace can be freed if no further FFTs of the same length will be
    needed.

The following functions compute the transforms of real and half-complex data,

int gsl_fft_real_transform (double *data*[], size_t *stride*,   Function
        size_t *n*, const gsl_fft_real_wavetable * *wavetable*,
        gsl_fft_real_workspace * *work*)
int gsl_fft_halfcomplex_transform (double *data*[], size_t   Function
        stride, size_t *n*, const gsl_fft_halfcomplex_wavetable *
        wavetable, gsl_fft_real_workspace * *work*)

These functions compute the FFT of *data*, a real or half-complex array of
length *n*, using a mixed radix decimation-in-frequency algorithm. For gsl_
fft_real_transform *data* is an array of time-ordered real data. For gsl_
fft_halfcomplex_transform *data* contains fourier coefficients in the half-
complex ordering described above. There is no restriction on the length *n*.
Efficient modules are provided for subtransforms of length 2, 3, 4 and 5.
Any remaining factors are computed with a slow, $O(n^2)$, general-n module.
The caller must supply a *wavetable* containing trigonometric lookup tables
and a workspace *work*.

int gsl_fft_real_unpack (const double *real_coefficient*[],   Function
        gsl_complex_packed_array *complex_coefficient*, size_t *stride*,
        size_t *n*)

This function converts a single real array, *real_coefficient* into an equivalent complex array, *complex_coefficient*, (with imaginary part set to zero),
suitable for gsl_fft_complex routines. The algorithm for the conversion is
simply,

```
for (i = 0; i < n; i++)
  {
    complex_coefficient[i*stride].real
      = real_coefficient[i*stride];
    complex_coefficient[i*stride].imag
      = 0.0;
  }
```

int gsl_fft_halfcomplex_unpack (const double   Function
        halfcomplex_coefficient[], gsl_complex_packed_array
        complex_coefficient, size_t *stride*, size_t *n*)

This function converts *halfcomplex_coefficient*, an array of half-complex coefficients as returned by gsl_fft_real_transform, into an ordinary complex
array, *complex_coefficient*. It fills in the complex array using the symmetry
$z_k = z_{N-k}^*$ to reconstruct the redundant elements. The algorithm for the
conversion is,

```
complex_coefficient[0].real
  = halfcomplex_coefficient[0];
complex_coefficient[0].imag
  = 0.0;

for (i = 1; i < n - i; i++)
  {
    double hc_real
```

```
          = halfcomplex_coefficient[(2 * i - 1)*stride];
        double hc_imag
          = halfcomplex_coefficient[(2 * i)*stride];
        complex_coefficient[i*stride].real = hc_real;
        complex_coefficient[i*stride].imag = hc_imag;
        complex_coefficient[(n - i)*stride].real = hc_real;
        complex_coefficient[(n - i)*stride].imag = -hc_imag;
      }

  if (i == n - i)
    {
      complex_coefficient[i*stride].real
        = halfcomplex_coefficient[(n - 1)*stride];
      complex_coefficient[i*stride].imag
        = 0.0;
    }
```

Here is an example program using `gsl_fft_real_transform` and `gsl_fft_halfcomplex_inverse`. It generates a real signal in the shape of a square pulse. The pulse is fourier transformed to frequency space, and all but the lowest ten frequency components are removed from the array of fourier coefficients returned by `gsl_fft_real_transform`.

The remaining fourier coefficients are transformed back to the time-domain, to give a filtered version of the square pulse. Since fourier coefficients are stored using the half-complex symmetry both positive and negative frequencies are removed and the final filtered signal is also real.

```
#include <stdio.h>
#include <math.h>
#include <gsl/gsl_errno.h>
#include <gsl/gsl_fft_real.h>
#include <gsl/gsl_fft_halfcomplex.h>

int
main (void)
{
  int i, n = 100;
  double data[n];

  gsl_fft_real_wavetable * real;
  gsl_fft_halfcomplex_wavetable * hc;
  gsl_fft_real_workspace * work;

  for (i = 0; i < n; i++)
    {
      data[i] = 0.0;
    }

  for (i = n / 3; i < 2 * n / 3; i++)
```

```
      {
        data[i] = 1.0;
      }

  for (i = 0; i < n; i++)
      {
        printf ("%d: %e\n", i, data[i]);
      }
  printf ("\n");

  work = gsl_fft_real_workspace_alloc (n);
  real = gsl_fft_real_wavetable_alloc (n);

  gsl_fft_real_transform (data, 1, n,
                          real, work);

  gsl_fft_real_wavetable_free (real);

  for (i = 11; i < n; i++)
      {
        data[i] = 0;
      }

  hc = gsl_fft_halfcomplex_wavetable_alloc (n);

  gsl_fft_halfcomplex_inverse (data, 1, n,
                               hc, work);
  gsl_fft_halfcomplex_wavetable_free (hc);

  for (i = 0; i < n; i++)
      {
        printf ("%d: %e\n", i, data[i]);
      }

  gsl_fft_real_workspace_free (work);
  return 0;
}
```

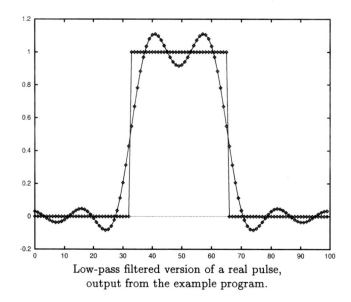

Low-pass filtered version of a real pulse,
output from the example program.

15.8 References and Further Reading

A good starting point for learning more about the FFT is the review article *Fast Fourier Transforms: A Tutorial Review and A State of the Art* by Duhamel and Vetterli,

> P. Duhamel and M. Vetterli. Fast fourier transforms: A tutorial review and a state of the art. *Signal Processing*, 19:259–299, 1990.

To find out about the algorithms used in the GSL routines you may want to consult the document *GSL FFT Algorithms* (it is included in GSL, as 'doc/fftalgorithms.tex'). This has general information on FFTs and explicit derivations of the implementation for each routine. There are also references to the relevant literature. For convenience some of the more important references are reproduced below.

There are several introductory books on the FFT with example programs, such as *The Fast Fourier Transform* by Brigham and *DFT/FFT and Convolution Algorithms* by Burrus and Parks,

> E. Oran Brigham. *The Fast Fourier Transform*. Prentice Hall, 1974.

> C. S. Burrus and T. W. Parks. *DFT/FFT and Convolution Algorithms*. Wiley, 1984.

Both these introductory books cover the radix-2 FFT in some detail. The mixed-radix algorithm at the heart of the FFTPACK routines is reviewed in Clive Temperton's paper,

> Clive Temperton. Self-sorting mixed-radix fast fourier transforms. *Journal of Computational Physics*, 52(1):1–23, 1983.

The derivation of FFTs for real-valued data is explained in the following two articles,

Henrik V. Sorenson, Douglas L. Jones, Michael T. Heideman, and C. Sidney Burrus. Real-valued fast fourier transform algorithms. *IEEE Transactions on Acoustics, Speech, and Signal Processing*, ASSP-35(6):849–863, 1987.

Clive Temperton. Fast mixed-radix real fourier transforms. *Journal of Computational Physics*, 52:340–350, 1983.

In 1979 the IEEE published a compendium of carefully-reviewed Fortran FFT programs in *Programs for Digital Signal Processing*. It is a useful reference for implementations of many different FFT algorithms,

Digital Signal Processing Committee and IEEE Acoustics, Speech, and Signal Processing Committee, editors. *Programs for Digital Signal Processing*. IEEE Press, 1979.

For large-scale FFT work we recommend the use of the dedicated FFTW library by Frigo and Johnson. The FFTW library is self-optimizing—it automatically tunes itself for each hardware platform in order to achieve maximum performance. It is available under the GNU GPL.

FFTW Website, http://www.fftw.org/

The source code for FFTPACK is available from Netlib,

FFTPACK, http://www.netlib.org/fftpack/

16 Numerical Integration

This chapter describes routines for performing numerical integration (quadrature) of a function in one dimension. There are routines for adaptive and non-adaptive integration of general functions, with specialised routines for specific cases. These include integration over infinite and semi-infinite ranges, singular integrals, including logarithmic singularities, computation of Cauchy principal values and oscillatory integrals. The library reimplements the algorithms used in QUADPACK, a numerical integration package written by Piessens, Doncker-Kapenga, Uberhuber and Kahaner. Fortran code for QUADPACK is available on Netlib.

The functions described in this chapter are declared in the header file 'gsl_integration.h'.

16.1 Introduction

Each algorithm computes an approximation to a definite integral of the form,

$$I = \int_a^b f(x)w(x)\,dx$$

where $w(x)$ is a weight function (for general integrands $w(x) = 1$). The user provides absolute and relative error bounds (*epsabs*, *epsrel*) which specify the following accuracy requirement,

$$|RESULT - I| \le \max(epsabs, epsrel\,|I|)$$

where *RESULT* is the numerical approximation obtained by the algorithm. The algorithms attempt to estimate the absolute error $ABSERR = |RESULT - I|$ in such a way that the following inequality holds,

$$|RESULT - I| \le ABSERR \le \max(epsabs, epsrel\,|I|)$$

In short, the routines return the first approximation which has an absolute error smaller than *epsabs* or a relative error smaller than *epsrel*.

Note that this is an *either-or* constraint, not simultaneous. To compute to a specified absolute error, set *epsrel* to zero. To compute to a specified relative error, set *epsabs* to zero. The routines will fail to converge if the error bounds are too stringent, but always return the best approximation obtained up to that stage.

The algorithms in QUADPACK use a naming convention based on the following letters,

    Q - quadrature routine

    N - non-adaptive integrator
    A - adaptive integrator

    G - general integrand (user-defined)

```
W - weight function with integrand

S - singularities can be more readily integrated
P - points of special difficulty can be supplied
I - infinite range of integration
O - oscillatory weight function, cos or sin
F - Fourier integral
C - Cauchy principal value
```

The algorithms are built on pairs of quadrature rules, a higher order rule and a lower order rule. The higher order rule is used to compute the best approximation to an integral over a small range. The difference between the results of the higher order rule and the lower order rule gives an estimate of the error in the approximation.

16.1.1 Integrands without weight functions

The algorithms for general functions (without a weight function) are based on Gauss-Kronrod rules.

A Gauss-Kronrod rule begins with a classical Gaussian quadrature rule of order m. This is extended with additional points between each of the abscissae to give a higher order Kronrod rule of order $2m+1$. The Kronrod rule is efficient because it reuses existing function evaluations from the Gaussian rule.

The higher order Kronrod rule is used as the best approximation to the integral, and the difference between the two rules is used as an estimate of the error in the approximation.

16.1.2 Integrands with weight functions

For integrands with weight functions the algorithms use Clenshaw-Curtis quadrature rules.

A Clenshaw-Curtis rule begins with an n-th order Chebyshev polynomial approximation to the integrand. This polynomial can be integrated exactly to give an approximation to the integral of the original function. The Chebyshev expansion can be extended to higher orders to improve the approximation and provide an estimate of the error.

16.1.3 Integrands with singular weight functions

The presence of singularities (or other behavior) in the integrand can cause slow convergence in the Chebyshev approximation. The modified Clenshaw-Curtis rules used in QUADPACK separate out several common weight functions which cause slow convergence.

These weight functions are integrated analytically against the Chebyshev polynomials to precompute *modified Chebyshev moments*. Combining the moments with the Chebyshev approximation to the function gives the desired integral. The use of analytic integration for the singular part of the function allows exact cancellations and substantially improves the overall convergence behavior of the integration.

16.2 QNG non-adaptive Gauss-Kronrod integration

The QNG algorithm is a non-adaptive procedure which uses fixed Gauss-Kronrod-Patterson abscissae to sample the integrand at a maximum of 87 points. It is provided for fast integration of smooth functions.

int gsl_integration_qng (const gsl_function * f, double a,   Function
          double b, double epsabs, double epsrel, double * result,
          double * abserr, size_t * neval)

This function applies the Gauss-Kronrod 10-point, 21-point, 43-point and 87-point integration rules in succession until an estimate of the integral of f over (a, b) is achieved within the desired absolute and relative error limits, epsabs and epsrel. The function returns the final approximation, result, an estimate of the absolute error, abserr and the number of function evaluations used, neval. The Gauss-Kronrod rules are designed in such a way that each rule uses all the results of its predecessors, in order to minimize the total number of function evaluations.

16.3 QAG adaptive integration

The QAG algorithm is a simple adaptive integration procedure. The integration region is divided into subintervals, and on each iteration the subinterval with the largest estimated error is bisected. This reduces the overall error rapidly, as the subintervals become concentrated around local difficulties in the integrand. These subintervals are managed by a gsl_integration_workspace struct, which handles the memory for the subinterval ranges, results and error estimates.

gsl_integration_workspace * gsl_integration_workspace_alloc   Function
          (size_t n)

This function allocates a workspace sufficient to hold n double precision intervals, their integration results and error estimates.

void gsl_integration_workspace_free   Function
          (gsl_integration_workspace * w)

This function frees the memory associated with the workspace w.

int gsl_integration_qag (const gsl_function * f, double a,   Function
          double b, double epsabs, double epsrel, size_t limit, int key,
          gsl_integration_workspace * workspace, double * result,
          double * abserr)

This function applies an integration rule adaptively until an estimate of the integral of f over (a, b) is achieved within the desired absolute and relative error limits, epsabs and epsrel. The function returns the final approximation, result, and an estimate of the absolute error, abserr. The integration rule is determined by the value of key, which should be chosen from the following symbolic names,

```
GSL_INTEG_GAUSS15   (key = 1)
GSL_INTEG_GAUSS21   (key = 2)
GSL_INTEG_GAUSS31   (key = 3)
GSL_INTEG_GAUSS41   (key = 4)
GSL_INTEG_GAUSS51   (key = 5)
GSL_INTEG_GAUSS61   (key = 6)
```

corresponding to the 15, 21, 31, 41, 51 and 61 point Gauss-Kronrod rules. The higher-order rules give better accuracy for smooth functions, while lower-order rules save time when the function contains local difficulties, such as discontinuities.

On each iteration the adaptive integration strategy bisects the interval with the largest error estimate. The subintervals and their results are stored in the memory provided by *workspace*. The maximum number of subintervals is given by *limit*, which may not exceed the allocated size of the workspace.

16.4 QAGS adaptive integration with singularities

The presence of an integrable singularity in the integration region causes an adaptive routine to concentrate new subintervals around the singularity. As the subintervals decrease in size the successive approximations to the integral converge in a limiting fashion. This approach to the limit can be accelerated using an extrapolation procedure. The QAGS algorithm combines adaptive bisection with the Wynn epsilon-algorithm to speed up the integration of many types of integrable singularities.

int gsl_integration_qags (const gsl_function * f, double a,   Function
        double b, double *epsabs*, double *epsrel*, size_t *limit*,
        gsl_integration_workspace * *workspace*, double * *result*,
        double * *abserr*)

This function applies the Gauss-Kronrod 21-point integration rule adaptively until an estimate of the integral of f over (a, b) is achieved within the desired absolute and relative error limits, *epsabs* and *epsrel*. The results are extrapolated using the epsilon-algorithm, which accelerates the convergence of the integral in the presence of discontinuities and integrable singularities. The function returns the final approximation from the extrapolation, *result*, and an estimate of the absolute error, *abserr*. The subintervals and their results are stored in the memory provided by *workspace*. The maximum number of subintervals is given by *limit*, which may not exceed the allocated size of the workspace.

16.5 QAGP adaptive integration with known singular points

int gsl_integration_qagp (const gsl_function * f, double * pts,   Function
        size_t npts, double epsabs, double epsrel, size_t limit,
        gsl_integration_workspace * workspace, double * result,
        double * abserr)

This function applies the adaptive integration algorithm QAGS taking account of the user-supplied locations of singular points. The array pts of
length npts should contain the endpoints of the integration ranges defined
by the integration region and locations of the singularities. For example,
to integrate over the region (a, b) with break-points at x_1, x_2, x_3 (where
$a < x_1 < x_2 < x_3 < b$) the following pts array should be used

        pts[0] = a
        pts[1] = x_1
        pts[2] = x_2
        pts[3] = x_3
        pts[4] = b

with $npts = 5$.

If you know the locations of the singular points in the integration region
then this routine will be faster than QAGS.

16.6 QAGI adaptive integration on infinite intervals

int gsl_integration_qagi (gsl_function * f, double epsabs,   Function
        double epsrel, size_t limit, gsl_integration_workspace *
        workspace, double * result, double * abserr)

This function computes the integral of the function f over the infinite interval
$(-\infty, +\infty)$. The integral is mapped onto the semi-open interval $(0, 1]$ using
the transformation $x = (1 - t)/t$,

$$\int_{-\infty}^{+\infty} dx \, f(x) = \int_0^1 dt \, (f((1-t)/t) + f(-(1-t)/t))/t^2.$$

It is then integrated using the QAGS algorithm. The normal 21-point Gauss-
Kronrod rule of QAGS is replaced by a 15-point rule, because the transformation can generate an integrable singularity at the origin. In this case a
lower-order rule is more efficient.

int gsl_integration_qagiu (gsl_function * f, double a, double   Function
        epsabs, double epsrel, size_t limit,
        gsl_integration_workspace * workspace, double * result,
        double * abserr)

This function computes the integral of the function f over the semi-infinite
interval $(a, +\infty)$. The integral is mapped onto the semi-open interval $(0, 1]$

using the transformation $x = a + (1 - t)/t$,

$$\int_a^{+\infty} dx\, f(x) = \int_0^1 dt\, f(a + (1 - t)/t)/t^2$$

and then integrated using the QAGS algorithm.

int gsl_integration_qagil (gsl_function * f, double b, double   Function
          epsabs, double epsrel, size_t limit,
          gsl_integration_workspace * workspace, double * result,
          double * abserr)

This function computes the integral of the function f over the semi-infinite
interval $(-\infty, b)$. The integral is mapped onto the semi-open interval $(0, 1]$
using the transformation $x = b - (1 - t)/t$,

$$\int_{-\infty}^b dx\, f(x) = \int_0^1 dt\, f(b - (1 - t)/t)/t^2$$

and then integrated using the QAGS algorithm.

16.7 QAWC adaptive integration for Cauchy principal values

int gsl_integration_qawc (gsl_function * f, double a, double   Function
          b, double c, double epsabs, double epsrel, size_t limit,
          gsl_integration_workspace * workspace, double * result,
          double * abserr)

This function computes the Cauchy principal value of the integral of f over
(a, b), with a singularity at c,

$$I = \int_a^b dx\, \frac{f(x)}{x - c} = \lim_{\epsilon \to 0} \left\{ \int_a^{c-\epsilon} dx\, \frac{f(x)}{x - c} + \int_{c+\epsilon}^b dx\, \frac{f(x)}{x - c} \right\}$$

The adaptive bisection algorithm of QAG is used, with modifications to
ensure that subdivisions do not occur at the singular point $x = c$. When a
subinterval contains the point $x = c$ or is close to it then a special 25-point
modified Clenshaw-Curtis rule is used to control the singularity. Further
away from the singularity the algorithm uses an ordinary 15-point Gauss-
Kronrod integration rule.

16.8 QAWS adaptive integration for singular functions

The QAWS algorithm is designed for integrands with algebraic-logarithmic singularities at the end-points of an integration region. In order to work efficiently the algorithm requires a precomputed table of Chebyshev moments.

gsl_integration_qaws_table *   Function
        gsl_integration_qaws_table_alloc (double alpha, double beta, int mu, int nu)
This function allocates space for a gsl_integration_qaws_table struct describing a singular weight function $W(x)$ with the parameters $(\alpha, \beta, \mu, \nu)$,

$$W(x) = (x - a)^\alpha (b - x)^\beta \log^\mu(x - a) \log^\nu(b - x)$$

where $\alpha > -1$, $\beta > -1$, and $\mu = 0, 1$, $\nu = 0, 1$. The weight function can take four different forms depending on the values of μ and ν,

$$
\begin{aligned}
W(x) &= (x - a)^\alpha (b - x)^\beta & (\mu = 0, \nu = 0) \\
W(x) &= (x - a)^\alpha (b - x)^\beta \log(x - a) & (\mu = 1, \nu = 0) \\
W(x) &= (x - a)^\alpha (b - x)^\beta \log(b - x) & (\mu = 0, \nu = 1) \\
W(x) &= (x - a)^\alpha (b - x)^\beta \log(x - a) \log(b - x) & (\mu = 1, \nu = 1)
\end{aligned}
$$

The singular points (a, b) do not have to be specified until the integral is computed, where they are the endpoints of the integration range.

The function returns a pointer to the newly allocated table gsl_integration_qaws_table if no errors were detected, and 0 in the case of error.

int gsl_integration_qaws_table_set   Function
        (gsl_integration_qaws_table * t, double alpha, double beta, int mu, int nu)
This function modifies the parameters $(\alpha, \beta, \mu, \nu)$ of an existing gsl_integration_qaws_table struct t.

void gsl_integration_qaws_table_free   Function
        (gsl_integration_qaws_table * t)
This function frees all the memory associated with the gsl_integration_qaws_table struct t.

int gsl_integration_qaws (gsl_function * f, const double a,   Function
        const double b, gsl_integration_qaws_table * t, const
        double epsabs, const double epsrel, const size_t limit,
        gsl_integration_workspace * workspace, double * result,
        double * abserr)
This function computes the integral of the function $f(x)$ over the interval (a, b) with the singular weight function $(x-a)^\alpha (b-x)^\beta \log^\mu(x-a) \log^\nu(b-x)$.

The parameters of the weight function $(\alpha, \beta, \mu, \nu)$ are taken from the table t. The integral is,

$$I = \int_a^b dx\, f(x)(x-a)^\alpha (b-x)^\beta \log^\mu(x-a) \log^\nu(b-x).$$

The adaptive bisection algorithm of QAG is used. When a subinterval contains one of the endpoints then a special 25-point modified Clenshaw-Curtis rule is used to control the singularities. For subintervals which do not include the endpoints an ordinary 15-point Gauss-Kronrod integration rule is used.

16.9 QAWO adaptive integration for oscillatory functions

The QAWO algorithm is designed for integrands with an oscillatory factor, $\sin(\omega x)$ or $\cos(\omega x)$. In order to work efficiently the algorithm requires a table of Chebyshev moments which must be pre-computed with calls to the functions below.

gsl_integration_qawo_table *   Function
      gsl_integration_qawo_table_alloc (double *omega*, double L,
      enum gsl_integration_qawo_enum *sine*, size_t n)
This function allocates space for a gsl_integration_qawo_table struct and its associated workspace describing a sine or cosine weight function $W(x)$ with the parameters (ω, L),

$$W(x) = \left\{ \begin{array}{c} \sin(\omega x) \\ \cos(\omega x) \end{array} \right\}$$

The parameter L must be the length of the interval over which the function will be integrated $L = b - a$. The choice of sine or cosine is made with the parameter *sine* which should be chosen from one of the two following symbolic values:

      GSL_INTEG_COSINE
      GSL_INTEG_SINE

The gsl_integration_qawo_table is a table of the trigonometric coefficients required in the integration process. The parameter n determines the number of levels of coefficients that are computed. Each level corresponds to one bisection of the interval L, so that n levels are sufficient for subintervals down to the length $L/2^n$. The integration routine gsl_integration_qawo returns the error GSL_ETABLE if the number of levels is insufficient for the requested accuracy.

int gsl_integration_qawo_table_set   Function
      (gsl_integration_qawo_table * t, double *omega*, double L,
      enum gsl_integration_qawo_enum *sine*)
This function changes the parameters *omega*, L and *sine* of the existing workspace t.

int gsl_integration_qawo_table_set_length   Function
        (gsl_integration_qawo_table * t, double L)
This function allows the length parameter L of the workspace t to be
changed.

void gsl_integration_qawo_table_free   Function
        (gsl_integration_qawo_table * t)
This function frees all the memory associated with the workspace t.

int gsl_integration_qawo (gsl_function * f, const double a,   Function
        const double epsabs, const double epsrel, const size_t limit,
        gsl_integration_workspace * workspace,
        gsl_integration_qawo_table * wf, double * result, double *
        abserr)
This function uses an adaptive algorithm to compute the integral of f over
(a, b) with the weight function $\sin(\omega x)$ or $\cos(\omega x)$ defined by the table wf,

$$I = \int_a^b dx\, f(x) \left\{ \begin{array}{c} \sin(\omega x) \\ \cos(\omega x) \end{array} \right\}$$

The results are extrapolated using the epsilon-algorithm to accelerate the
convergence of the integral. The function returns the final approximation
from the extrapolation, result, and an estimate of the absolute error, abserr.
The subintervals and their results are stored in the memory provided by
workspace. The maximum number of subintervals is given by limit, which
may not exceed the allocated size of the workspace.

Those subintervals with "large" widths d where $d\omega > 4$ are computed using
a 25-point Clenshaw-Curtis integration rule, which handles the oscillatory
behavior. Subintervals with a "small" widths where $d\omega < 4$ are computed
using a 15-point Gauss-Kronrod integration.

16.10 QAWF adaptive integration for Fourier integrals

int gsl_integration_qawf (gsl_function * f, const double a,   Function
        const double epsabs, const size_t limit,
        gsl_integration_workspace * workspace,
        gsl_integration_workspace * cycle_workspace,
        gsl_integration_qawo_table * wf, double * result, double *
        abserr)
This function attempts to compute a Fourier integral of the function f over
the semi-infinite interval $[a, +\infty)$.

$$I = \int_a^{+\infty} dx\, f(x) \left\{ \begin{array}{c} \sin(\omega x) \\ \cos(\omega x) \end{array} \right\}$$

The parameter ω and choice of sin or cos is taken from the table wf (the
length L can take any value, since it is overridden by this function to a value

appropriate for the fourier integration). The integral is computed using the QAWO algorithm over each of the subintervals,

$$C_1 = [a, a + c]$$
$$C_2 = [a + c, a + 2c]$$
$$\ldots = \ldots$$
$$C_k = [a + (k - 1)c, a + kc]$$

where $c = (2\,\mathrm{floor}(|\omega|) + 1)\pi/|\omega|$. The width c is chosen to cover an odd number of periods so that the contributions from the intervals alternate in sign and are monotonically decreasing when f is positive and monotonically decreasing. The sum of this sequence of contributions is accelerated using the epsilon-algorithm.

This function works to an overall absolute tolerance of *abserr*. The following strategy is used: on each interval C_k the algorithm tries to achieve the tolerance

$$TOL_k = u_k\, abserr$$

where $u_k = (1 - p)p^{k-1}$ and $p = 9/10$. The sum of the geometric series of contributions from each interval gives an overall tolerance of *abserr*.

If the integration of a subinterval leads to difficulties then the accuracy requirement for subsequent intervals is relaxed,

$$TOL_k = u_k\, \max(abserr, \max_{i<k}\{E_i\})$$

where E_k is the estimated error on the interval C_k.

The subintervals and their results are stored in the memory provided by *workspace*. The maximum number of subintervals is given by *limit*, which may not exceed the allocated size of the workspace. The integration over each subinterval uses the memory provided by *cycle_workspace* as workspace for the QAWO algorithm.

16.11 Error codes

In addition to the standard error codes for invalid arguments the functions can return the following values,

GSL_EMAXITER
:   the maximum number of subdivisions was exceeded.

GSL_EROUND
:   cannot reach tolerance because of roundoff error, or roundoff error was detected in the extrapolation table.

GSL_ESING
:   a non-integrable singularity or other bad integrand behavior was found in the integration interval.

GSL_EDIVERGE
:   the integral is divergent, or too slowly convergent to be integrated numerically.

16.12 Examples

The integrator QAGS will handle a large class of definite integrals. For example, consider the following integral, which has an algebraic-logarithmic singularity at the origin,

$$\int_0^1 x^{-1/2} \log(x)\, dx = -4$$

The program below computes this integral to a relative accuracy bound of 1e-7.

```
#include <stdio.h>
#include <math.h>
#include <gsl/gsl_integration.h>

double f (double x, void * params) {
  double alpha = *(double *) params;
  double f = log(alpha*x) / sqrt(x);
  return f;
}

int
main (void)
{
  gsl_integration_workspace * w
    = gsl_integration_workspace_alloc (1000);

  double result, error;
  double expected = -4.0;
  double alpha = 1.0;

  gsl_function F;
  F.function = &f;
  F.params = &alpha;

  gsl_integration_qags (&F, 0, 1, 0, 1e-7, 1000,
                        w, &result, &error);

  printf ("result          = % .18f\n", result);
  printf ("exact result    = % .18f\n", expected);
  printf ("estimated error = % .18f\n", error);
  printf ("actual error    = % .18f\n", result - expected);
  printf ("intervals =  %d\n", w->size);

  gsl_integration_workspace_free (w);

  return 0;
}
```

The results below show that the desired accuracy is achieved after 8 subdivisions.

```
$ ./a.out
  result            = -3.999999999999973799
  exact result      = -4.000000000000000000
  estimated error =   0.000000000000246025
  actual error      =   0.000000000000026201
  intervals =   8
```

In fact, the extrapolation procedure used by QAGS produces an accuracy of almost twice as many digits. The error estimate returned by the extrapolation procedure is larger than the actual error, giving a margin of safety of one order of magnitude.

16.13 References and Further Reading

The following book is the definitive reference for QUADPACK, and was written by the original authors. It provides descriptions of the algorithms, program listings, test programs and examples. It also includes useful advice on numerical integration and many references to the numerical integration literature used in developing QUADPACK.

R. Piessens, E. de Doncker-Kapenga, C.W. Uberhuber, D.K. Kahaner. QUADPACK *A subroutine package for automatic integration* Springer Verlag, 1983.

17 Random Number Generation

The library provides a large collection of random number generators which can be accessed through a uniform interface. Environment variables allow you to select different generators and seeds at runtime, so that you can easily switch between generators without needing to recompile your program. Each instance of a generator keeps track of its own state, allowing the generators to be used in multi-threaded programs. Additional functions are available for transforming uniform random numbers into samples from continuous or discrete probability distributions such as the Gaussian, log-normal or Poisson distributions.

These functions are declared in the header file 'gsl_rng.h'.

17.1 General comments on random numbers

In 1988, Park and Miller wrote a paper entitled "Random number generators: good ones are hard to find." [Commun. ACM, 31, 1192–1201]. Fortunately, some excellent random number generators are available, though poor ones are still in common use. You may be happy with the system-supplied random number generator on your computer, but you should be aware that as computers get faster, requirements on random number generators increase. Nowadays, a simulation that calls a random number generator millions of times can often finish before you can make it down the hall to the coffee machine and back.

A very nice review of random number generators was written by Pierre L'Ecuyer, as Chapter 4 of the book: Handbook on Simulation, Jerry Banks, ed. (Wiley, 1997). The chapter is available in postscript from L'Ecuyer's ftp site (see references). Knuth's volume on Seminumerical Algorithms (originally published in 1968) devotes 170 pages to random number generators, and has recently been updated in its 3rd edition (1997). It is brilliant, a classic. If you don't own it, you should stop reading right now, run to the nearest bookstore, and buy it.

A good random number generator will satisfy both theoretical and statistical properties. Theoretical properties are often hard to obtain (they require real math!), but one prefers a random number generator with a long period, low serial correlation, and a tendency *not* to "fall mainly on the planes." Statistical tests are performed with numerical simulations. Generally, a random number generator is used to estimate some quantity for which the theory of probability provides an exact answer. Comparison to this exact answer provides a measure of "randomness".

17.2 The Random Number Generator Interface

It is important to remember that a random number generator is not a "real" function like sine or cosine. Unlike real functions, successive calls to a random number generator yield different return values. Of course that is just what you want for a random number generator, but to achieve this effect, the generator must keep track of some kind of "state" variable. Sometimes this state is just an integer (sometimes just the value of the previously generated random number), but often it is more complicated than that and may involve a whole array of numbers, possibly with some indices thrown in. To use the random number generators, you do not need to know the details of what comprises the state, and besides that varies from algorithm to algorithm.

The random number generator library uses two special structs, `gsl_rng_type` which holds static information about each type of generator and `gsl_rng` which describes an instance of a generator created from a given `gsl_rng_type`.

The functions described in this section are declared in the header file 'gsl_rng.h'.

17.3 Random number generator initialization

`gsl_rng * gsl_rng_alloc (const gsl_rng_type * T)`   Function
This function returns a pointer to a newly-created instance of a random number generator of type T. For example, the following code creates an instance of the Tausworthe generator,

        gsl_rng * r = gsl_rng_alloc (gsl_rng_taus);

If there is insufficient memory to create the generator then the function returns a null pointer and the error handler is invoked with an error code of `GSL_ENOMEM`.

The generator is automatically initialized with the default seed, `gsl_rng_default_seed`. This is zero by default but can be changed either directly or by using the environment variable `GSL_RNG_SEED` (see Section 17.6 [Random number environment variables], page 221).

The details of the available generator types are described later in this chapter.

`void gsl_rng_set (const gsl_rng * r, unsigned long int s)`   Function
This function initializes (or 'seeds') the random number generator. If the generator is seeded with the same value of s on two different runs, the same stream of random numbers will be generated by successive calls to the routines below. If different values of $s \geq 1$ are supplied, then the generated streams of random numbers should be completely different. If the seed s is zero then the standard seed from the original implementation is used instead. For example, the original Fortran source code for the `ranlux` generator used a seed of 314159265, and so choosing s equal to zero reproduces this when using `gsl_rng_ranlux`.

When using multiple seeds with the same generator, choose seed values greater than zero to avoid collisions with the default setting.

Note that the most generators only accept 32-bit seeds, with higher values being reduced modulo 2^{32}. For generators with smaller ranges the maximum seed value will typically be lower.

void gsl_rng_free (gsl_rng * r)   *Function*

This function frees all the memory associated with the generator r.

17.4 Sampling from a random number generator

The following functions return uniformly distributed random numbers, either as integers or double precision floating point numbers. Inline versions of these functions are used when HAVE_INLINE is defined. To obtain non-uniform distributions see Chapter 19 [Random Number Distributions], page 239.

unsigned long int gsl_rng_get (const gsl_rng * r)   *Function*

This function returns a random integer from the generator r. The minimum and maximum values depend on the algorithm used, but all integers in the range [min,max] are equally likely. The values of min and max can determined using the auxiliary functions gsl_rng_max (r) and gsl_rng_min (r).

double gsl_rng_uniform (const gsl_rng * r)   *Function*

This function returns a double precision floating point number uniformly distributed in the range [0,1). The range includes 0.0 but excludes 1.0. The value is typically obtained by dividing the result of gsl_rng_get(r) by gsl_rng_max(r) + 1.0 in double precision. Some generators compute this ratio internally so that they can provide floating point numbers with more than 32 bits of randomness (the maximum number of bits that can be portably represented in a single unsigned long int).

double gsl_rng_uniform_pos (const gsl_rng * r)   *Function*

This function returns a positive double precision floating point number uniformly distributed in the range (0,1), excluding both 0.0 and 1.0. The number is obtained by sampling the generator with the algorithm of gsl_rng_uniform until a non-zero value is obtained. You can use this function if you need to avoid a singularity at 0.0.

unsigned long int gsl_rng_uniform_int (const gsl_rng * r,   *Function*
            unsigned long int n)

This function returns a random integer from 0 to $n - 1$ inclusive by scaling down and/or discarding samples from the generator r. All integers in the range $[0, n - 1]$ are produced with equal probability. For generators with a non-zero minimum value an offset is applied so that zero is returned with the correct probability.

Note that this function is designed for sampling from ranges smaller than the range of the underlying generator. The parameter n must be less than or equal to the range of the generator r. If n is larger than the range of the generator then the function calls the error handler with an error code of GSL_EINVAL and returns zero.

In particular, this function is not intended for generating the full range of unsigned integer values $[0, 2^{32} - 1]$. Instead choose a generator with the maximal integer range and zero mimimum value, such as gsl_rng_ranlxd1, gsl_rng_mt19937 or gsl_rng_taus, and sample it directly using gsl_rng_get. The range of each generator can be found using the auxiliary functions described in the next section.

17.5 Auxiliary random number generator functions

The following functions provide information about an existing generator. You should use them in preference to hard-coding the generator parameters into your own code.

const char * gsl_rng_name (const gsl_rng * r)   Function
    This function returns a pointer to the name of the generator. For example,

        printf ("r is a '%s' generator\n",
                gsl_rng_name (r));

    would print something like r is a 'taus' generator.

unsigned long int gsl_rng_max (const gsl_rng * r)   Function
    gsl_rng_max returns the largest value that gsl_rng_get can return.

unsigned long int gsl_rng_min (const gsl_rng * r)   Function
    gsl_rng_min returns the smallest value that gsl_rng_get can return. Usu-
    ally this value is zero. There are some generators with algorithms that
    cannot return zero, and for these generators the minimum value is 1.

void * gsl_rng_state (const gsl_rng * r)   Function
size_t gsl_rng_size (const gsl_rng * r)   Function
    These functions return a pointer to the state of generator r and its size.
    You can use this information to access the state directly. For example, the
    following code will write the state of a generator to a stream,

        void * state = gsl_rng_state (r);
        size_t n = gsl_rng_size (r);
        fwrite (state, n, 1, stream);

const gsl_rng_type ** gsl_rng_types_setup (void)   Function
    This function returns a pointer to an array of all the available generator
    types, terminated by a null pointer. The function should be called once
    at the start of the program, if needed. The following code fragment shows
    how to iterate over the array of generator types to print the names of the
    available algorithms,

        const gsl_rng_type **t, **t0;

        t0 = gsl_rng_types_setup ();

        printf ("Available generators:\n");

        for (t = t0; *t != 0; t++)

```
    {
      printf ("%s\n", (*t)->name);
    }
```

17.6 Random number environment variables

The library allows you to choose a default generator and seed from the environment variables GSL_RNG_TYPE and GSL_RNG_SEED and the function gsl_rng_env_setup. This makes it easy try out different generators and seeds without having to recompile your program.

const gsl_rng_type * gsl_rng_env_setup (void)   Function
    This function reads the environment variables GSL_RNG_TYPE and GSL_RNG_SEED and uses their values to set the corresponding library variables gsl_rng_default and gsl_rng_default_seed. These global variables are defined as follows,

```
        extern const gsl_rng_type *gsl_rng_default
        extern unsigned long int gsl_rng_default_seed
```

    The environment variable GSL_RNG_TYPE should be the name of a generator, such as taus or mt19937. The environment variable GSL_RNG_SEED should contain the desired seed value. It is converted to an unsigned long int using the C library function strtoul.

    If you don't specify a generator for GSL_RNG_TYPE then gsl_rng_mt19937 is used as the default. The initial value of gsl_rng_default_seed is zero.

Here is a short program which shows how to create a global generator using the environment variables GSL_RNG_TYPE and GSL_RNG_SEED,

```
        #include <stdio.h>
        #include <gsl/gsl_rng.h>

        gsl_rng * r;  /* global generator */

        int
        main (void)
        {
          const gsl_rng_type * T;

          gsl_rng_env_setup();

          T = gsl_rng_default;
          r = gsl_rng_alloc (T);

          printf ("generator type: %s\n", gsl_rng_name (r));
          printf ("seed = %lu\n", gsl_rng_default_seed);
          printf ("first value = %lu\n", gsl_rng_get (r));

          gsl_rng_free (r);
          return 0;
```

```
}
```

Running the program without any environment variables uses the initial defaults, an mt19937 generator with a seed of 0,

```
$ ./a.out
  generator type: mt19937
  seed = 0
  first value = 4293858116
```

By setting the two variables on the command line we can change the default generator and the seed,

```
$ GSL_RNG_TYPE="taus" GSL_RNG_SEED=123 ./a.out
GSL_RNG_TYPE=taus
GSL_RNG_SEED=123
generator type: taus
seed = 123
first value = 2720986350
```

17.7 Copying random number generator state

The above methods do not expose the random number 'state' which changes from call to call. It is often useful to be able to save and restore the state. To permit these practices, a few somewhat more advanced functions are supplied. These include:

int gsl_rng_memcpy (gsl_rng * dest, const gsl_rng * src)   Function
This function copies the random number generator src into the pre-existing generator dest, making dest into an exact copy of src. The two generators must be of the same type.

gsl_rng * gsl_rng_clone (const gsl_rng * r)   Function
This function returns a pointer to a newly created generator which is an exact copy of the generator r.

17.8 Reading and writing random number generator state

The library provides functions for reading and writing the random number state to a file as binary data.

int gsl_rng_fwrite (FILE * stream, const gsl_rng * r)   Function
This function writes the random number state of the random number generator r to the stream stream in binary format. The return value is 0 for success and GSL_EFAILED if there was a problem writing to the file. Since the data is written in the native binary format it may not be portable between different architectures.

int gsl_rng_fread (FILE * *stream*, gsl_rng * *r*)   Function
This function reads the random number state into the random number generator *r* from the open stream *stream* in binary format. The random number
generator *r* must be preinitialized with the correct random number generator type since type information is not saved. The return value is 0 for
success and GSL_EFAILED if there was a problem reading from the file. The
data is assumed to have been written in the native binary format on the
same architecture.

17.9 Random number generator algorithms

The functions described above make no reference to the actual algorithm used.
This is deliberate so that you can switch algorithms without having to change
any of your application source code. The library provides a large number of
generators of different types, including simulation quality generators, generators
provided for compatibility with other libraries and historical generators from the
past.

The following generators are recommended for use in simulation. They have
extremely long periods, low correlation and pass most statistical tests. For the
most reliable source of uncorrelated numbers, the second-generation RANLUX
generators have the strongest proof of randomness.

gsl_rng_mt19937   Generator
The MT19937 generator of Makoto Matsumoto and Takuji Nishimura is a
variant of the twisted generalized feedback shift-register algorithm, and is
known as the "Mersenne Twister" generator. It has a Mersenne prime period
of $2^{19937} - 1$ (about 10^{6000}) and is equi-distributed in 623 dimensions. It has
passed the DIEHARD statistical tests. It uses 624 words of state per generator
and is comparable in speed to the other generators. The original generator
used a default seed of 4357 and choosing *s* equal to zero in gsl_rng_set
reproduces this. Later versions switched to 5489 as the default seed, you
can choose this explicitly via gsl_rng_set instead if you require it.

For more information see,

> Makoto Matsumoto and Takuji Nishimura, "Mersenne Twister: A 623-
> dimensionally equidistributed uniform pseudorandom number genera
> tor". *ACM Transactions on Modeling and Computer Simulation*, Vol.
> 8, No. 1 (Jan. 1998), Pages 3–30

The generator gsl_rng_mt19937 uses the second revision of the seeding procedure published by the two authors above in 2002. The original seeding
procedures could cause spurious artifacts for some seed values. They are
still available through the alternative generators gsl_rng_mt19937_1999 and
gsl_rng_mt19937_1998.

gsl_rng_ranlxs0   Generator
gsl_rng_ranlxs1   Generator
gsl_rng_ranlxs2   Generator

The generator ranlxs0 is a second-generation version of the RANLUX algorithm of Lüscher, which produces "luxury random numbers". This generator provides single precision output (24 bits) at three luxury levels ranlxs0, ranlxs1 and ranlxs2, in increasing order of strength. It uses double-precision floating point arithmetic internally and can be significantly faster than the integer version of ranlux, particularly on 64-bit architectures. The period of the generator is about 10^{171}. The algorithm has mathematically proven properties and can provide truly decorrelated numbers at a known level of randomness. The higher luxury levels provide increased decorrelation between samples as an additional safety margin.

gsl_rng_ranlxd1   Generator
gsl_rng_ranlxd2   Generator

These generators produce double precision output (48 bits) from the RANLXS generator. The library provides two luxury levels ranlxd1 and ranlxd2, in increasing order of strength.

gsl_rng_ranlux   Generator
gsl_rng_ranlux389   Generator

The ranlux generator is an implementation of the original algorithm developed by Lüscher. It uses a lagged-fibonacci-with-skipping algorithm to produce "luxury random numbers". It is a 24-bit generator, originally designed for single-precision IEEE floating point numbers. This implementation is based on integer arithmetic, while the second-generation versions RANLXS and RANLXD described above provide floating-point implementations which will be faster on many platforms. The period of the generator is about 10^{171}. The algorithm has mathematically proven properties and it can provide truly decorrelated numbers at a known level of randomness. The default level of decorrelation recommended by Lüscher is provided by gsl_rng_ranlux, while gsl_rng_ranlux389 gives the highest level of randomness, with all 24 bits decorrelated. Both types of generator use 24 words of state per generator.

For more information see,

> M. Lüscher, "A portable high-quality random number generator for lattice field theory calculations", *Computer Physics Communications*, 79 (1994) 100–110.

> F. James, "RANLUX: A Fortran implementation of the high-quality pseudo-random number generator of Lüscher", *Computer Physics Communications*, 79 (1994) 111–114

gsl_rng_cmrg Generator

This is a combined multiple recursive generator by L'Ecuyer. Its sequence is,

$$z_n = (x_n - y_n) \bmod m_1$$

where the two underlying generators x_n and y_n are,

$$x_n = (a_1 x_{n-1} + a_2 x_{n-2} + a_3 x_{n-3}) \bmod m_1$$
$$y_n = (b_1 y_{n-1} + b_2 y_{n-2} + b_3 y_{n-3}) \bmod m_2$$

with coefficients $a_1 = 0$, $a_2 = 63308$, $a_3 = -183326$, $b_1 = 86098$, $b_2 = 0$, $b_3 = -539608$, and moduli $m_1 = 2^{31} - 1 = 2147483647$ and $m_2 = 2145483479$.

The period of this generator is $\mathrm{lcm}(m_1^3 - 1, m_2^3 - 1)$, which is approximately 2^{185} (about 10^{56}). It uses 6 words of state per generator. For more information see,

> P. L'Ecuyer, "Combined Multiple Recursive Random Number Generators", *Operations Research*, 44, 5 (1996), 816–822.

gsl_rng_mrg Generator

This is a fifth-order multiple recursive generator by L'Ecuyer, Blouin and Coutre. Its sequence is,

$$x_n = (a_1 x_{n-1} + a_5 x_{n-5}) \bmod m$$

with $a_1 = 107374182$, $a_2 = a_3 = a_4 = 0$, $a_5 = 104480$ and $m = 2^{31} - 1$.

The period of this generator is about 10^{46}. It uses 5 words of state per generator. More information can be found in the following paper,

> P. L'Ecuyer, F. Blouin, and R. Coutre, "A search for good multiple recursive random number generators", *ACM Transactions on Modeling and Computer Simulation* 3, 87–98 (1993).

gsl_rng_taus Generator
gsl_rng_taus2 Generator

This is a maximally equidistributed combined Tausworthe generator by L'Ecuyer. The sequence is,

$$x_n = (s_n^1 \oplus s_n^2 \oplus s_n^3)$$

where,

$$s_{n+1}^1 = (((s_n^1 \& 4294967294) \ll 12) \oplus (((s_n^1 \ll 13) \oplus s_n^1) \gg 19))$$
$$s_{n+1}^2 = (((s_n^2 \& 4294967288) \ll 4) \oplus (((s_n^2 \ll 2) \oplus s_n^2) \gg 25))$$
$$s_{n+1}^3 = (((s_n^3 \& 4294967280) \ll 17) \oplus (((s_n^3 \ll 3) \oplus s_n^3) \gg 11))$$

computed modulo 2^{32}. In the formulas above $\oplus$ denotes "exclusive-or". Note that the algorithm relies on the properties of 32-bit unsigned integers and has been implemented using a bitmask of 0xFFFFFFFF to make it work on 64 bit machines.

The period of this generator is 2^{88} (about 10^{26}). It uses 3 words of state per generator. For more information see,

P. L'Ecuyer, "Maximally Equidistributed Combined Tausworthe Generators", *Mathematics of Computation*, 65, 213 (1996), 203–213.

The generator gsl_rng_taus2 uses the same algorithm as gsl_rng_taus but with an improved seeding procedure described in the paper,

P. L'Ecuyer, "Tables of Maximally Equidistributed Combined LFSR Generators", *Mathematics of Computation*, 68, 225 (1999), 261–269

The generator gsl_rng_taus2 should now be used in preference to gsl_rng_taus.

gsl_rng_gfsr4   Generator

The gfsr4 generator is like a lagged-fibonacci generator, and produces each number as an xor'd sum of four previous values.

$$r_n = r_{n-A} \oplus r_{n-B} \oplus r_{n-C} \oplus r_{n-D}$$

Ziff (ref below) notes that "it is now widely known" that two-tap registers (such as R250, which is described below) have serious flaws, the most obvious one being the three-point correlation that comes from the definition of the generator. Nice mathematical properties can be derived for GFSR's, and numerics bears out the claim that 4-tap GFSR's with appropriately chosen offsets are as random as can be measured, using the author's test.

This implementation uses the values suggested the example on p392 of Ziff's article: $A = 471$, $B = 1586$, $C = 6988$, $D = 9689$.

If the offsets are appropriately chosen (such as the one ones in this implementation), then the sequence is said to be maximal; that means that the period is $2^D - 1$, where D is the longest lag. (It is one less than 2^D because it is not permitted to have all zeros in the ra[] array.) For this implementation with $D = 9689$ that works out to about 10^{2917}.

Note that the implementation of this generator using a 32-bit integer amounts to 32 parallel implementations of one-bit generators. One consequence of this is that the period of this 32-bit generator is the same as for the one-bit generator. Moreover, this independence means that all 32-bit patterns are equally likely, and in particular that 0 is an allowed random value. (We are grateful to Heiko Bauke for clarifying for us these properties of GFSR random number generators.)

For more information see,

Robert M. Ziff, "Four-tap shift-register-sequence random-number generators", *Computers in Physics*, 12(4), Jul/Aug 1998, pp 385–392.

17.10 Unix random number generators

The standard Unix random number generators rand, random and rand48 are provided as part of GSL. Although these generators are widely available individually often they aren't all available on the same platform. This makes it difficult to write portable code using them and so we have included the complete set of Unix generators in GSL for convenience. Note that these generators don't produce high-quality randomness and aren't suitable for work requiring accurate statistics. However, if you won't be measuring statistical quantities and just want to introduce some variation into your program then these generators are quite acceptable.

gsl_rng_rand   Generator

This is the BSD rand generator. Its sequence is

$$x_{n+1} = (ax_n + c) \bmod m$$

with $a = 1103515245$, $c = 12345$ and $m = 2^{31}$. The seed specifies the initial value, x_1. The period of this generator is 2^{31}, and it uses 1 word of storage per generator.

gsl_rng_random_bsd   Generator
gsl_rng_random_libc5   Generator
gsl_rng_random_glibc2   Generator

These generators implement the random family of functions, a set of linear feedback shift register generators originally used in BSD Unix. There are several versions of random in use today: the original BSD version (e.g. on SunOS4), a libc5 version (found on older GNU/Linux systems) and a glibc2 version. Each version uses a different seeding procedure, and thus produces different sequences.

The original BSD routines accepted a variable length buffer for the generator state, with longer buffers providing higher-quality randomness. The random function implemented algorithms for buffer lengths of 8, 32, 64, 128 and 256 bytes, and the algorithm with the largest length that would fit into the user-supplied buffer was used. To support these algorithms additional generators are available with the following names,

    gsl_rng_random8_bsd
    gsl_rng_random32_bsd
    gsl_rng_random64_bsd
    gsl_rng_random128_bsd
    gsl_rng_random256_bsd

where the numeric suffix indicates the buffer length. The original BSD random function used a 128-byte default buffer and so gsl_rng_random_bsd has been made equivalent to gsl_rng_random128_bsd. Corresponding versions of the libc5 and glibc2 generators are also available, with the names gsl_rng_random8_libc5, gsl_rng_random8_glibc2, etc.

gsl_rng_rand48   Generator

This is the Unix rand48 generator. Its sequence is

$$x_{n+1} = (ax_n + c) \bmod m$$

defined on 48-bit unsigned integers with $a = 25214903917$, $c = 11$ and $m = 2^{48}$. The seed specifies the upper 32 bits of the initial value, x_1, with the lower 16 bits set to 0x330E. The function gsl_rng_get returns the upper 32 bits from each term of the sequence. This does not have a direct parallel in the original rand48 functions, but forcing the result to type long int reproduces the output of mrand48. The function gsl_rng_uniform uses the full 48 bits of internal state to return the double precision number x_n/m, which is equivalent to the function drand48. Note that some versions of the GNU C Library contained a bug in mrand48 function which caused it to produce different results (only the lower 16-bits of the return value were set).

17.11 Other random number generators

The generators in this section are provided for compatibility with existing libraries. If you are converting an existing program to use GSL then you can select these generators to check your new implementation against the original one, using the same random number generator. After verifying that your new program reproduces the original results you can then switch to a higher-quality generator.

Note that most of the generators in this section are based on single linear congruence relations, which are the least sophisticated type of generator. In particular, linear congruences have poor properties when used with a non-prime modulus, as several of these routines do (e.g. with a power of two modulus, 2^{31} or 2^{32}). This leads to periodicity in the least significant bits of each number, with only the higher bits having any randomness. Thus if you want to produce a random bitstream it is best to avoid using the least significant bits.

gsl_rng_ranf   Generator

This is the CRAY random number generator RANF. Its sequence is

$$x_{n+1} = (ax_n) \bmod m$$

defined on 48-bit unsigned integers with $a = 44485709377909$ and $m = 2^{48}$. The seed specifies the lower 32 bits of the initial value, x_1, with the lowest bit set to prevent the seed taking an even value. The upper 16 bits of x_1 are set to 0. A consequence of this procedure is that the pairs of seeds 2 and 3, 4 and 5, etc produce the same sequences.

The generator compatible with the CRAY MATHLIB routine RANF. It produces double precision floating point numbers which should be identical to those from the original RANF.

There is a subtlety in the implementation of the seeding. The initial state is reversed through one step, by multiplying by the modular inverse of a mod m. This is done for compatibility with the original CRAY implementation.

Note that you can only seed the generator with integers up to 2^{32}, while the original CRAY implementation uses non-portable wide integers which can cover all 2^{48} states of the generator.

The function `gsl_rng_get` returns the upper 32 bits from each term of the sequence. The function `gsl_rng_uniform` uses the full 48 bits to return the double precision number x_n/m.

The period of this generator is 2^{46}.

`gsl_rng_ranmar`   Generator

This is the RANMAR lagged-fibonacci generator of Marsaglia, Zaman and Tsang. It is a 24-bit generator, originally designed for single-precision IEEE floating point numbers. It was included in the CERNLIB high-energy physics library.

`gsl_rng_r250`   Generator

This is the shift-register generator of Kirkpatrick and Stoll. The sequence is based on the recurrence

$$x_n = x_{n-103} \oplus x_{n-250}$$

where $\oplus$ denotes "exclusive-or", defined on 32-bit words. The period of this generator is about 2^{250} and it uses 250 words of state per generator.

For more information see,

> S. Kirkpatrick and E. Stoll, "A very fast shift-register sequence random number generator", *Journal of Computational Physics*, 40, 517–526 (1981)

`gsl_rng_tt800`   Generator

This is an earlier version of the twisted generalized feedback shift-register generator, and has been superseded by the development of MT19937. However, it is still an acceptable generator in its own right. It has a period of 2^{800} and uses 33 words of storage per generator.

For more information see,

> Makoto Matsumoto and Yoshiharu Kurita, "Twisted GFSR Generators II", *ACM Transactions on Modelling and Computer Simulation*, Vol. 4, No. 3, 1994, pages 254–266.

`gsl_rng_vax`   Generator

This is the VAX generator `MTH$RANDOM`. Its sequence is,

$$x_{n+1} = (a x_n + c) \bmod m$$

with $a = 69069$, $c = 1$ and $m = 2^{32}$. The seed specifies the initial value, x_1. The period of this generator is 2^{32} and it uses 1 word of storage per generator.

`gsl_rng_transputer`   Generator

This is the random number generator from the INMOS Transputer Development system. Its sequence is,

$$x_{n+1} = (a x_n) \bmod m$$

with $a = 1664525$ and $m = 2^{32}$. The seed specifies the initial value, x_1.

`gsl_rng_randu`   Generator

This is the IBM RANDU generator. Its sequence is

$$x_{n+1} = (a x_n) \bmod m$$

with $a = 65539$ and $m = 2^{31}$. The seed specifies the initial value, x_1. The period of this generator was only 2^{29}. It has become a textbook example of a poor generator.

`gsl_rng_minstd`   Generator

This is Park and Miller's "minimal standard" MINSTD generator, a simple linear congruence which takes care to avoid the major pitfalls of such algorithms. Its sequence is,

$$x_{n+1} = (a x_n) \bmod m$$

with $a = 16807$ and $m = 2^{31} - 1 = 2147483647$. The seed specifies the initial value, x_1. The period of this generator is about 2^{31}.

This generator is used in the IMSL Library (subroutine RNUN) and in MATLAB (the RAND function). It is also sometimes known by the acronym "GGL" (I'm not sure what that stands for).

For more information see,

> Park and Miller, "Random Number Generators: Good ones are hard to find", *Communications of the ACM*, October 1988, Volume 31, No 10, pages 1192–1201.

`gsl_rng_uni`   Generator
`gsl_rng_uni32`   Generator

This is a reimplementation of the 16-bit SLATEC random number generator RUNIF. A generalization of the generator to 32 bits is provided by `gsl_rng_uni32`. The original source code is available from NETLIB.

`gsl_rng_slatec`   Generator

This is the SLATEC random number generator RAND. It is ancient. The original source code is available from NETLIB.

`gsl_rng_zuf`   Generator

This is the ZUFALL lagged Fibonacci series generator of Peterson. Its sequence is,

$$t = u_{n-273} + u_{n-607}$$
$$u_n = t - \mathrm{floor}(t)$$

The original source code is available from NETLIB. For more information see,

W. Petersen, "Lagged Fibonacci Random Number Generators for the
NEC SX-3", *International Journal of High Speed Computing* (1994).

`gsl_rng_knuthran2`   Generator
This is a second-order multiple recursive generator described by Knuth in
Seminumerical Algorithms, 3rd Ed., page 108. Its sequence is,

$$x_n = (a_1 x_{n-1} + a_2 x_{n-2}) \bmod m$$

with $a_1 = 271828183$, $a_2 = 314159269$, and $m = 2^{31} - 1$.

`gsl_rng_knuthran2002`   Generator
`gsl_rng_knuthran`   Generator
This is a second-order multiple recursive generator described by Knuth in
Seminumerical Algorithms, 3rd Ed., Section 3.6. Knuth provides its C code.
The updated routine `gsl_rng_knuthran2002` is from the revised 9th printing
and corrects some weaknesses in the earlier version, which is implemented
as `gsl_rng_knuthran`.

`gsl_rng_borosh13`   Generator
`gsl_rng_fishman18`   Generator
`gsl_rng_fishman20`   Generator
`gsl_rng_lecuyer21`   Generator
`gsl_rng_waterman14`   Generator
These multiplicative generators are taken from Knuth's *Seminumerical Al-
gorithms*, 3rd Ed., pages 106–108. Their sequence is,

$$x_{n+1} = (a x_n) \bmod m$$

where the seed specifies the initial value, x_1. The parameters a and m
are as follows, Borosh-Niederreiter: $a = 1812433253$, $m = 2^{32}$, Fishman18:
$a = 62089911$, $m = 2^{31} - 1$, Fishman20: $a = 48271$, $m = 2^{31} - 1$, L'Ecuyer:
$a = 40692$, $m = 2^{31} - 249$, Waterman: $a = 1566083941$, $m = 2^{32}$.

`gsl_rng_fishman2x`   Generator
This is the L'Ecuyer–Fishman random number generator. It is taken from
Knuth's *Seminumerical Algorithms*, 3rd Ed., page 108. Its sequence is,

$$z_{n+1} = (x_n - y_n) \bmod m$$

with $m = 2^{31} - 1$. x_n and y_n are given by the `fishman20` and `lecuyer21`
algorithms. The seed specifies the initial value, x_1.

`gsl_rng_coveyou`   Generator
This is the Coveyou random number generator. It is taken from Knuth's
Seminumerical Algorithms, 3rd Ed., Section 3.2.2. Its sequence is,

$$x_{n+1} = (x_n(x_n + 1)) \bmod m$$

with $m = 2^{32}$. The seed specifies the initial value, x_1.

17.12 Performance

The following table shows the relative performance of a selection the available random number generators. The fastest simulation quality generators are taus, gfsr4 and mt19937. The generators which offer the best mathematically-proven quality are those based on the RANLUX algorithm.

```
1754 k ints/sec,    870 k doubles/sec, taus
1613 k ints/sec,    855 k doubles/sec, gfsr4
1370 k ints/sec,    769 k doubles/sec, mt19937
 565 k ints/sec,    571 k doubles/sec, ranlxs0
 400 k ints/sec,    405 k doubles/sec, ranlxs1
 490 k ints/sec,    389 k doubles/sec, mrg
 407 k ints/sec,    297 k doubles/sec, ranlux
 243 k ints/sec,    254 k doubles/sec, ranlxd1
 251 k ints/sec,    253 k doubles/sec, ranlxs2
 238 k ints/sec,    215 k doubles/sec, cmrg
 247 k ints/sec,    198 k doubles/sec, ranlux389
 141 k ints/sec,    140 k doubles/sec, ranlxd2

1852 k ints/sec,    935 k doubles/sec, ran3
 813 k ints/sec,    575 k doubles/sec, ran0
 787 k ints/sec,    476 k doubles/sec, ran1
 379 k ints/sec,    292 k doubles/sec, ran2
```

17.13 Examples

The following program demonstrates the use of a random number generator to produce uniform random numbers in the range [0.0, 1.0),

```c
#include <stdio.h>
#include <gsl/gsl_rng.h>

int
main (void)
{
  const gsl_rng_type * T;
  gsl_rng * r;

  int i, n = 10;

  gsl_rng_env_setup ();

  T = gsl_rng_default;
  r = gsl_rng_alloc (T);

  for (i = 0; i < n; i++)
    {
      double u = gsl_rng_uniform (r);
      printf ("%.5f\n", u);
```

```
            }

        gsl_rng_free (r);

        return 0;
      }
```

Here is the output of the program,

```
    $ ./a.out
      0.99974
      0.16291
      0.28262
      0.94720
      0.23166
      0.48497
      0.95748
      0.74431
      0.54004
      0.73995
```

The numbers depend on the seed used by the generator. The default seed can be changed with the GSL_RNG_SEED environment variable to produce a different stream of numbers. The generator itself can be changed using the environment variable GSL_RNG_TYPE. Here is the output of the program using a seed value of 123 and the multiple-recursive generator mrg,

```
    $ GSL_RNG_SEED=123 GSL_RNG_TYPE=mrg ./a.out
      GSL_RNG_TYPE=mrg
      GSL_RNG_SEED=123
      0.33050
      0.86631
      0.32982
      0.67620
      0.53391
      0.06457
      0.16847
      0.70229
      0.04371
      0.86374
```

17.14 References and Further Reading

The subject of random number generation and testing is reviewed extensively in Knuth's *Seminumerical Algorithms*.

> Donald E. Knuth, *The Art of Computer Programming: Seminumerical Algorithms* (Vol 2, 3rd Ed, 1997), Addison-Wesley, ISBN 0201896842.

Further information is available in the review paper written by Pierre L'Ecuyer,

> P. L'Ecuyer, "Random Number Generation", Chapter 4 of the Handbook on Simulation, Jerry Banks Ed., Wiley, 1998, 93–137.
>
> http://www.iro.umontreal.ca/~lecuyer/papers.html

The source code for the DIEHARD random number generator tests is also available online,

> *DIEHARD source code* G. Marsaglia,
>
> http://stat.fsu.edu/pub/diehard/

A comprehensive set of random number generator tests is available from NIST,

> NIST Special Publication 800-22, "A Statistical Test Suite for the Validation of Random Number Generators and Pseudo Random Number Generators for Cryptographic Applications".
>
> http://csrc.nist.gov/rng/

17.15 Acknowledgements

Thanks to Makoto Matsumoto, Takuji Nishimura and Yoshiharu Kurita for making the source code to their generators (MT19937, MM&TN; TT800, MM&YK) available under the GNU General Public License. Thanks to Martin Lüscher for providing notes and source code for the RANLXS and RANLXD generators.

18 Quasi-Random Sequences

This chapter describes functions for generating quasi-random sequences in arbitrary dimensions. A quasi-random sequence progressively covers a d-dimensional space with a set of points that are uniformly distributed. Quasi-random sequences are also known as low-discrepancy sequences. The quasi-random sequence generators use an interface that is similar to the interface for random number generators, except that seeding is not required—each generator produces a single sequence.

The functions described in this section are declared in the header file 'gsl_qrng.h'.

18.1 Quasi-random number generator initialization

gsl_qrng * gsl_qrng_alloc (const gsl_qrng_type * T, unsigned   *Function*
        int d)

This function returns a pointer to a newly-created instance of a quasi-random sequence generator of type T and dimension d. If there is insufficient memory to create the generator then the function returns a null pointer and the error handler is invoked with an error code of GSL_ENOMEM.

void gsl_qrng_free (gsl_qrng * q)   *Function*

This function frees all the memory associated with the generator q.

void gsl_qrng_init (gsl_qrng * q)   *Function*

This function reinitializes the generator q to its starting point. Note that quasi-random sequences do not use a seed and always produce the same set of values.

18.2 Sampling from a quasi-random number generator

int gsl_qrng_get (const gsl_qrng * q, double x[])   *Function*

This function stores the next point from the sequence generator q in the array x. The space available for x must match the dimension of the generator. The point x will lie in the range $0 < x_i < 1$ for each x_i. An inline version of this function is used when HAVE_INLINE is defined.

18.3 Auxiliary quasi-random number generator functions

const char * gsl_qrng_name (const gsl_qrng * q)   Function
    This function returns a pointer to the name of the generator.

size_t gsl_qrng_size (const gsl_qrng * q)   Function
void * gsl_qrng_state (const gsl_qrng * q)   Function
    These functions return a pointer to the state of generator r and its size.
    You can use this information to access the state directly. For example, the
    following code will write the state of a generator to a stream,

```
void * state = gsl_qrng_state (q);
size_t n = gsl_qrng_size (q);
fwrite (state, n, 1, stream);
```

18.4 Saving and resorting quasi-random number generator state

int gsl_qrng_memcpy (gsl_qrng * dest, const gsl_qrng * src)   Function
    This function copies the quasi-random sequence generator src into the pre-
    existing generator dest, making dest into an exact copy of src. The two
    generators must be of the same type.

gsl_qrng * gsl_qrng_clone (const gsl_qrng * q)   Function
    This function returns a pointer to a newly created generator which is an
    exact copy of the generator q.

18.5 Quasi-random number generator algorithms

The following quasi-random sequence algorithms are available,

gsl_qrng_niederreiter_2   Generator
    This generator uses the algorithm described in Bratley, Fox, Niederreiter,
    ACM Trans. Model. Comp. Sim. 2, 195 (1992). It is valid up to 12
    dimensions.

gsl_qrng_sobol   Generator
    This generator uses the Sobol sequence described in Antonov, Saleev, USSR
    Comput. Maths. Math. Phys. 19, 252 (1980). It is valid up to 40 dimen-
    sions.

gsl_qrng_halton   Generator
gsl_qrng_reversehalton   Generator
    These generators use the Halton and reverse Halton sequences described in
    J.H. Halton, Numerische Mathematik 2, 84-90 (1960) and B. Vandewoestyne
    and R. Cools Computational and Applied Mathematics 189, 1&2, 341-361
    (2006). They are valid up to 1229 dimensions.

18.6 Examples

The following program prints the first 1024 points of the 2-dimensional Sobol sequence.

```
#include <stdio.h>
#include <gsl/gsl_qrng.h>

int
main (void)
{
  int i;
  gsl_qrng * q = gsl_qrng_alloc (gsl_qrng_sobol, 2);

  for (i = 0; i < 1024; i++)
    {
      double v[2];
      gsl_qrng_get (q, v);
      printf ("%.5f %.5f\n", v[0], v[1]);
    }

  gsl_qrng_free (q);
  return 0;
}
```

Here is the output from the program,

```
$ ./a.out
0.50000 0.50000
0.75000 0.25000
0.25000 0.75000
0.37500 0.37500
0.87500 0.87500
0.62500 0.12500
0.12500 0.62500
....
```

It can be seen that successive points progressively fill-in the spaces between previous points.

The following plot shows the distribution in the x-y plane of the first 1024 points from the Sobol sequence,

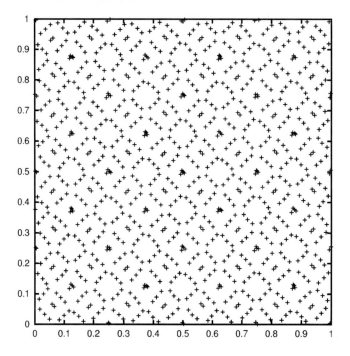

Distribution of the first 1024 points
from the quasi-random Sobol sequence

18.7 References

The implementations of the quasi-random sequence routines are based on the algorithms described in the following paper,

P. Bratley and B.L. Fox and H. Niederreiter, "Algorithm 738: Programs to Generate Niederreiter's Low-discrepancy Sequences", *ACM Transactions on Mathematical Software*, Vol. 20, No. 4, December, 1994, p. 494–495.

19 Random Number Distributions

This chapter describes functions for generating random variates and computing their probability distributions. Samples from the distributions described in this chapter can be obtained using any of the random number generators in the library as an underlying source of randomness.

In the simplest cases a non-uniform distribution can be obtained analytically from the uniform distribution of a random number generator by applying an appropriate transformation. This method uses one call to the random number generator. More complicated distributions are created by the *acceptance-rejection* method, which compares the desired distribution against a distribution which is similar and known analytically. This usually requires several samples from the generator.

The library also provides cumulative distribution functions and inverse cumulative distribution functions, sometimes referred to as quantile functions. The cumulative distribution functions and their inverses are computed separately for the upper and lower tails of the distribution, allowing full accuracy to be retained for small results.

The functions for random variates and probability density functions described in this section are declared in 'gsl_randist.h'. The corresponding cumulative distribution functions are declared in 'gsl_cdf.h'.

Note that the discrete random variate functions always return a value of type unsigned int, and on most platforms this has a maximum value of $2^{32} - 1 \approx 4.29 \times 10^9$. They should only be called with a safe range of parameters (where there is a negligible probability of a variate exceeding this limit) to prevent incorrect results due to overflow.

19.1 Introduction

Continuous random number distributions are defined by a probability density function, $p(x)$, such that the probability of x occurring in the infinitesimal range x to $x + dx$ is $p\,dx$.

The cumulative distribution function for the lower tail $P(x)$ is defined by the integral,

$$P(x) = \int_{-\infty}^{x} dx' p(x')$$

and gives the probability of a variate taking a value less than x.

The cumulative distribution function for the upper tail $Q(x)$ is defined by the integral,

$$Q(x) = \int_{x}^{+\infty} dx' p(x')$$

and gives the probability of a variate taking a value greater than x.

The upper and lower cumulative distribution functions are related by $P(x) + Q(x) = 1$ and satisfy $0 \le P(x) \le 1$, $0 \le Q(x) \le 1$.

The inverse cumulative distributions, $x = P^{-1}(P)$ and $x = Q^{-1}(Q)$ give the values of x which correspond to a specific value of P or Q. They can be used to find confidence limits from probability values.

For discrete distributions the probability of sampling the integer value k is given by $p(k)$, where $\sum_k p(k) = 1$. The cumulative distribution for the lower tail $P(k)$ of a discrete distribution is defined as,

$$P(k) = \sum_{i \le k} p(i)$$

where the sum is over the allowed range of the distribution less than or equal to k.

The cumulative distribution for the upper tail of a discrete distribution $Q(k)$ is defined as

$$Q(k) = \sum_{i > k} p(i)$$

giving the sum of probabilities for all values greater than k. These two definitions satisfy the identity $P(k) + Q(k) = 1$.

If the range of the distribution is 1 to n inclusive then $P(n) = 1$, $Q(n) = 0$ while $P(1) = p(1)$, $Q(1) = 1 - p(1)$.

19.2 The Gaussian Distribution

double gsl_ran_gaussian (const gsl_rng * r, double *sigma*)   Function
   This function returns a Gaussian random variate, with mean zero and standard deviation *sigma*. The probability distribution for Gaussian random variates is,

$$p(x)dx = \frac{1}{\sqrt{2\pi\sigma^2}} \exp(-x^2/2\sigma^2)dx$$

   for x in the range $-\infty$ to $+\infty$. Use the transformation $z = \mu + x$ on the numbers returned by gsl_ran_gaussian to obtain a Gaussian distribution with mean μ. This function uses the Box-Mueller algorithm which requires two calls to the random number generator r.

double gsl_ran_gaussian_pdf (double x, double *sigma*)   Function
   This function computes the probability density $p(x)$ at x for a Gaussian distribution with standard deviation *sigma*, using the formula given above.

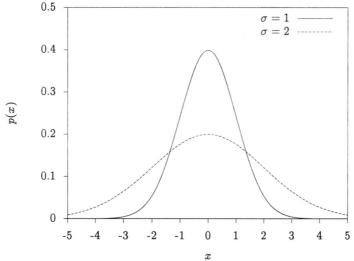

Gaussian Distribution

double gsl_ran_gaussian_ziggurat (const gsl_rng * r, double   Function
         sigma)
double gsl_ran_gaussian_ratio_method (const gsl_rng * r,   Function
         double *sigma*)
   This function computes a Gaussian random variate using the alternative Marsaglia-Tsang ziggurat and Kinderman-Monahan-Leva ratio methods. The Ziggurat algorithm is the fastest available algorithm in most cases.

double gsl_ran_ugaussian (const gsl_rng * r)   Function
double gsl_ran_ugaussian_pdf (double x)   Function
double gsl_ran_ugaussian_ratio_method (const gsl_rng * r)   Function
  These functions compute results for the unit Gaussian distribution. They
  are equivalent to the functions above with a standard deviation of one, *sigma*
  = 1.

double gsl_cdf_gaussian_P (double x, double *sigma*)   Function
double gsl_cdf_gaussian_Q (double x, double *sigma*)   Function
double gsl_cdf_gaussian_Pinv (double P, double *sigma*)   Function
double gsl_cdf_gaussian_Qinv (double Q, double *sigma*)   Function
  These functions compute the cumulative distribution functions $P(x)$, $Q(x)$
  and their inverses for the Gaussian distribution with standard deviation
  sigma.

double gsl_cdf_ugaussian_P (double x)   Function
double gsl_cdf_ugaussian_Q (double x)   Function
double gsl_cdf_ugaussian_Pinv (double P)   Function
double gsl_cdf_ugaussian_Qinv (double Q)   Function
  These functions compute the cumulative distribution functions $P(x)$, $Q(x)$
  and their inverses for the unit Gaussian distribution.

19.3 The Gaussian Tail Distribution

double gsl_ran_gaussian_tail (const gsl_rng * r, double a,   Function
        double *sigma*)
   This function provides random variates from the upper tail of a Gaussian
   distribution with standard deviation *sigma*. The values returned are larger
   than the lower limit a, which must be positive. The method is based on
   Marsaglia's famous rectangle-wedge-tail algorithm (Ann. Math. Stat. 32,
   894–899 (1961)), with this aspect explained in Knuth, v2, 3rd ed, p139,586
   (exercise 11).

   The probability distribution for Gaussian tail random variates is,

   $$p(x)dx = \frac{1}{N(a;\sigma)\sqrt{2\pi\sigma^2}} \exp(-x^2/2\sigma^2)dx$$

   for $x > a$ where $N(a;\sigma)$ is the normalization constant,

   $$N(a;\sigma) = \frac{1}{2}\text{erfc}\left(\frac{a}{\sqrt{2\sigma^2}}\right).$$

double gsl_ran_gaussian_tail_pdf (double x, double a, double   Function
        sigma)
   This function computes the probability density $p(x)$ at x for a Gaussian
   tail distribution with standard deviation *sigma* and lower limit a, using the
   formula given above.

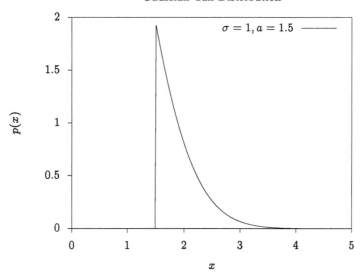

Gaussian Tail Distribution

double `gsl_ran_ugaussian_tail` (const `gsl_rng` * r, double a)   Function
double `gsl_ran_ugaussian_tail_pdf` (double x, double a)   Function
  These functions compute results for the tail of a unit Gaussian distribution.
  They are equivalent to the functions above with a standard deviation of one,
  sigma = 1.

19.4 The Bivariate Gaussian Distribution

void gsl_ran_bivariate_gaussian (const gsl_rng * r, double   Function
         sigma_x, double sigma_y, double rho, double * x, double * y)
    This function generates a pair of correlated Gaussian variates, with mean
    zero, correlation coefficient rho and standard deviations sigma_x and
    sigma_y in the x and y directions. The probability distribution for bivariate
    Gaussian random variates is,

$$p(x,y)dxdy = \frac{1}{2\pi\sigma_x\sigma_y\sqrt{1-\rho^2}} \exp\left(-\frac{(x^2/\sigma_x^2 + y^2/\sigma_y^2 - 2\rho xy/(\sigma_x\sigma_y))}{2(1-\rho^2)}\right) dxdy$$

    for x,y in the range $-\infty$ to $+\infty$. The correlation coefficient rho should lie
    between 1 and -1.

double gsl_ran_bivariate_gaussian_pdf (double x, double y,   Function
         double sigma_x, double sigma_y, double rho)
    This function computes the probability density $p(x,y)$ at (x,y) for a bivari-
    ate Gaussian distribution with standard deviations sigma_x, sigma_y and
    correlation coefficient rho, using the formula given above.

Bivariate Gaussian Distribution

$$\sigma_x = 1, \sigma_y = 1, \rho = 0.9$$

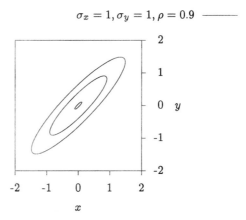

19.5 The Exponential Distribution

double gsl_ran_exponential (const gsl_rng * r, double *mu*)   Function
  This function returns a random variate from the exponential distribution
  with mean *mu*. The distribution is,

$$p(x)dx = \frac{1}{\mu} \exp(-x/\mu)dx$$

  for $x \geq 0$.

double gsl_ran_exponential_pdf (double x, double *mu*)   Function
  This function computes the probability density $p(x)$ at x for an exponential
  distribution with mean *mu*, using the formula given above.

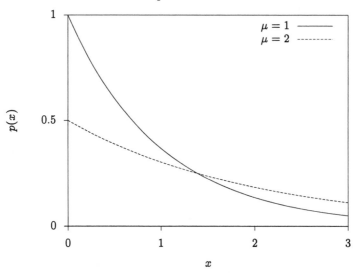

Exponential Distribution

double gsl_cdf_exponential_P (double x, double *mu*)   Function
double gsl_cdf_exponential_Q (double x, double *mu*)   Function
double gsl_cdf_exponential_Pinv (double P, double *mu*)   Function
double gsl_cdf_exponential_Qinv (double Q, double *mu*)   Function
  These functions compute the cumulative distribution functions $P(x)$, $Q(x)$
  and their inverses for the exponential distribution with mean *mu*.

19.6 The Laplace Distribution

double gsl_ran_laplace (const gsl_rng * r, double a)   Function
  This function returns a random variate from the Laplace distribution with
  width a. The distribution is,

$$p(x)dx = \frac{1}{2a} \exp(-|x/a|)dx$$

  for $-\infty < x < \infty$.

double gsl_ran_laplace_pdf (double x, double a)   Function
  This function computes the probability density $p(x)$ at x for a Laplace distribution with width a, using the formula given above.

Laplace Distribution (Two-sided Exponential)

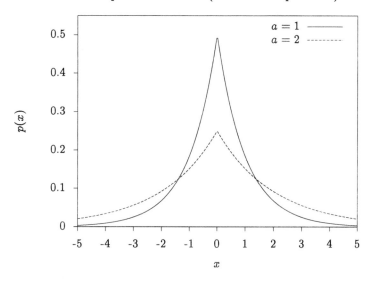

double gsl_cdf_laplace_P (double x, double a)   Function
double gsl_cdf_laplace_Q (double x, double a)   Function
double gsl_cdf_laplace_Pinv (double P, double a)   Function
double gsl_cdf_laplace_Qinv (double Q, double a)   Function
  These functions compute the cumulative distribution functions $P(x)$, $Q(x)$
  and their inverses for the Laplace distribution with width a.

19.7 The Exponential Power Distribution

double gsl_ran_exppow (const gsl_rng * r, double a, double b)   *Function*
This function returns a random variate from the exponential power distribution with scale parameter a and exponent b. The distribution is,

$$p(x)dx = \frac{1}{2a\Gamma(1 + 1/b)} \exp(-|x/a|^b)dx$$

for $x \geq 0$. For $b = 1$ this reduces to the Laplace distribution. For $b = 2$ it has the same form as a gaussian distribution, but with $a = \sqrt{2}\sigma$.

double gsl_ran_exppow_pdf (double x, double a, double b)   *Function*
This function computes the probability density $p(x)$ at x for an exponential power distribution with scale parameter a and exponent b, using the formula given above.

Exponential Power Distribution

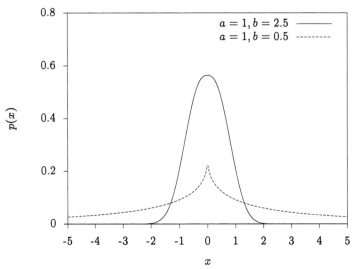

double gsl_cdf_exppow_P (double x, double a, double b)   *Function*
double gsl_cdf_exppow_Q (double x, double a, double b)   *Function*
These functions compute the cumulative distribution functions $P(x)$, $Q(x)$ for the exponential power distribution with parameters a and b.

19.8 The Cauchy Distribution

double gsl_ran_cauchy (const gsl_rng * r, double a)   Function
This function returns a random variate from the Cauchy distribution with
scale parameter a. The probability distribution for Cauchy random variates
is,

$$p(x)dx = \frac{1}{a\pi(1 + (x/a)^2)}dx$$

for x in the range $-\infty$ to $+\infty$. The Cauchy distribution is also known as
the Lorentz distribution.

double gsl_ran_cauchy_pdf (double x, double a)   Function
This function computes the probability density $p(x)$ at x for a Cauchy distribution with scale parameter a, using the formula given above.

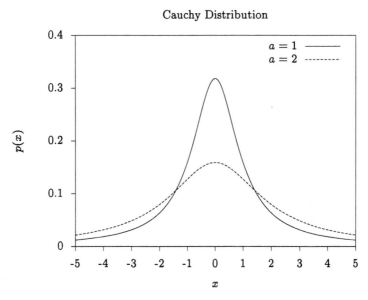

double gsl_cdf_cauchy_P (double x, double a)   Function
double gsl_cdf_cauchy_Q (double x, double a)   Function
double gsl_cdf_cauchy_Pinv (double P, double a)   Function
double gsl_cdf_cauchy_Qinv (double Q, double a)   Function
These functions compute the cumulative distribution functions $P(x)$, $Q(x)$
and their inverses for the Cauchy distribution with scale parameter a.

19.9 The Rayleigh Distribution

double gsl_ran_rayleigh (const gsl_rng * r, double *sigma*)   Function
This function returns a random variate from the Rayleigh distribution with
scale parameter *sigma*. The distribution is,

$$p(x)dx = \frac{x}{\sigma^2} \exp(-x^2/(2\sigma^2))dx$$

for $x > 0$.

double gsl_ran_rayleigh_pdf (double *x*, double *sigma*)   Function
This function computes the probability density $p(x)$ at x for a Rayleigh
distribution with scale parameter *sigma*, using the formula given above.

Rayleigh Distribution

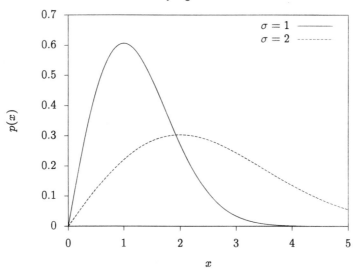

double gsl_cdf_rayleigh_P (double *x*, double *sigma*)   Function
double gsl_cdf_rayleigh_Q (double *x*, double *sigma*)   Function
double gsl_cdf_rayleigh_Pinv (double *P*, double *sigma*)   Function
double gsl_cdf_rayleigh_Qinv (double *Q*, double *sigma*)   Function
These functions compute the cumulative distribution functions $P(x)$, $Q(x)$
and their inverses for the Rayleigh distribution with scale parameter *sigma*.

19.10 The Rayleigh Tail Distribution

double gsl_ran_rayleigh_tail (const gsl_rng * r, double a,   *Function*
    double *sigma*)

This function returns a random variate from the tail of the Rayleigh distribution with scale parameter *sigma* and a lower limit of a. The distribution is,

$$p(x)dx = \frac{x}{\sigma^2} \exp((a^2 - x^2)/(2\sigma^2))dx$$

for $x > a$.

double gsl_ran_rayleigh_tail_pdf (double x, double a, double   *Function*
    sigma)

This function computes the probability density $p(x)$ at x for a Rayleigh tail distribution with scale parameter *sigma* and lower limit a, using the formula given above.

Rayleigh Tail Distribution

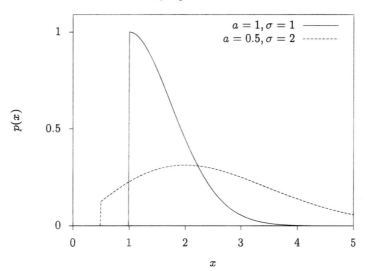

19.11 The Landau Distribution

double gsl_ran_landau (const gsl_rng * r)   Function
  This function returns a random variate from the Landau distribution. The
  probability distribution for Landau random variates is defined analytically
  by the complex integral,

$$p(x) = \frac{1}{2\pi i} \int_{c-i\infty}^{c+i\infty} ds\ \exp(s \log(s) + xs)$$

For numerical purposes it is more convenient to use the following equivalent
form of the integral,

$$p(x) = (1/\pi) \int_{0}^{\infty} dt\ \exp(-t \log(t) - xt) \sin(\pi t).$$

double gsl_ran_landau_pdf (double x)   Function
  This function computes the probability density $p(x)$ at x for the Landau
  distribution using an approximation to the formula given above.

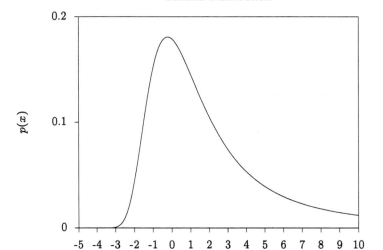

Landau Distribution

19.12 The Levy alpha-Stable Distributions

double gsl_ran_levy (const gsl_rng * r, double c, double *alpha*)   Function
   This function returns a random variate from the Levy symmetric stable distribution with scale c and exponent *alpha*. The symmetric stable probability distribution is defined by a fourier transform,

$$p(x) = \frac{1}{2\pi} \int_{-\infty}^{+\infty} dt \exp(-itx - |ct|^{\alpha})$$

There is no explicit solution for the form of $p(x)$ and the library does not define a corresponding pdf function. For $\alpha = 1$ the distribution reduces to the Cauchy distribution. For $\alpha = 2$ it is a Gaussian distribution with $\sigma = \sqrt{2}c$. For $\alpha < 1$ the tails of the distribution become extremely wide. The algorithm only works for $0 < \alpha \leq 2$.

Levy Distribution

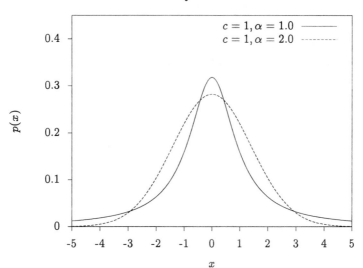

19.13 The Levy skew alpha-Stable Distribution

double gsl_ran_levy_skew (const gsl_rng * r, double c, double   *Function*
        alpha, double beta)
    This function returns a random variate from the Levy skew stable distri-
    bution with scale c, exponent alpha and skewness parameter beta. The
    skewness parameter must lie in the range $[-1, 1]$. The Levy skew stable
    probability distribution is defined by a fourier transform,

$$p(x) = \frac{1}{2\pi} \int_{-\infty}^{+\infty} dt \, \exp(-itx - |ct|^\alpha (1 - i\beta \text{sign}(t) \tan(\pi\alpha/2)))$$

When $\alpha = 1$ the term $\tan(\pi\alpha/2)$ is replaced by $-(2/\pi) \log |t|$. There is
no explicit solution for the form of $p(x)$ and the library does not define
a corresponding pdf function. For $\alpha = 2$ the distribution reduces to a
Gaussian distribution with $\sigma = \sqrt{2}c$ and the skewness parameter has no
effect. For $\alpha < 1$ the tails of the distribution become extremely wide. The
symmetric distribution corresponds to $\beta = 0$.

The algorithm only works for $0 < \alpha \le 2$.

The Levy alpha-stable distributions have the property that if N alpha-stable
variates are drawn from the distribution $p(c, \alpha, \beta)$ then the sum $Y = X_1 + X_2 + \ldots + X_N$ will also be distributed as an alpha-stable variate, $p(N^{1/\alpha}c, \alpha, \beta)$.

Levy Skew Distribution

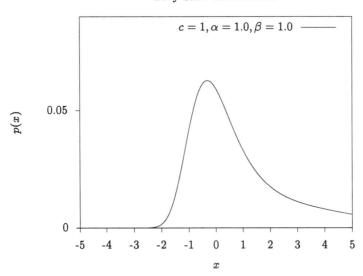

19.14 The Gamma Distribution

double gsl_ran_gamma (const gsl_rng * r, double a, double b)   Function
This function returns a random variate from the gamma distribution. The distribution function is,

$$p(x)dx = \frac{1}{\Gamma(a)b^a} x^{a-1} e^{-x/b} dx$$

for $x > 0$.

The gamma distribution with an integer parameter a is known as the Erlang distribution.

The variates are computed using the Marsaglia-Tsang fast gamma method. This function for this method was previously called gsl_ran_gamma_mt and can still be accessed using this name.

double gsl_ran_gamma_knuth (const gsl_rng * r, double a,   Function
        double b)
This function returns a gamma variate using the algorithms from Knuth (vol 2).

double gsl_ran_gamma_pdf (double x, double a, double b)   Function
This function computes the probability density $p(x)$ at x for a gamma distribution with parameters a and b, using the formula given above.

Gamma Distribution

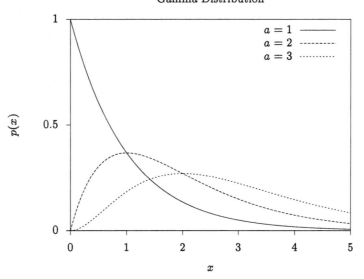

double `gsl_cdf_gamma_P` (double *x*, double *a*, double *b*)   Function
double `gsl_cdf_gamma_Q` (double *x*, double *a*, double *b*)   Function
double `gsl_cdf_gamma_Pinv` (double *P*, double *a*, double *b*)   Function
double `gsl_cdf_gamma_Qinv` (double *Q*, double *a*, double *b*)   Function

These functions compute the cumulative distribution functions $P(x)$, $Q(x)$ and their inverses for the gamma distribution with parameters a and b.

19.15 The Flat (Uniform) Distribution

double gsl_ran_flat (const gsl_rng * r, double a, double b)   Function
    This function returns a random variate from the flat (uniform) distribution
    from a to b. The distribution is,

$$p(x)dx = \frac{1}{(b-a)}dx$$

if $a \le x < b$ and 0 otherwise.

double gsl_ran_flat_pdf (double x, double a, double b)   Function
    This function computes the probability density $p(x)$ at x for a uniform dis-
    tribution from a to b, using the formula given above.

Flat Distribution

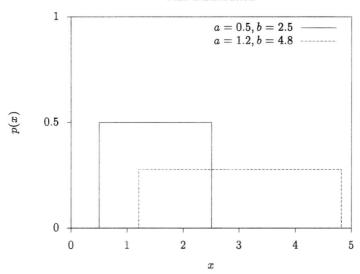

double gsl_cdf_flat_P (double x, double a, double b)   Function
double gsl_cdf_flat_Q (double x, double a, double b)   Function
double gsl_cdf_flat_Pinv (double P, double a, double b)   Function
double gsl_cdf_flat_Qinv (double Q, double a, double b)   Function
    These functions compute the cumulative distribution functions $P(x)$, $Q(x)$
    and their inverses for a uniform distribution from a to b.

19.16 The Lognormal Distribution

double gsl_ran_lognormal (const gsl_rng * r, double *zeta*,   Function
        double *sigma*)

This function returns a random variate from the lognormal distribution. The distribution function is,

$$p(x)dx = \frac{1}{x\sqrt{2\pi\sigma^2}} \exp(-(\ln(x) - \zeta)^2/2\sigma^2)dx$$

for $x > 0$.

double gsl_ran_lognormal_pdf (double *x*, double *zeta*, double   Function
        sigma)

This function computes the probability density $p(x)$ at x for a lognormal distribution with parameters *zeta* and *sigma*, using the formula given above.

Lognormal Distribution

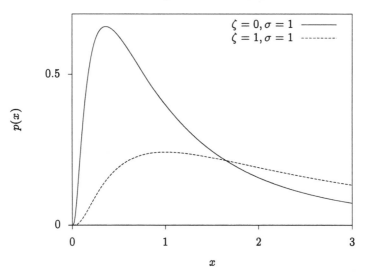

double gsl_cdf_lognormal_P (double *x*, double *zeta*, double   Function
        sigma)

double gsl_cdf_lognormal_Q (double *x*, double *zeta*, double   Function
        sigma)

double gsl_cdf_lognormal_Pinv (double *P*, double *zeta*, double   Function
        sigma)

double gsl_cdf_lognormal_Qinv (double *Q*, double *zeta*, double   Function
        sigma)

These functions compute the cumulative distribution functions $P(x)$, $Q(x)$ and their inverses for the lognormal distribution with parameters *zeta* and *sigma*.

19.17 The Chi-squared Distribution

The chi-squared distribution arises in statistics. If Y_i are n independent gaussian random variates with unit variance then the sum-of-squares,

$$X_i = \sum_i Y_i^2$$

has a chi-squared distribution with n degrees of freedom.

double gsl_ran_chisq (const gsl_rng * r, double nu)   *Function*
    This function returns a random variate from the chi-squared distribution with *nu* degrees of freedom. The distribution function is,

$$p(x)dx = \frac{1}{2\Gamma(\nu/2)}(x/2)^{\nu/2-1}\exp(-x/2)dx$$

    for $x \geq 0$.

double gsl_ran_chisq_pdf (double x, double nu)   *Function*
    This function computes the probability density $p(x)$ at x for a chi-squared distribution with *nu* degrees of freedom, using the formula given above.

Chi-squared Distribution

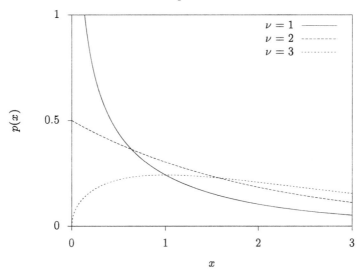

double gsl_cdf_chisq_P (double x, double nu)   *Function*
double gsl_cdf_chisq_Q (double x, double nu)   *Function*
double gsl_cdf_chisq_Pinv (double P, double nu)   *Function*
double gsl_cdf_chisq_Qinv (double Q, double nu)   *Function*
    These functions compute the cumulative distribution functions $P(x)$, $Q(x)$ and their inverses for the chi-squared distribution with *nu* degrees of freedom.

19.18 The F-distribution

The F-distribution arises in statistics. If Y_1 and Y_2 are chi-squared deviates with ν_1 and ν_2 degrees of freedom then the ratio,

$$X = \frac{(Y_1/\nu_1)}{(Y_2/\nu_2)}$$

has an F-distribution $F(x; \nu_1, \nu_2)$.

double gsl_ran_fdist (const gsl_rng * r, double *nu1*, double   *Function*
        nu2)

This function returns a random variate from the F-distribution with degrees of freedom *nu1* and *nu2*. The distribution function is,

$$p(x)dx = \frac{\Gamma((\nu_1 + \nu_2)/2)}{\Gamma(\nu_1/2)\Gamma(\nu_2/2)} \nu_1^{\nu_1/2} \nu_2^{\nu_2/2} x^{\nu_1/2-1} (\nu_2 + \nu_1 x)^{-\nu_1/2-\nu_2/2}$$

for $x \geq 0$.

double gsl_ran_fdist_pdf (double *x*, double *nu1*, double *nu2*)   *Function*

This function computes the probability density $p(x)$ at x for an F-distribution with *nu1* and *nu2* degrees of freedom, using the formula given above.

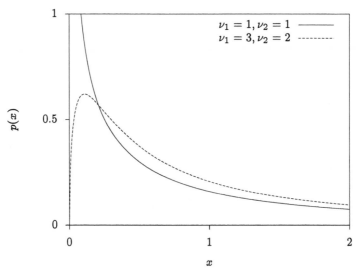

F-Distribution

double gsl_cdf_fdist_P (double x, double *nu1*, double *nu2*)   Function
double gsl_cdf_fdist_Q (double x, double *nu1*, double *nu2*)   Function
double gsl_cdf_fdist_Pinv (double P, double *nu1*, double *nu2*)   Function
double gsl_cdf_fdist_Qinv (double Q, double *nu1*, double *nu2*)   Function

These functions compute the cumulative distribution functions $P(x)$, $Q(x)$ and their inverses for the F-distribution with *nu1* and *nu2* degrees of freedom.

19.19 The t-distribution

The t-distribution arises in statistics. If Y_1 has a normal distribution and Y_2 has a chi-squared distribution with ν degrees of freedom then the ratio,

$$X = \frac{Y_1}{\sqrt{Y_2/\nu}}$$

has a t-distribution $t(x; \nu)$ with ν degrees of freedom.

double gsl_ran_tdist (const gsl_rng * r, double nu)   Function
    This function returns a random variate from the t-distribution. The distribution function is,

$$p(x)dx = \frac{\Gamma((\nu+1)/2)}{\sqrt{\pi\nu}\Gamma(\nu/2)}(1+x^2/\nu)^{-(\nu+1)/2}dx$$

    for $-\infty < x < +\infty$.

double gsl_ran_tdist_pdf (double x, double nu)   Function
    This function computes the probability density $p(x)$ at x for a t-distribution with nu degrees of freedom, using the formula given above.

Student's t distribution

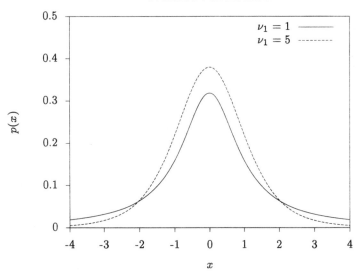

double gsl_cdf_tdist_P (double x, double nu)   Function
double gsl_cdf_tdist_Q (double x, double nu)   Function
double gsl_cdf_tdist_Pinv (double P, double nu)   Function
double gsl_cdf_tdist_Qinv (double Q, double nu)   Function
    These functions compute the cumulative distribution functions $P(x)$, $Q(x)$ and their inverses for the t-distribution with nu degrees of freedom.

19.20 The Beta Distribution

double gsl_ran_beta (const gsl_rng * r, double a, double b)   Function
  This function returns a random variate from the beta distribution. The
  distribution function is,

$$p(x)dx = \frac{\Gamma(a+b)}{\Gamma(a)\Gamma(b)} x^{a-1}(1-x)^{b-1} dx$$

  for $0 \leq x \leq 1$.

double gsl_ran_beta_pdf (double x, double a, double b)   Function
  This function computes the probability density $p(x)$ at x for a beta distribution with parameters a and b, using the formula given above.

Beta Distribution

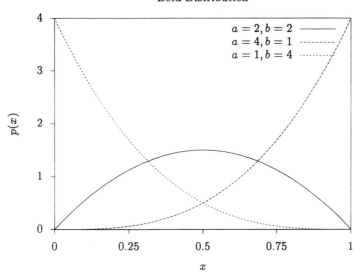

double gsl_cdf_beta_P (double x, double a, double b)   Function
double gsl_cdf_beta_Q (double x, double a, double b)   Function
double gsl_cdf_beta_Pinv (double P, double a, double b)   Function
double gsl_cdf_beta_Qinv (double Q, double a, double b)   Function
  These functions compute the cumulative distribution functions $P(x)$, $Q(x)$
  and their inverses for the beta distribution with parameters a and b.

19.21 The Logistic Distribution

double gsl_ran_logistic (const gsl_rng * r, double a)   Function
    This function returns a random variate from the logistic distribution. The
    distribution function is,

$$p(x)dx = \frac{\exp(-x/a)}{a(1 + \exp(-x/a))^2} dx$$

for $-\infty < x < +\infty$.

double gsl_ran_logistic_pdf (double x, double a)   Function
    This function computes the probability density $p(x)$ at x for a logistic dis-
    tribution with scale parameter a, using the formula given above.

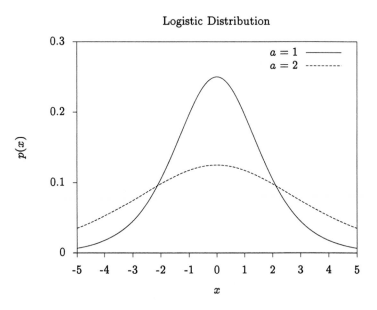

double gsl_cdf_logistic_P (double x, double a)   Function
double gsl_cdf_logistic_Q (double x, double a)   Function
double gsl_cdf_logistic_Pinv (double P, double a)   Function
double gsl_cdf_logistic_Qinv (double Q, double a)   Function
    These functions compute the cumulative distribution functions $P(x)$, $Q(x)$
    and their inverses for the logistic distribution with scale parameter a.

19.22 The Pareto Distribution

double gsl_ran_pareto (const gsl_rng * r, double a, double b)   Function
This function returns a random variate from the Pareto distribution of order a. The distribution function is,

$$p(x)dx = (a/b)/(x/b)^{a+1}dx$$

for $x \geq b$.

double gsl_ran_pareto_pdf (double x, double a, double b)   Function
This function computes the probability density $p(x)$ at x for a Pareto distribution with exponent a and scale b, using the formula given above.

Pareto Distribution

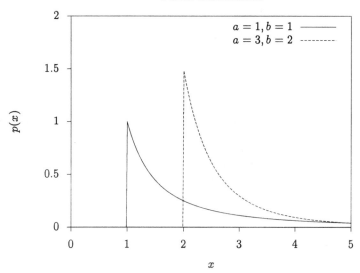

double gsl_cdf_pareto_P (double x, double a, double b)   Function
double gsl_cdf_pareto_Q (double x, double a, double b)   Function
double gsl_cdf_pareto_Pinv (double P, double a, double b)   Function
double gsl_cdf_pareto_Qinv (double Q, double a, double b)   Function
These functions compute the cumulative distribution functions $P(x)$, $Q(x)$ and their inverses for the Pareto distribution with exponent a and scale b.

19.23 Spherical Vector Distributions

The spherical distributions generate random vectors, located on a spherical surface. They can be used as random directions, for example in the steps of a random walk.

void **gsl_ran_dir_2d** (const gsl_rng * r, double * x, double * y)   *Function*
void **gsl_ran_dir_2d_trig_method** (const gsl_rng * r, double *   *Function*
        x, double * y)

This function returns a random direction vector $v = (x,y)$ in two dimensions. The vector is normalized such that $|v|^2 = x^2 + y^2 = 1$. The obvious way to do this is to take a uniform random number between 0 and 2π and let x and y be the sine and cosine respectively. Two trig functions would have been expensive in the old days, but with modern hardware implementations, this is sometimes the fastest way to go. This is the case for the Pentium (but not the case for the Sun Sparcstation). One can avoid the trig evaluations by choosing x and y in the interior of a unit circle (choose them at random from the interior of the enclosing square, and then reject those that are outside the unit circle), and then dividing by $\sqrt{x^2 + y^2}$. A much cleverer approach, attributed to von Neumann (See Knuth, v2, 3rd ed, p140, exercise 23), requires neither trig nor a square root. In this approach, u and v are chosen at random from the interior of a unit circle, and then $x = (u^2 - v^2)/(u^2 + v^2)$ and $y = 2uv/(u^2 + v^2)$.

void **gsl_ran_dir_3d** (const gsl_rng * r, double * x, double * y,   *Function*
        double * z)

This function returns a random direction vector $v = (x,y,z)$ in three dimensions. The vector is normalized such that $|v|^2 = x^2 + y^2 + z^2 = 1$. The method employed is due to Robert E. Knop (CACM 13, 326 (1970)), and explained in Knuth, v2, 3rd ed, p136. It uses the surprising fact that the distribution projected along any axis is actually uniform (this is only true for 3 dimensions).

void **gsl_ran_dir_nd** (const gsl_rng * r, size_t n, double * x)   *Function*

This function returns a random direction vector $v = (x_1, x_2, \ldots, x_n)$ in n dimensions. The vector is normalized such that $|v|^2 = x_1^2 + x_2^2 + \cdots + x_n^2 = 1$. The method uses the fact that a multivariate gaussian distribution is spherically symmetric. Each component is generated to have a gaussian distribution, and then the components are normalized. The method is described by Knuth, v2, 3rd ed, p135–136, and attributed to G. W. Brown, Modern Mathematics for the Engineer (1956).

19.24 The Weibull Distribution

double gsl_ran_weibull (const gsl_rng * r, double a, double b)   Function
   This function returns a random variate from the Weibull distribution. The
   distribution function is,

$$p(x)dx = \frac{b}{a^b} x^{b-1} \exp(-(x/a)^b)dx$$

for $x \geq 0$.

double gsl_ran_weibull_pdf (double x, double a, double b)   Function
   This function computes the probability density $p(x)$ at x for a Weibull distribution with scale a and exponent b, using the formula given above.

Weibull Distribution

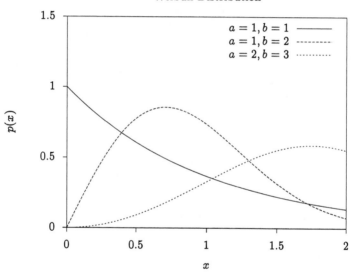

double gsl_cdf_weibull_P (double x, double a, double b)   Function
double gsl_cdf_weibull_Q (double x, double a, double b)   Function
double gsl_cdf_weibull_Pinv (double P, double a, double b)   Function
double gsl_cdf_weibull_Qinv (double Q, double a, double b)   Function
   These functions compute the cumulative distribution functions $P(x)$, $Q(x)$
   and their inverses for the Weibull distribution with scale a and exponent b.

19.25 The Type-1 Gumbel Distribution

double gsl_ran_gumbel1 (const gsl_rng * r, double a, double b)   Function
  This function returns a random variate from the Type-1 Gumbel distribution. The Type-1 Gumbel distribution function is,

$$p(x)dx = ab \exp(-(b \exp(-ax) + ax))dx$$

  for $-\infty < x < \infty$.

double gsl_ran_gumbel1_pdf (double x, double a, double b)   Function
  This function computes the probability density $p(x)$ at x for a Type-1 Gumbel distribution with parameters a and b, using the formula given above.

Type 1 Gumbel Distribution

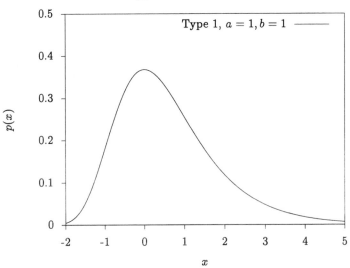

double gsl_cdf_gumbel1_P (double x, double a, double b)   Function
double gsl_cdf_gumbel1_Q (double x, double a, double b)   Function
double gsl_cdf_gumbel1_Pinv (double P, double a, double b)   Function
double gsl_cdf_gumbel1_Qinv (double Q, double a, double b)   Function
  These functions compute the cumulative distribution functions $P(x)$, $Q(x)$ and their inverses for the Type-1 Gumbel distribution with parameters a and b.

19.26 The Type-2 Gumbel Distribution

double gsl_ran_gumbel2 (const gsl_rng * r, double a, double b)   Function
    This function returns a random variate from the Type-2 Gumbel distribution. The Type-2 Gumbel distribution function is,

$$p(x)dx = abx^{-a-1}\exp(-bx^{-a})dx$$

for $0 < x < \infty$.

double gsl_ran_gumbel2_pdf (double x, double a, double b)   Function
    This function computes the probability density $p(x)$ at x for a Type-2 Gumbel distribution with parameters a and b, using the formula given above.

Type 2 Gumbel Distribution

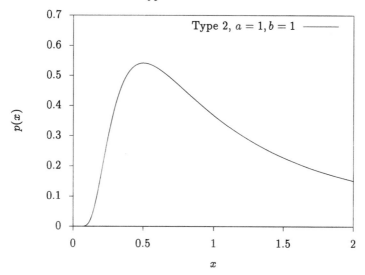

double gsl_cdf_gumbel2_P (double x, double a, double b)   Function
double gsl_cdf_gumbel2_Q (double x, double a, double b)   Function
double gsl_cdf_gumbel2_Pinv (double P, double a, double b)   Function
double gsl_cdf_gumbel2_Qinv (double Q, double a, double b)   Function
    These functions compute the cumulative distribution functions $P(x)$, $Q(x)$ and their inverses for the Type-2 Gumbel distribution with parameters a and b.

19.27 The Dirichlet Distribution

void gsl_ran_dirichlet (const gsl_rng * r, size_t K, const   *Function*
      double alpha[], double theta[])
This function returns an array of K random variates from a Dirichlet distribution of order K-1. The distribution function is

$$p(\theta_1,\ldots,\theta_K)\,d\theta_1\cdots d\theta_K = \frac{1}{Z}\prod_{i=1}^{K}\theta_i^{\alpha_i-1}\,\delta(1-\sum_{i=1}^{K}\theta_i)d\theta_1\cdots d\theta_K$$

for $\theta_i \geq 0$ and $\alpha_i \geq 0$. The delta function ensures that $\sum \theta_i = 1$. The normalization factor Z is

$$Z = \frac{\prod_{i=1}^{K}\Gamma(\alpha_i)}{\Gamma(\sum_{i=1}^{K}\alpha_i)}$$

The random variates are generated by sampling K values from gamma distributions with parameters $a = \alpha_i$, $b = 1$, and renormalizing. See A.M. Law, W.D. Kelton, *Simulation Modeling and Analysis* (1991).

double gsl_ran_dirichlet_pdf (size_t K, const double alpha[],   *Function*
      const double theta[])
This function computes the probability density $p(\theta_1,\ldots,\theta_K)$ at theta[K] for a Dirichlet distribution with parameters alpha[K], using the formula given above.

double gsl_ran_dirichlet_lnpdf (size_t K, const double   *Function*
      alpha[], const double theta[])
This function computes the logarithm of the probability density $p(\theta_1,\ldots,\theta_K)$ for a Dirichlet distribution with parameters alpha[K].

19.28 General Discrete Distributions

Given K discrete events with different probabilities $P[k]$, produce a random value k consistent with its probability.

The obvious way to do this is to preprocess the probability list by generating a cumulative probability array with $K + 1$ elements:

$$C[0] = 0$$
$$C[k + 1] = C[k] + P[k].$$

Note that this construction produces $C[K] = 1$. Now choose a uniform deviate u between 0 and 1, and find the value of k such that $C[k] \leq u < C[k + 1]$. Although this in principle requires of order $\log K$ steps per random number generation, they are fast steps, and if you use something like $\lfloor uK \rfloor$ as a starting point, you can often do pretty well.

But faster methods have been devised. Again, the idea is to preprocess the probability list, and save the result in some form of lookup table; then the individual calls for a random discrete event can go rapidly. An approach invented by G. Marsaglia (Generating discrete random numbers in a computer, Comm ACM 6, 37–38 (1963)) is very clever, and readers interested in examples of good algorithm design are directed to this short and well-written paper. Unfortunately, for large K, Marsaglia's lookup table can be quite large.

A much better approach is due to Alastair J. Walker (An efficient method for generating discrete random variables with general distributions, ACM Trans on Mathematical Software 3, 253–256 (1977); see also Knuth, v2, 3rd ed, p120–121,139). This requires two lookup tables, one floating point and one integer, but both only of size K. After preprocessing, the random numbers are generated in O(1) time, even for large K. The preprocessing suggested by Walker requires $O(K^2)$ effort, but that is not actually necessary, and the implementation provided here only takes $O(K)$ effort. In general, more preprocessing leads to faster generation of the individual random numbers, but a diminishing return is reached pretty early. Knuth points out that the optimal preprocessing is combinatorially difficult for large K.

This method can be used to speed up some of the discrete random number generators below, such as the binomial distribution. To use it for something like the Poisson Distribution, a modification would have to be made, since it only takes a finite set of K outcomes.

gsl_ran_discrete_t * gsl_ran_discrete_preproc (size_t K,   Function
          const double * P)

    This function returns a pointer to a structure that contains the lookup table for the discrete random number generator. The array $P[]$ contains the probabilities of the discrete events; these array elements must all be positive, but they needn't add up to one (so you can think of them more generally as "weights")—the preprocessor will normalize appropriately. This return value is used as an argument for the gsl_ran_discrete function below.

size_t gsl_ran_discrete (const gsl_rng * r, const   Function
        gsl_ran_discrete_t * g)
    After the preprocessor, above, has been called, you use this function to get
    the discrete random numbers.

double gsl_ran_discrete_pdf (size_t k, const   Function
        gsl_ran_discrete_t * g)
    Returns the probability $P[k]$ of observing the variable k. Since $P[k]$ is not
    stored as part of the lookup table, it must be recomputed; this computation
    takes $O(K)$, so if K is large and you care about the original array $P[k]$ used
    to create the lookup table, then you should just keep this original array $P[k]$
    around.

void gsl_ran_discrete_free (gsl_ran_discrete_t * g)   Function
    De-allocates the lookup table pointed to by g.

19.29 The Poisson Distribution

unsigned int gsl_ran_poisson (const gsl_rng * r, double mu)   Function
    This function returns a random integer from the Poisson distribution with
    mean mu. The probability distribution for Poisson variates is,

$$p(k) = \frac{\mu^k}{k!} \exp(-\mu)$$

for $k \geq 0$.

double gsl_ran_poisson_pdf (unsigned int k, double mu)   Function
    This function computes the probability $p(k)$ of obtaining k from a Poisson
    distribution with mean mu, using the formula given above.

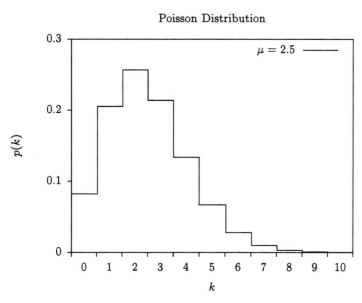

double gsl_cdf_poisson_P (unsigned int k, double mu)   Function
double gsl_cdf_poisson_Q (unsigned int k, double mu)   Function
    These functions compute the cumulative distribution functions $P(k)$, $Q(k)$
    for the Poisson distribution with parameter mu.

19.30 The Bernoulli Distribution

unsigned int **gsl_ran_bernoulli** (const gsl_rng * r, double p)   *Function*
    This function returns either 0 or 1, the result of a Bernoulli trial with prob-
    ability p. The probability distribution for a Bernoulli trial is,

$$p(0) = 1 - p$$
$$p(1) = p$$

double **gsl_ran_bernoulli_pdf** (unsigned int k, double p)   *Function*
    This function computes the probability $p(k)$ of obtaining k from a Bernoulli
    distribution with probability parameter p, using the formula given above.

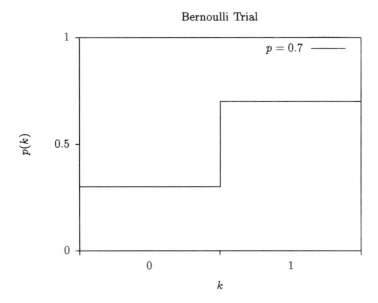

19.31 The Binomial Distribution

unsigned int gsl_ran_binomial (const gsl_rng * r, double p,   Function
        unsigned int n)
  This function returns a random integer from the binomial distribution, the
  number of successes in n independent trials with probability p. The probability distribution for binomial variates is,

$$p(k) = \frac{n!}{k!(n-k)!}p^k(1-p)^{n-k}$$

  for $0 \le k \le n$.

double gsl_ran_binomial_pdf (unsigned int k, double p,   Function
        unsigned int n)
  This function computes the probability $p(k)$ of obtaining k from a binomial
  distribution with parameters p and n, using the formula given above.

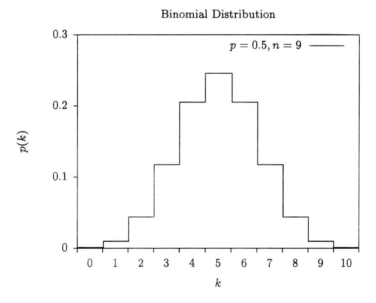

double gsl_cdf_binomial_P (unsigned int k, double p, unsigned   Function
        int n)
double gsl_cdf_binomial_Q (unsigned int k, double p, unsigned   Function
        int n)
  These functions compute the cumulative distribution functions $P(k)$, $Q(k)$
  for the binomial distribution with parameters p and n.

19.32 The Multinomial Distribution

void gsl_ran_multinomial (const gsl_rng * r, size_t K,   Function
        unsigned int N, const double p[], unsigned int n[])
This function computes a random sample n[] from the multinomial distribution formed by N trials from an underlying distribution p[K]. The distribution function for n[] is,

$$P(n_1, n_2, \cdots, n_K) = \frac{N!}{n_1! n_2! \cdots n_K!}\, p_1^{n_1} p_2^{n_2} \cdots p_K^{n_K}$$

where $(n_1, n_2, ..., n_K)$ are nonnegative integers with $\sum_{k=1}^{K} n_k = N$, and
$(p_1, p_2, \ldots, p_K)$ is a probability distribution with $\sum p_i = 1$. If the array $p[K]$
is not normalized then its entries will be treated as weights and normalized
appropriately. The arrays n[] and p[] must both be of length K.

Random variates are generated using the conditional binomial method (see
C.S. David, *The computer generation of multinomial random variates*,
Comp. Stat. Data Anal. 16 (1993) 205–217 for details).

double gsl_ran_multinomial_pdf (size_t K, const double p[],   Function
        const unsigned int n[])
This function computes the probability $P(n_1, n_2, \ldots, n_K)$ of sampling $n[K]$
from a multinomial distribution with parameters $p[K]$, using the formula
given above.

double gsl_ran_multinomial_lnpdf (size_t K, const double   Function
        p[], const unsigned int n[])
This function returns the logarithm of the probability for the multinomial
distribution $P(n_1, n_2, \ldots, n_K)$ with parameters $p[K]$.

19.33 The Negative Binomial Distribution

unsigned int gsl_ran_negative_binomial (const gsl_rng * r,   Function
        double p, double n)
  This function returns a random integer from the negative binomial distribution, the number of failures occurring before n successes in independent
  trials with probability p of success. The probability distribution for negative
  binomial variates is,

$$p(k) = \frac{\Gamma(n+k)}{\Gamma(k+1)\Gamma(n)} p^n (1-p)^k$$

  Note that n is not required to be an integer.

double gsl_ran_negative_binomial_pdf (unsigned int k, double   Function
        p, double n)
  This function computes the probability $p(k)$ of obtaining k from a negative
  binomial distribution with parameters p and n, using the formula given
  above.

Negative Binomial Distribution

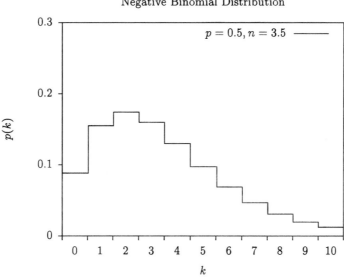

double gsl_cdf_negative_binomial_P (unsigned int k, double p,   Function
        double n)
double gsl_cdf_negative_binomial_Q (unsigned int k, double p,   Function
        double n)
  These functions compute the cumulative distribution functions $P(k)$, $Q(k)$
  for the negative binomial distribution with parameters p and n.

19.34 The Pascal Distribution

unsigned int gsl_ran_pascal (const gsl_rng * r, double p,   *Function*
        unsigned int n)

This function returns a random integer from the Pascal distribution. The
Pascal distribution is simply a negative binomial distribution with an integer
value of n.

$$p(k) = \frac{(n+k-1)!}{k!(n-1)!} p^n (1-p)^k$$

for $k \geq 0$

double gsl_ran_pascal_pdf (unsigned int k, double p, unsigned   *Function*
        int n)

This function computes the probability $p(k)$ of obtaining k from a Pascal
distribution with parameters p and n, using the formula given above.

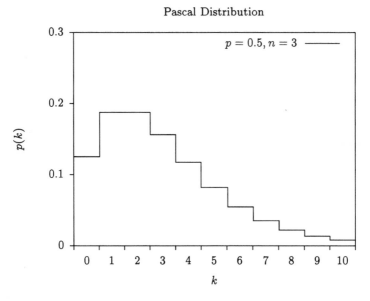

Pascal Distribution

double gsl_cdf_pascal_P (unsigned int k, double p, unsigned   *Function*
        int n)
double gsl_cdf_pascal_Q (unsigned int k, double p, unsigned   *Function*
        int n)

These functions compute the cumulative distribution functions $P(k)$, $Q(k)$
for the Pascal distribution with parameters p and n.

19.35 The Geometric Distribution

unsigned int gsl_ran_geometric (const gsl_rng * r, double p)   Function
  This function returns a random integer from the geometric distribution, the
  number of independent trials with probability p until the first success. The
  probability distribution for geometric variates is,

$$p(k) = p(1 - p)^{k-1}$$

  for $k \geq 1$. Note that the distribution begins with $k = 1$ with this definition.
  There is another convention in which the exponent $k - 1$ is replaced by k.

double gsl_ran_geometric_pdf (unsigned int k, double p)   Function
  This function computes the probability $p(k)$ of obtaining k from a geometric
  distribution with probability parameter p, using the formula given above.

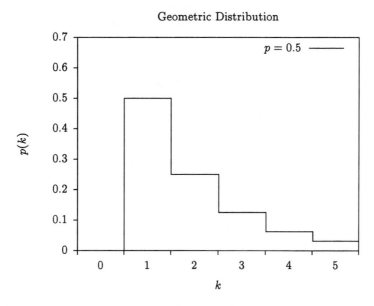

Geometric Distribution

double gsl_cdf_geometric_P (unsigned int k, double p)   Function
double gsl_cdf_geometric_Q (unsigned int k, double p)   Function
  These functions compute the cumulative distribution functions $P(k)$, $Q(k)$
  for the geometric distribution with parameter p.

19.36 The Hypergeometric Distribution

unsigned int gsl_ran_hypergeometric (const gsl_rng * r,   *Function*
          unsigned int n1, unsigned int n2, unsigned int t)
This function returns a random integer from the hypergeometric distribution. The probability distribution for hypergeometric random variates is,

$$p(k) = C(n_1, k)C(n_2, t - k)/C(n_1 + n_2, t)$$

where $C(a, b) = a!/(b!(a-b)!)$ and $t \leq n_1+n_2$. The domain of k is $\max(0, t - n_2), \ldots, \min(t, n_1)$.

If a population contains n_1 elements of "type 1" and n_2 elements of "type 2" then the hypergeometric distribution gives the probability of obtaining k elements of "type 1" in t samples from the population without replacement.

double gsl_ran_hypergeometric_pdf (unsigned int k, unsigned   *Function*
          int n1, unsigned int n2, unsigned int t)
This function computes the probability $p(k)$ of obtaining k from a hypergeometric distribution with parameters n1, n2, t, using the formula given above.

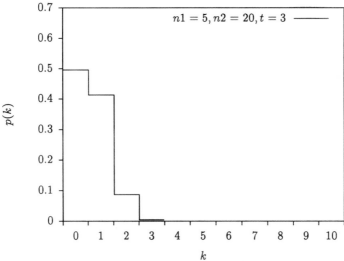

Hypergeometric Distribution

double gsl_cdf_hypergeometric_P (unsigned int k, unsigned int   *Function*
          n1, unsigned int n2, unsigned int t)
double gsl_cdf_hypergeometric_Q (unsigned int k, unsigned int   *Function*
          n1, unsigned int n2, unsigned int t)
These functions compute the cumulative distribution functions $P(k)$, $Q(k)$ for the hypergeometric distribution with parameters n1, n2 and t.

19.37 The Logarithmic Distribution

unsigned int gsl_ran_logarithmic (const gsl_rng * r, double p)   Function
   This function returns a random integer from the logarithmic distribution.
   The probability distribution for logarithmic random variates is,

$$p(k) = \frac{-1}{\log(1-p)} \left(\frac{p^k}{k} \right)$$

   for $k \geq 1$.

double gsl_ran_logarithmic_pdf (unsigned int k, double p)   Function
   This function computes the probability $p(k)$ of obtaining k from a logarithmic distribution with probability parameter p, using the formula given
   above.

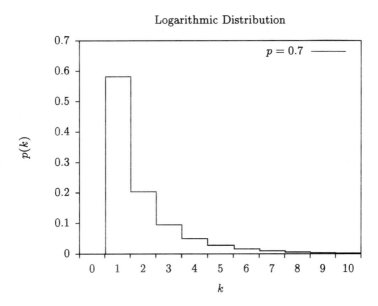

Logarithmic Distribution

19.38 Shuffling and Sampling

The following functions allow the shuffling and sampling of a set of objects. The algorithms rely on a random number generator as a source of randomness and a poor quality generator can lead to correlations in the output. In particular it is important to avoid generators with a short period. For more information see Knuth, v2, 3rd ed, Section 3.4.2, "Random Sampling and Shuffling".

void gsl_ran_shuffle (const gsl_rng * r, void * base, size_t n,   Function
          size_t size)

This function randomly shuffles the order of n objects, each of size size, stored in the array base[0..n-1]. The output of the random number generator r is used to produce the permutation. The algorithm generates all possible n! permutations with equal probability, assuming a perfect source of random numbers.

The following code shows how to shuffle the numbers from 0 to 51,

```
int a[52];

for (i = 0; i < 52; i++)
  {
    a[i] = i;
  }

gsl_ran_shuffle (r, a, 52, sizeof (int));
```

int gsl_ran_choose (const gsl_rng * r, void * dest, size_t k,   Function
          void * src, size_t n, size_t size)

This function fills the array dest[k] with k objects taken randomly from the n elements of the array src[0..n-1]. The objects are each of size size. The output of the random number generator r is used to make the selection. The algorithm ensures all possible samples are equally likely, assuming a perfect source of randomness.

The objects are sampled *without* replacement, thus each object can only appear once in dest[k]. It is required that k be less than or equal to n. The objects in dest will be in the same relative order as those in src. You will need to call gsl_ran_shuffle(r, dest, n, size) if you want to randomize the order.

The following code shows how to select a random sample of three unique numbers from the set 0 to 99,

```
double a[3], b[100];

for (i = 0; i < 100; i++)
  {
    b[i] = (double) i;
  }

gsl_ran_choose (r, a, 3, b, 100, sizeof (double));
```

void gsl_ran_sample (const gsl_rng * r, void * *dest*, size_t k,   Function
        void * *src*, size_t n, size_t *size*)
  This function is like gsl_ran_choose but samples k items from the original
  array of n items *src* with replacement, so the same object can appear more
  than once in the output sequence *dest*. There is no requirement that k be
  less than n in this case.

19.39 Examples

  The following program demonstrates the use of a random number generator
to produce variates from a distribution. It prints 10 samples from the Poisson
distribution with a mean of 3.

```
#include <stdio.h>
#include <gsl/gsl_rng.h>
#include <gsl/gsl_randist.h>

int
main (void)
{
  const gsl_rng_type * T;
  gsl_rng * r;

  int i, n = 10;
  double mu = 3.0;

  /* create a generator chosen by the
     environment variable GSL_RNG_TYPE */

  gsl_rng_env_setup();

  T = gsl_rng_default;
  r = gsl_rng_alloc (T);

  /* print n random variates chosen from
     the poisson distribution with mean
     parameter mu */

  for (i = 0; i < n; i++)
    {
      unsigned int k = gsl_ran_poisson (r, mu);
      printf (" %u", k);
    }

  printf ("\n");
  gsl_rng_free (r);
  return 0;
}
```

If the library and header files are installed under '/usr/local' (the default location) then the program can be compiled with these options,

```
$ gcc -Wall demo.c -lgsl -lgslcblas -lm
```

Here is the output of the program,

```
$ ./a.out
  2 5 5 2 1 0 3 4 1 1
```

The variates depend on the seed used by the generator. The seed for the default generator type `gsl_rng_default` can be changed with the `GSL_RNG_SEED` environment variable to produce a different stream of variates,

```
$ GSL_RNG_SEED=123 ./a.out
GSL_RNG_SEED=123
  4 5 6 3 3 1 4 2 5 5
```

The following program generates a random walk in two dimensions.

```
#include <stdio.h>
#include <gsl/gsl_rng.h>
#include <gsl/gsl_randist.h>

int
main (void)
{
  int i;
  double x = 0, y = 0, dx, dy;

  const gsl_rng_type * T;
  gsl_rng * r;

  gsl_rng_env_setup();
  T = gsl_rng_default;
  r = gsl_rng_alloc (T);

  printf ("%g %g\n", x, y);

  for (i = 0; i < 10; i++)
    {
      gsl_ran_dir_2d (r, &dx, &dy);
      x += dx; y += dy;
      printf ("%g %g\n", x, y);
    }

  gsl_rng_free (r);
  return 0;
}
```

Here is the output from the program, three 10-step random walks from the origin,

Random walk

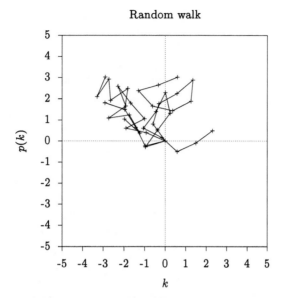

The following program computes the upper and lower cumulative distribution functions for the standard normal distribution at $x = 2$.

```
#include <stdio.h>
#include <gsl/gsl_cdf.h>

int
main (void)
{
  double P, Q;
  double x = 2.0;

  P = gsl_cdf_ugaussian_P (x);
  printf ("prob(x < %f) = %f\n", x, P);

  Q = gsl_cdf_ugaussian_Q (x);
  printf ("prob(x > %f) = %f\n", x, Q);

  x = gsl_cdf_ugaussian_Pinv (P);
  printf ("Pinv(%f) = %f\n", P, x);

  x = gsl_cdf_ugaussian_Qinv (Q);
  printf ("Qinv(%f) = %f\n", Q, x);

  return 0;
}
```

Here is the output of the program,

```
prob(x < 2.000000) = 0.977250
prob(x > 2.000000) = 0.022750
Pinv(0.977250) = 2.000000
Qinv(0.022750) = 2.000000
```

19.40 References and Further Reading

For an encyclopaedic coverage of the subject readers are advised to consult the book *Non-Uniform Random Variate Generation* by Luc Devroye. It covers every imaginable distribution and provides hundreds of algorithms.

> Luc Devroye, *Non-Uniform Random Variate Generation*, Springer-Verlag, ISBN 0-387-96305-7.

The subject of random variate generation is also reviewed by Knuth, who describes algorithms for all the major distributions.

> Donald E. Knuth, *The Art of Computer Programming: Seminumerical Algorithms* (Vol 2, 3rd Ed, 1997), Addison-Wesley, ISBN 0201896842.

The Particle Data Group provides a short review of techniques for generating distributions of random numbers in the "Monte Carlo" section of its Annual Review of Particle Physics.

> *Review of Particle Properties* R.M. Barnett et al., Physical Review D54, 1 (1996) http://pdg.lbl.gov/.

The Review of Particle Physics is available online in postscript and pdf format.

An overview of methods used to compute cumulative distribution functions can be found in *Statistical Computing* by W.J. Kennedy and J.E. Gentle. Another general reference is *Elements of Statistical Computing* by R.A. Thisted.

> William E. Kennedy and James E. Gentle, *Statistical Computing* (1980), Marcel Dekker, ISBN 0-8247-6898-1.

> Ronald A. Thisted, *Elements of Statistical Computing* (1988), Chapman & Hall, ISBN 0-412-01371-1.

The cumulative distribution functions for the Gaussian distribution are based on the following papers,

> *Rational Chebyshev Approximations Using Linear Equations*, W.J. Cody, W. Fraser, J.F. Hart. Numerische Mathematik 12, 242–251 (1968).

> *Rational Chebyshev Approximations for the Error Function*, W.J. Cody. Mathematics of Computation 23, n107, 631–637 (July 1969).

20 Statistics

This chapter describes the statistical functions in the library. The basic statistical functions include routines to compute the mean, variance and standard deviation. More advanced functions allow you to calculate absolute deviations, skewness, and kurtosis as well as the median and arbitrary percentiles. The algorithms use recurrence relations to compute average quantities in a stable way, without large intermediate values that might overflow.

The functions are available in versions for datasets in the standard floating-point and integer types. The versions for double precision floating-point data have the prefix gsl_stats and are declared in the header file 'gsl_statistics_double.h'. The versions for integer data have the prefix gsl_stats_int and are declared in the header file 'gsl_statistics_int.h'.

20.1 Mean, Standard Deviation and Variance

double gsl_stats_mean (const double *data[]*, size_t *stride*,   Function
        size_t *n*)
This function returns the arithmetic mean of *data*, a dataset of length n with stride *stride*. The arithmetic mean, or *sample mean*, is denoted by $\hat{\mu}$ and defined as,

$$\hat{\mu} = \frac{1}{N} \sum x_i$$

where x_i are the elements of the dataset *data*. For samples drawn from a gaussian distribution the variance of $\hat{\mu}$ is σ^2/N.

double gsl_stats_variance (const double *data[]*, size_t *stride*,   Function
        size_t *n*)
This function returns the estimated, or *sample*, variance of *data*, a dataset of length n with stride *stride*. The estimated variance is denoted by $\hat{\sigma}^2$ and is defined by,

$$\hat{\sigma}^2 = \frac{1}{(N-1)} \sum (x_i - \hat{\mu})^2$$

where x_i are the elements of the dataset *data*. Note that the normalization factor of $1/(N-1)$ results from the derivation of $\hat{\sigma}^2$ as an unbiased estimator of the population variance σ^2. For samples drawn from a gaussian distribution the variance of $\hat{\sigma}^2$ itself is $2\sigma^4/N$.

This function computes the mean via a call to gsl_stats_mean. If you have already computed the mean then you can pass it directly to gsl_stats_variance_m.

double gsl_stats_variance_m (const double data[], size_t   Function
        stride, size_t n, double mean)
This function returns the sample variance of data relative to the given value
of mean. The function is computed with $\hat{\mu}$ replaced by the value of mean
that you supply,

$$\hat{\sigma}^2 = \frac{1}{(N-1)} \sum (x_i - mean)^2$$

double gsl_stats_sd (const double data[], size_t stride, size_t   Function
        n)
double gsl_stats_sd_m (const double data[], size_t stride,   Function
        size_t n, double mean)
The standard deviation is defined as the square root of the variance. These
functions return the square root of the corresponding variance functions
above.

double gsl_stats_tss (const double data[], size_t stride,   Function
        size_t n)
double gsl_stats_tss_m (const double data[], size_t stride,   Function
        size_t n, double mean)
These functions return the total sum of squares (TSS) of data about the
mean. For gsl_stats_tss_m the user-supplied value of mean is used, and
for gsl_stats_tss it is computed using gsl_stats_mean.

$$\text{TSS} = \sum (x_i - mean)^2$$

double gsl_stats_variance_with_fixed_mean (const double   Function
        data[], size_t stride, size_t n, double mean)
This function computes an unbiased estimate of the variance of data when
the population mean mean of the underlying distribution is known a priori.
In this case the estimator for the variance uses the factor $1/N$ and the sample
mean $\hat{\mu}$ is replaced by the known population mean μ,

$$\hat{\sigma}^2 = \frac{1}{N} \sum (x_i - \mu)^2$$

double gsl_stats_sd_with_fixed_mean (const double data[],   Function
        size_t stride, size_t n, double mean)
This function calculates the standard deviation of data for a fixed population
mean mean. The result is the square root of the corresponding variance
function.

20.2 Absolute deviation

double gsl_stats_absdev (const double *data*[], size_t *stride*,   *Function*
        size_t *n*)
    This function computes the absolute deviation from the mean of *data*, a
    dataset of length *n* with stride *stride*. The absolute deviation from the
    mean is defined as,

$$absdev = \frac{1}{N} \sum |x_i - \hat{\mu}|$$

    where x_i are the elements of the dataset *data*. The absolute deviation from
    the mean provides a more robust measure of the width of a distribution than
    the variance. This function computes the mean of *data* via a call to gsl_
    stats_mean.

double gsl_stats_absdev_m (const double *data*[], size_t *stride*,   *Function*
        size_t *n*, double *mean*)
    This function computes the absolute deviation of the dataset *data* relative
    to the given value of *mean*,

$$absdev = \frac{1}{N} \sum |x_i - mean|$$

    This function is useful if you have already computed the mean of *data* (and
    want to avoid recomputing it), or wish to calculate the absolute deviation
    relative to another value (such as zero, or the median).

20.3 Higher moments (skewness and kurtosis)

double gsl_stats_skew (const double *data*[], size_t *stride*,   *Function*
        size_t *n*)
    This function computes the skewness of *data*, a dataset of length *n* with
    stride *stride*. The skewness is defined as,

$$skew = \frac{1}{N} \sum \left(\frac{x_i - \hat{\mu}}{\hat{\sigma}}\right)^3$$

    where x_i are the elements of the dataset *data*. The skewness measures the
    asymmetry of the tails of a distribution.

    The function computes the mean and estimated standard deviation of *data*
    via calls to gsl_stats_mean and gsl_stats_sd.

double gsl_stats_skew_m_sd (const double *data*[], size_t   *Function*
        stride, size_t *n*, double *mean*, double *sd*)
    This function computes the skewness of the dataset *data* using the given
    values of the mean *mean* and standard deviation *sd*,

$$skew = \frac{1}{N} \sum \left(\frac{x_i - mean}{sd}\right)^3$$

    These functions are useful if you have already computed the mean and stan-
    dard deviation of *data* and want to avoid recomputing them.

double gsl_stats_kurtosis (const double *data*[], size_t *stride*,   Function
        size_t *n*)
This function computes the kurtosis of *data*, a dataset of length *n* with stride
stride. The kurtosis is defined as,

$$kurtosis = \left(\frac{1}{N} \sum \left(\frac{x_i - \hat{\mu}}{\hat{\sigma}} \right)^4 \right) - 3$$

The kurtosis measures how sharply peaked a distribution is, relative to its
width. The kurtosis is normalized to zero for a gaussian distribution.

double gsl_stats_kurtosis_m_sd (const double *data*[], size_t   Function
        stride, size_t *n*, double *mean*, double *sd*)
This function computes the kurtosis of the dataset *data* using the given
values of the mean *mean* and standard deviation *sd*,

$$kurtosis = \frac{1}{N} \left(\sum \left(\frac{x_i - mean}{sd} \right)^4 \right) - 3$$

This function is useful if you have already computed the mean and standard
deviation of *data* and want to avoid recomputing them.

20.4 Autocorrelation

double gsl_stats_lag1_autocorrelation (const double *data*[],   Function
        const size_t *stride*, const size_t *n*)
This function computes the lag-1 autocorrelation of the dataset *data*.

$$a_1 = \frac{\sum_{i=1}^{n} (x_i - \hat{\mu})(x_{i-1} - \hat{\mu})}{\sum_{i=1}^{n} (x_i - \hat{\mu})(x_i - \hat{\mu})}$$

double gsl_stats_lag1_autocorrelation_m (const double *data*[],   Function
        const size_t *stride*, const size_t *n*, const double *mean*)
This function computes the lag-1 autocorrelation of the dataset *data* using
the given value of the mean *mean*.

20.5 Covariance

double gsl_stats_covariance (const double *data1*[], const   Function
        size_t *stride1*, const double *data2*[], const size_t *stride2*,
        const size_t *n*)
This function computes the covariance of the datasets *data1* and *data2*
which must both be of the same length *n*.

$$covar = \frac{1}{(n-1)} \sum_{i=1}^{n} (x_i - \hat{x})(y_i - \hat{y})$$

double gsl_stats_covariance_m (const double *data1*[], const   Function
        size_t *stride1*, const double *data2*[], const size_t *stride2*,
        const size_t *n*, const double *mean1*, const double *mean2*)
This function computes the covariance of the datasets *data1* and *data2* using
the given values of the means, *mean1* and *mean2*. This is useful if you
have already computed the means of *data1* and *data2* and want to avoid
recomputing them.

20.6 Correlation

double gsl_stats_correlation (const double *data1*[], const   Function
        size_t *stride1*, const double *data2*[], const size_t *stride2*,
        const size_t *n*)
This function efficiently computes the Pearson correlation coefficient between the datasets *data1* and *data2* which must both be of the same length
n.

$$r = \frac{cov(x,y)}{\hat{\sigma}_x \hat{\sigma}_y} = \frac{\frac{1}{n-1}\sum(x_i - \hat{x})(y_i - \hat{y})}{\sqrt{\frac{1}{n-1}\sum(x_i - \hat{x})^2}\sqrt{\frac{1}{n-1}\sum(y_i - \hat{y})^2}}$$

20.7 Weighted Samples

The functions described in this section allow the computation of statistics for
weighted samples. The functions accept an array of samples, x_i, with associated
weights, w_i. Each sample x_i is considered as having been drawn from a Gaussian
distribution with variance σ_i^2. The sample weight w_i is defined as the reciprocal
of this variance, $w_i = 1/\sigma_i^2$. Setting a weight to zero corresponds to removing
a sample from a dataset.

double gsl_stats_wmean (const double *w*[], size_t *wstride*,   Function
        const double *data*[], size_t *stride*, size_t *n*)
This function returns the weighted mean of the dataset *data* with stride
stride and length *n*, using the set of weights *w* with stride *wstride* and
length *n*. The weighted mean is defined as,

$$\hat{\mu} = \frac{\sum w_i x_i}{\sum w_i}$$

double gsl_stats_wvariance (const double w[], size_t *wstride*,   Function
            const double *data*[], size_t *stride*, size_t n)

This function returns the estimated variance of the dataset *data* with stride *stride* and length n, using the set of weights w with stride *wstride* and length n. The estimated variance of a weighted dataset is calculated as,

$$\hat{\sigma}^2 = \frac{\sum w_i}{(\sum w_i)^2 - \sum(w_i^2)} \sum w_i (x_i - \hat{\mu})^2$$

Note that this expression reduces to an unweighted variance with the familiar $1/(N-1)$ factor when there are N equal non-zero weights.

double gsl_stats_wvariance_m (const double w[], size_t   Function
            wstride, const double *data*[], size_t *stride*, size_t n, double
            wmean)

This function returns the estimated variance of the weighted dataset *data* using the given weighted mean *wmean*.

double gsl_stats_wsd (const double w[], size_t *wstride*, const   Function
            double *data*[], size_t *stride*, size_t n)

The standard deviation is defined as the square root of the variance. This function returns the square root of the corresponding variance function gsl_stats_wvariance above.

double gsl_stats_wsd_m (const double w[], size_t *wstride*,   Function
            const double *data*[], size_t *stride*, size_t n, double *wmean*)

This function returns the square root of the corresponding variance function gsl_stats_wvariance_m above.

double gsl_stats_wvariance_with_fixed_mean (const double   Function
            w[], size_t *wstride*, const double *data*[], size_t *stride*,
            size_t n, const double *mean*)

This function computes an unbiased estimate of the variance of the weighted dataset *data* when the population mean *mean* of the underlying distribution is known *a priori*. In this case the estimator for the variance replaces the sample mean $\hat{\mu}$ by the known population mean μ,

$$\hat{\sigma}^2 = \frac{\sum w_i (x_i - \mu)^2}{\sum w_i}$$

double gsl_stats_wsd_with_fixed_mean (const double w[],   Function
            size_t *wstride*, const double *data*[], size_t *stride*, size_t n,
            const double *mean*)

The standard deviation is defined as the square root of the variance. This function returns the square root of the corresponding variance function above.

double gsl_stats_wtss (const double w[], const size_t *wstride*,   Function
      const double *data*[], size_t *stride*, size_t n)

double gsl_stats_wtss_m (const double w[], const size_t   Function
      wstride, const double *data*[], size_t *stride*, size_t n, double
      wmean)

These functions return the weighted total sum of squares (TSS) of *data*
about the weighted mean. For gsl_stats_wtss_m the user-supplied value of
wmean is used, and for gsl_stats_wtss it is computed using gsl_stats_
wmean.

$$\mathrm{TSS} = \sum w_i(x_i - wmean)^2$$

double gsl_stats_wabsdev (const double w[], size_t *wstride*,   Function
      const double *data*[], size_t *stride*, size_t n)

This function computes the weighted absolute deviation from the weighted
mean of *data*. The absolute deviation from the mean is defined as,

$$absdev = \frac{\sum w_i |x_i - \hat{\mu}|}{\sum w_i}$$

double gsl_stats_wabsdev_m (const double w[], size_t *wstride*,   Function
      const double *data*[], size_t *stride*, size_t n, double *wmean*)

This function computes the absolute deviation of the weighted dataset *data*
about the given weighted mean *wmean*.

double gsl_stats_wskew (const double w[], size_t *wstride*,   Function
      const double *data*[], size_t *stride*, size_t n)

This function computes the weighted skewness of the dataset *data*.

$$skew = \frac{\sum w_i((x_i - xbar)/\sigma)^3}{\sum w_i}$$

double gsl_stats_wskew_m_sd (const double w[], size_t   Function
      wstride, const double *data*[], size_t *stride*, size_t n, double
      wmean, double *wsd*)

This function computes the weighted skewness of the dataset *data* using the
given values of the weighted mean and weighted standard deviation, *wmean*
and *wsd*.

double gsl_stats_wkurtosis (const double w[], size_t *wstride*,   Function
      const double *data*[], size_t *stride*, size_t n)

This function computes the weighted kurtosis of the dataset *data*.

$$kurtosis = \frac{\sum w_i((x_i - xbar)/sigma)^4}{\sum w_i} - 3$$

double gsl_stats_wkurtosis_m_sd (const double w[], size_t   Function
        wstride, const double data[], size_t stride, size_t n, double
        wmean, double wsd)

This function computes the weighted kurtosis of the dataset data using the given values of the weighted mean and weighted standard deviation, wmean and wsd.

20.8 Maximum and Minimum values

The following functions find the maximum and minimum values of a dataset (or their indices). If the data contains NaNs then a NaN will be returned, since the maximum or minimum value is undefined. For functions which return an index, the location of the first NaN in the array is returned.

double gsl_stats_max (const double data[], size_t stride,   Function
        size_t n)

This function returns the maximum value in data, a dataset of length n with stride stride. The maximum value is defined as the value of the element x_i which satisfies $x_i \geq x_j$ for all j.

If you want instead to find the element with the largest absolute magnitude you will need to apply fabs or abs to your data before calling this function.

double gsl_stats_min (const double data[], size_t stride,   Function
        size_t n)

This function returns the minimum value in data, a dataset of length n with stride stride. The minimum value is defined as the value of the element x_i which satisfies $x_i \leq x_j$ for all j.

If you want instead to find the element with the smallest absolute magnitude you will need to apply fabs or abs to your data before calling this function.

void gsl_stats_minmax (double * min, double * max, const   Function
        double data[], size_t stride, size_t n)

This function finds both the minimum and maximum values min, max in data in a single pass.

size_t gsl_stats_max_index (const double data[], size_t   Function
        stride, size_t n)

This function returns the index of the maximum value in data, a dataset of length n with stride stride. The maximum value is defined as the value of the element x_i which satisfies $x_i \geq x_j$ for all j. When there are several equal maximum elements then the first one is chosen.

size_t gsl_stats_min_index (const double data[], size_t   Function
        stride, size_t n)

This function returns the index of the minimum value in data, a dataset of length n with stride stride. The minimum value is defined as the value of the element x_i which satisfies $x_i \geq x_j$ for all j. When there are several equal minimum elements then the first one is chosen.

void gsl_stats_minmax_index (size_t * min_index, size_t *   Function
        max_index, const double data[], size_t stride, size_t n)
This function returns the indexes min_index, max_index of the minimum
and maximum values in data in a single pass.

20.9 Median and Percentiles

The median and percentile functions described in this section operate on
sorted data. For convenience we use quantiles, measured on a scale of 0 to 1,
instead of percentiles (which use a scale of 0 to 100).

double gsl_stats_median_from_sorted_data (const double   Function
        sorted_data[], size_t stride, size_t n)
This function returns the median value of sorted_data, a dataset of length n
with stride stride. The elements of the array must be in ascending numerical
order. There are no checks to see whether the data are sorted, so the function
gsl_sort should always be used first.

When the dataset has an odd number of elements the median is the value
of element $(n-1)/2$. When the dataset has an even number of elements the
median is the mean of the two nearest middle values, elements $(n-1)/2$ and
$n/2$. Since the algorithm for computing the median involves interpolation
this function always returns a floating-point number, even for integer data
types.

double gsl_stats_quantile_from_sorted_data (const double   Function
        sorted_data[], size_t stride, size_t n, double f)
This function returns a quantile value of sorted_data, a double-precision
array of length n with stride stride. The elements of the array must be in
ascending numerical order. The quantile is determined by the f, a fraction
between 0 and 1. For example, to compute the value of the 75th percentile
f should have the value 0.75.

There are no checks to see whether the data are sorted, so the function gsl_
sort should always be used first.

The quantile is found by interpolation, using the formula

$$\text{quantile} = (1-\delta)x_i + \delta x_{i+1}$$

where i is floor($(n-1)f$) and δ is $(n-1)f - i$.

Thus the minimum value of the array (data[0*stride]) is given by f equal
to zero, the maximum value (data[(n-1)*stride]) is given by f equal to
one and the median value is given by f equal to 0.5. Since the algorithm
for computing quantiles involves interpolation this function always returns
a floating-point number, even for integer data types.

20.10 Examples

Here is a basic example of how to use the statistical functions:

```
#include <stdio.h>
#include <gsl/gsl_statistics.h>

int
main(void)
{
  double data[5] = {17.2, 18.1, 16.5, 18.3, 12.6};
  double mean, variance, largest, smallest;

  mean     = gsl_stats_mean(data, 1, 5);
  variance = gsl_stats_variance(data, 1, 5);
  largest  = gsl_stats_max(data, 1, 5);
  smallest = gsl_stats_min(data, 1, 5);

  printf ("The dataset is %g, %g, %g, %g, %g\n",
          data[0], data[1], data[2], data[3], data[4]);

  printf ("The sample mean is %g\n", mean);
  printf ("The estimated variance is %g\n", variance);
  printf ("The largest value is %g\n", largest);
  printf ("The smallest value is %g\n", smallest);
  return 0;
}
```

The program should produce the following output,

```
The dataset is 17.2, 18.1, 16.5, 18.3, 12.6
The sample mean is 16.54
The estimated variance is 4.2984
The largest value is 18.3
The smallest value is 12.6
```

Here is an example using sorted data,

```
#include <stdio.h>
#include <gsl/gsl_sort.h>
#include <gsl/gsl_statistics.h>

int
main(void)
{
  double data[5] = {17.2, 18.1, 16.5, 18.3, 12.6};
  double median, upperq, lowerq;

  printf ("Original dataset:  %g, %g, %g, %g, %g\n",
          data[0], data[1], data[2], data[3], data[4]);

  gsl_sort (data, 1, 5);
```

```
printf ("Sorted dataset: %g, %g, %g, %g, %g\n",
        data[0], data[1], data[2], data[3], data[4]);

median
  = gsl_stats_median_from_sorted_data (data,
                                       1, 5);

upperq
  = gsl_stats_quantile_from_sorted_data (data,
                                         1, 5,
                                         0.75);
lowerq
  = gsl_stats_quantile_from_sorted_data (data,
                                         1, 5,
                                         0.25);

printf ("The median is %g\n", median);
printf ("The upper quartile is %g\n", upperq);
printf ("The lower quartile is %g\n", lowerq);
return 0;
}
```

This program should produce the following output,

```
Original dataset: 17.2, 18.1, 16.5, 18.3, 12.6
Sorted dataset: 12.6, 16.5, 17.2, 18.1, 18.3
The median is 17.2
The upper quartile is 18.1
The lower quartile is 16.5
```

20.11 References and Further Reading

The standard reference for almost any topic in statistics is the multi-volume *Advanced Theory of Statistics* by Kendall and Stuart.

> Maurice Kendall, Alan Stuart, and J. Keith Ord. *The Advanced Theory of Statistics* (multiple volumes) reprinted as *Kendall's Advanced Theory of Statistics*. Wiley, ISBN 047023380X.

Many statistical concepts can be more easily understood by a Bayesian approach. The following book by Gelman, Carlin, Stern and Rubin gives a comprehensive coverage of the subject.

> Andrew Gelman, John B. Carlin, Hal S. Stern, Donald B. Rubin. *Bayesian Data Analysis*. Chapman & Hall, ISBN 0412039915.

For physicists the Particle Data Group provides useful reviews of Probability and Statistics in the "Mathematical Tools" section of its Annual Review of Particle Physics.

> *Review of Particle Properties* R.M. Barnett et al., Physical Review D54, 1 (1996)

The Review of Particle Physics is available online at the website pdg.lbl.gov.

21 Histograms

This chapter describes functions for creating histograms. Histograms provide a convenient way of summarizing the distribution of a set of data. A histogram consists of a set of *bins* which count the number of events falling into a given range of a continuous variable x. In GSL the bins of a histogram contain floating-point numbers, so they can be used to record both integer and non-integer distributions. The bins can use arbitrary sets of ranges (uniformly spaced bins are the default). Both one and two-dimensional histograms are supported.

Once a histogram has been created it can also be converted into a probability distribution function. The library provides efficient routines for selecting random samples from probability distributions. This can be useful for generating simulations based on real data.

The functions are declared in the header files 'gsl_histogram.h' and 'gsl_histogram2d.h'.

21.1 The histogram struct

A histogram is defined by the following struct,

gsl_histogram   Data Type

> size_t n
>> This is the number of histogram bins
>
> double * range
>> The ranges of the bins are stored in an array of $n + 1$ elements pointed to by *range*.
>
> double * bin
>> The counts for each bin are stored in an array of n elements pointed to by *bin*. The bins are floating-point numbers, so you can increment them by non-integer values if necessary.

The range for *bin*[i] is given by *range*[i] to *range*[i+1]. For n bins there are $n + 1$ entries in the array *range*. Each bin is inclusive at the lower end and exclusive at the upper end. Mathematically this means that the bins are defined by the following inequality,

$$\text{bin[i] corresponds to range[i]} \leq x < \text{range[i+1]}$$

Here is a diagram of the correspondence between ranges and bins on the number-line for x,

```
     [ bin[0] )[ bin[1] )[ bin[2] )[ bin[3] )[ bin[4] )
 ---|---------|---------|---------|---------|---------|---   x
    r[0]      r[1]      r[2]      r[3]      r[4]      r[5]
```

In this picture the values of the *range* array are denoted by r. On the left-hand side of each bin the square bracket '[' denotes an inclusive lower bound ($r \leq x$), and the round parentheses ')' on the right-hand side denote an exclusive upper bound ($x < r$). Thus any samples which fall on the upper end of the histogram are excluded. If you want to include this value for the last bin you will need to add an extra bin to your histogram.

The gsl_histogram struct and its associated functions are defined in the header file 'gsl_histogram.h'.

21.2 Histogram allocation

The functions for allocating memory to a histogram follow the style of malloc and free. In addition they also perform their own error checking. If there is insufficient memory available to allocate a histogram then the functions call the error handler (with an error number of GSL_ENOMEM) in addition to returning a null pointer. Thus if you use the library error handler to abort your program then it isn't necessary to check every histogram alloc.

gsl_histogram * gsl_histogram_alloc (size_t n)   Function
    This function allocates memory for a histogram with n bins, and returns a pointer to a newly created gsl_histogram struct. If insufficient memory is available a null pointer is returned and the error handler is invoked with an error code of GSL_ENOMEM. The bins and ranges are not initialized, and should be prepared using one of the range-setting functions below in order to make the histogram ready for use.

int gsl_histogram_set_ranges (gsl_histogram * h, const double   Function
        range[], size_t size)
    This function sets the ranges of the existing histogram h using the array range of size size. The values of the histogram bins are reset to zero. The range array should contain the desired bin limits. The ranges can be arbitrary, subject to the restriction that they are monotonically increasing.

    The following example shows how to create a histogram with logarithmic bins with ranges [1,10), [10,100) and [100,1000).

        gsl_histogram * h = gsl_histogram_alloc (3);

        /* bin[0] covers the range 1 <= x < 10 */
        /* bin[1] covers the range 10 <= x < 100 */
        /* bin[2] covers the range 100 <= x < 1000 */

        double range[4] = { 1.0, 10.0, 100.0, 1000.0 };

        gsl_histogram_set_ranges (h, range, 4);

    Note that the size of the range array should be defined to be one element bigger than the number of bins. The additional element is required for the upper value of the final bin.

int gsl_histogram_set_ranges_uniform (gsl_histogram * h,   Function
        double *xmin*, double *xmax*)

This function sets the ranges of the existing histogram h to cover the range *xmin* to *xmax* uniformly. The values of the histogram bins are reset to zero. The bin ranges are shown in the table below,

bin[0]	corresponds to	$xmin \leq x < xmin + d$
bin[1]	corresponds to	$xmin + d \leq x < xmin + 2d$
...	...	...
bin[n-1]	corresponds to	$xmin + (n-1)d \leq x < xmax$

where d is the bin spacing, $d = (xmax - xmin)/n$.

void gsl_histogram_free (gsl_histogram * h)   Function

This function frees the histogram h and all of the memory associated with it.

21.3 Copying Histograms

int gsl_histogram_memcpy (gsl_histogram * *dest*, const   Function
        gsl_histogram * *src*)

This function copies the histogram *src* into the pre-existing histogram *dest*, making *dest* into an exact copy of *src*. The two histograms must be of the same size.

gsl_histogram * gsl_histogram_clone (const gsl_histogram *   Function
        src)

This function returns a pointer to a newly created histogram which is an exact copy of the histogram *src*.

21.4 Updating and accessing histogram elements

There are two ways to access histogram bins, either by specifying an x coordinate or by using the bin-index directly. The functions for accessing the histogram through x coordinates use a binary search to identify the bin which covers the appropriate range.

int gsl_histogram_increment (gsl_histogram * h, double x)   Function

This function updates the histogram h by adding one (1.0) to the bin whose range contains the coordinate x.

If x lies in the valid range of the histogram then the function returns zero to indicate success. If x is less than the lower limit of the histogram then the function returns GSL_EDOM, and none of bins are modified. Similarly, if the value of x is greater than or equal to the upper limit of the histogram then the function returns GSL_EDOM, and none of the bins are modified. The error handler is not called, however, since it is often necessary to compute histograms for a small range of a larger dataset, ignoring the values outside the range of interest.

int gsl_histogram_accumulate (gsl_histogram * *h*, double *x*,   Function
        double *weight*)
    This function is similar to gsl_histogram_increment but increases the value
    of the appropriate bin in the histogram *h* by the floating-point number
    weight.

double gsl_histogram_get (const gsl_histogram * *h*, size_t *i*)   Function
    This function returns the contents of the *i*-th bin of the histogram *h*. If *i* lies
    outside the valid range of indices for the histogram then the error handler
    is called with an error code of GSL_EDOM and the function returns 0.

int gsl_histogram_get_range (const gsl_histogram * *h*, size_t   Function
        i, double * *lower*, double * *upper*)
    This function finds the upper and lower range limits of the *i*-th bin of the
    histogram *h*. If the index *i* is valid then the corresponding range limits are
    stored in *lower* and *upper*. The lower limit is inclusive (i.e. events with
    this coordinate are included in the bin) and the upper limit is exclusive
    (i.e. events with the coordinate of the upper limit are excluded and fall in
    the neighboring higher bin, if it exists). The function returns 0 to indicate
    success. If *i* lies outside the valid range of indices for the histogram then the
    error handler is called and the function returns an error code of GSL_EDOM.

double gsl_histogram_max (const gsl_histogram * *h*)   Function
double gsl_histogram_min (const gsl_histogram * *h*)   Function
size_t gsl_histogram_bins (const gsl_histogram * *h*)   Function
    These functions return the maximum upper and minimum lower range limits
    and the number of bins of the histogram *h*. They provide a way of deter-
    mining these values without accessing the gsl_histogram struct directly.

void gsl_histogram_reset (gsl_histogram * *h*)   Function
    This function resets all the bins in the histogram *h* to zero.

21.5 Searching histogram ranges

The following functions are used by the access and update routines to locate
the bin which corresponds to a given *x* coordinate.

int gsl_histogram_find (const gsl_histogram * *h*, double *x*,   Function
        size_t * *i*)
    This function finds and sets the index *i* to the bin number which covers the
    coordinate *x* in the histogram *h*. The bin is located using a binary search.
    The search includes an optimization for histograms with uniform range, and
    will return the correct bin immediately in this case. If *x* is found in the
    range of the histogram then the function sets the index *i* and returns GSL_
    SUCCESS. If *x* lies outside the valid range of the histogram then the function
    returns GSL_EDOM and the error handler is invoked.

21.6 Histogram Statistics

double gsl_histogram_max_val (const gsl_histogram * h)   *Function*
    This function returns the maximum value contained in the histogram bins.

size_t gsl_histogram_max_bin (const gsl_histogram * h)   *Function*
    This function returns the index of the bin containing the maximum value.
    In the case where several bins contain the same maximum value the smallest
    index is returned.

double gsl_histogram_min_val (const gsl_histogram * h)   *Function*
    This function returns the minimum value contained in the histogram bins.

size_t gsl_histogram_min_bin (const gsl_histogram * h)   *Function*
    This function returns the index of the bin containing the minimum value. In
    the case where several bins contain the same maximum value the smallest
    index is returned.

double gsl_histogram_mean (const gsl_histogram * h)   *Function*
    This function returns the mean of the histogrammed variable, where the
    histogram is regarded as a probability distribution. Negative bin values are
    ignored for the purposes of this calculation. The accuracy of the result is
    limited by the bin width.

double gsl_histogram_sigma (const gsl_histogram * h)   *Function*
    This function returns the standard deviation of the histogrammed variable,
    where the histogram is regarded as a probability distribution. Negative bin
    values are ignored for the purposes of this calculation. The accuracy of the
    result is limited by the bin width.

double gsl_histogram_sum (const gsl_histogram * h)   *Function*
    This function returns the sum of all bin values. Negative bin values are
    included in the sum.

21.7 Histogram Operations

int gsl_histogram_equal_bins_p (const gsl_histogram * h1,   *Function*
        const gsl_histogram * h2)
    This function returns 1 if the all of the individual bin ranges of the two
    histograms are identical, and 0 otherwise.

int gsl_histogram_add (gsl_histogram * h1, const   *Function*
        gsl_histogram * h2)
    This function adds the contents of the bins in histogram h2 to the corre-
    sponding bins of histogram h1, i.e. $h'_1(i) = h_1(i) + h_2(i)$. The two histograms
    must have identical bin ranges.

int gsl_histogram_sub (gsl_histogram * *h1*, const   *Function*
        gsl_histogram * *h2*)
This function subtracts the contents of the bins in histogram *h2* from the
corresponding bins of histogram *h1*, i.e. $h'_1(i) = h_1(i) - h_2(i)$. The two
histograms must have identical bin ranges.

int gsl_histogram_mul (gsl_histogram * *h1*, const   *Function*
        gsl_histogram * *h2*)
This function multiplies the contents of the bins of histogram *h1* by the
contents of the corresponding bins in histogram *h2*, i.e. $h'_1(i) = h_1(i) * h_2(i)$.
The two histograms must have identical bin ranges.

int gsl_histogram_div (gsl_histogram * *h1*, const   *Function*
        gsl_histogram * *h2*)
This function divides the contents of the bins of histogram *h1* by the contents
of the corresponding bins in histogram *h2*, i.e. $h'_1(i) = h_1(i)/h_2(i)$. The two
histograms must have identical bin ranges.

int gsl_histogram_scale (gsl_histogram * *h*, double *scale*)   *Function*
This function multiplies the contents of the bins of histogram *h* by the
constant *scale*, i.e. $h'_1(i) = h_1(i) * scale$.

int gsl_histogram_shift (gsl_histogram * *h*, double *offset*)   *Function*
This function shifts the contents of the bins of histogram *h* by the constant
offset, i.e. $h'_1(i) = h_1(i) + offset$.

21.8 Reading and writing histograms

The library provides functions for reading and writing histograms to a file as
binary data or formatted text.

int gsl_histogram_fwrite (FILE * *stream*, const gsl_histogram *   *Function*
        h)
This function writes the ranges and bins of the histogram *h* to the stream
stream in binary format. The return value is 0 for success and GSL_EFAILED
if there was a problem writing to the file. Since the data is written in the
native binary format it may not be portable between different architectures.

int gsl_histogram_fread (FILE * *stream*, gsl_histogram * *h*)   *Function*
This function reads into the histogram *h* from the open stream *stream* in
binary format. The histogram *h* must be preallocated with the correct size
since the function uses the number of bins in *h* to determine how many bytes
to read. The return value is 0 for success and GSL_EFAILED if there was a
problem reading from the file. The data is assumed to have been written in
the native binary format on the same architecture.

int gsl_histogram_fprintf (FILE * *stream*, const gsl_histogram   Function
     * *h*, const char * *range_format*, const char * *bin_format*)
This function writes the ranges and bins of the histogram *h* line-by-line to
the stream *stream* using the format specifiers *range_format* and *bin_format*.
These should be one of the %g, %e or %f formats for floating point numbers.
The function returns 0 for success and GSL_EFAILED if there was a problem
writing to the file. The histogram output is formatted in three columns, and
the columns are separated by spaces, like this,

     range[0] range[1] bin[0]
     range[1] range[2] bin[1]
     range[2] range[3] bin[2]
     ....
     range[n-1] range[n] bin[n-1]

The values of the ranges are formatted using *range_format* and the value of
the bins are formatted using *bin_format*. Each line contains the lower and
upper limit of the range of the bins and the value of the bin itself. Since the
upper limit of one bin is the lower limit of the next there is duplication of
these values between lines but this allows the histogram to be manipulated
with line-oriented tools.

int gsl_histogram_fscanf (FILE * *stream*, gsl_histogram * *h*)   Function
This function reads formatted data from the stream *stream* into the histogram *h*. The data is assumed to be in the three-column format used by
gsl_histogram_fprintf. The histogram *h* must be preallocated with the
correct length since the function uses the size of *h* to determine how many
numbers to read. The function returns 0 for success and GSL_EFAILED if
there was a problem reading from the file.

21.9 Resampling from histograms

A histogram made by counting events can be regarded as a measurement of
a probability distribution. Allowing for statistical error, the height of each bin
represents the probability of an event where the value of x falls in the range of
that bin. The probability distribution function has the one-dimensional form
$p(x)dx$ where,

$$p(x) = n_i/(N w_i)$$

In this equation n_i is the number of events in the bin which contains x, w_i is
the width of the bin and N is the total number of events. The distribution of
events within each bin is assumed to be uniform.

21.10 The histogram probability distribution struct

The probability distribution function for a histogram consists of a set of *bins* which measure the probability of an event falling into a given range of a continuous variable x. A probability distribution function is defined by the following struct, which actually stores the cumulative probability distribution function. This is the natural quantity for generating samples via the inverse transform method, because there is a one-to-one mapping between the cumulative probability distribution and the range $[0,1]$. It can be shown that by taking a uniform random number in this range and finding its corresponding coordinate in the cumulative probability distribution we obtain samples with the desired probability distribution.

gsl_histogram_pdf   Data Type

> size_t n
>> This is the number of bins used to approximate the probability distribution function.
>
> double * range
>> The ranges of the bins are stored in an array of $n+1$ elements pointed to by *range*.
>
> double * sum
>> The cumulative probability for the bins is stored in an array of n elements pointed to by *sum*.

The following functions allow you to create a gsl_histogram_pdf struct which represents this probability distribution and generate random samples from it.

gsl_histogram_pdf * gsl_histogram_pdf_alloc (size_t n)   Function
This function allocates memory for a probability distribution with n bins and returns a pointer to a newly initialized gsl_histogram_pdf struct. If insufficient memory is available a null pointer is returned and the error handler is invoked with an error code of GSL_ENOMEM.

int gsl_histogram_pdf_init (gsl_histogram_pdf * p, const   Function
        gsl_histogram * h)
This function initializes the probability distribution p with the contents of the histogram h. If any of the bins of h are negative then the error handler is invoked with an error code of GSL_EDOM because a probability distribution cannot contain negative values.

void gsl_histogram_pdf_free (gsl_histogram_pdf * p)   Function
This function frees the probability distribution function p and all of the memory associated with it.

double gsl_histogram_pdf_sample (const gsl_histogram_pdf * p, Function
    double r)
    This function uses r, a uniform random number between zero and one, to
compute a single random sample from the probability distribution p. The
algorithm used to compute the sample s is given by the following formula,

$$s = \text{range}[i] + \delta * (\text{range}[i + 1] - \text{range}[i])$$

where i is the index which satisfies $sum[i] \leq r < sum[i + 1]$ and *delta* is
$(r - sum[i])/(sum[i + 1] - sum[i])$.

21.11 Example programs for histograms

    The following program shows how to make a simple histogram of a column
of numerical data supplied on stdin. The program takes three arguments,
specifying the upper and lower bounds of the histogram and the number of
bins. It then reads numbers from stdin, one line at a time, and adds them to
the histogram. When there is no more data to read it prints out the accumulated
histogram using gsl_histogram_fprintf.

```
#include <stdio.h>
#include <stdlib.h>
#include <gsl/gsl_histogram.h>

int
main (int argc, char **argv)
{
  double a, b;
  size_t n;

  if (argc != 4)
    {
      printf ("Usage: gsl-histogram xmin xmax n\n"
              "Computes a histogram of the data "
              "on stdin using n bins from xmin "
              "to xmax\n");
      exit (0);
    }

  a = atof (argv[1]);
  b = atof (argv[2]);
  n = atoi (argv[3]);

  {
    double x;
    gsl_histogram * h = gsl_histogram_alloc (n);
    gsl_histogram_set_ranges_uniform (h, a, b);

    while (fscanf (stdin, "%lg", &x) == 1)
      {
```

```
            gsl_histogram_increment (h, x);
        }
      gsl_histogram_fprintf (stdout, h, "%g", "%g");
      gsl_histogram_free (h);
    }
  exit (0);
}
```

Here is an example of the program in use. We generate 10000 random samples from a Cauchy distribution with a width of 30 and histogram them over the range -100 to 100, using 200 bins.

```
$ gsl-randist 0 10000 cauchy 30
  | gsl-histogram -100 100 200 > histogram.dat
```

A plot of the resulting histogram shows the familiar shape of the Cauchy distribution and the fluctuations caused by the finite sample size.

```
$ awk '{print $1, $3 ; print $2, $3}' histogram.dat
  | graph -T X
```

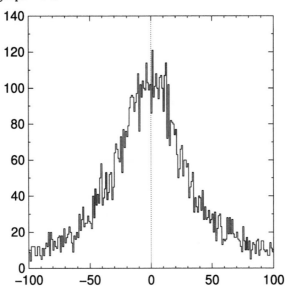

21.12 Two dimensional histograms

A two dimensional histogram consists of a set of *bins* which count the number of events falling in a given area of the (x, y) plane. The simplest way to use a two dimensional histogram is to record two-dimensional position information, $n(x, y)$. Another possibility is to form a *joint distribution* by recording related variables. For example a detector might record both the position of an event (x) and the amount of energy it deposited E. These could be histogrammed as the joint distribution $n(x, E)$.

21.13 The 2D histogram struct

Two dimensional histograms are defined by the following struct,

gsl_histogram2d   Data Type

>   size_t nx, ny
>   This is the number of histogram bins in the x and y directions.
>
>   double * xrange
>   The ranges of the bins in the x-direction are stored in an array of $nx+1$ elements pointed to by *xrange*.
>
>   double * yrange
>   The ranges of the bins in the y-direction are stored in an array of $ny+1$ elements pointed to by *yrange*.
>
>   double * bin
>   The counts for each bin are stored in an array pointed to by *bin*. The bins are floating-point numbers, so you can increment them by non-integer values if necessary. The array *bin* stores the two dimensional array of bins in a single block of memory according to the mapping bin(i,j) = bin[i * ny + j].

The range for bin(i,j) is given by xrange[i] to xrange[i+1] in the x-direction and yrange[j] to yrange[j+1] in the y-direction. Each bin is inclusive at the lower end and exclusive at the upper end. Mathematically this means that the bins are defined by the following inequality,

$$\begin{aligned} \text{bin(i,j) corresponds to} \quad & xrange[i] \leq x < xrange[i+1] \\ \text{and} \quad & yrange[j] \leq y < yrange[j+1] \end{aligned}$$

Note that any samples which fall on the upper sides of the histogram are excluded. If you want to include these values for the side bins you will need to add an extra row or column to your histogram.

The gsl_histogram2d struct and its associated functions are defined in the header file 'gsl_histogram2d.h'.

21.14 2D Histogram allocation

The functions for allocating memory to a 2D histogram follow the style of malloc and free. In addition they also perform their own error checking. If there is insufficient memory available to allocate a histogram then the functions call the error handler (with an error number of GSL_ENOMEM) in addition to returning a null pointer. Thus if you use the library error handler to abort your program then it isn't necessary to check every 2D histogram alloc.

gsl_histogram2d * gsl_histogram2d_alloc (size_t *nx*, size_t   Function
        ny)

This function allocates memory for a two-dimensional histogram with *nx* bins in the x direction and *ny* bins in the y direction. The function returns a pointer to a newly created gsl_histogram2d struct. If insufficient memory is available a null pointer is returned and the error handler is invoked with an error code of GSL_ENOMEM. The bins and ranges must be initialized with one of the functions below before the histogram is ready for use.

int gsl_histogram2d_set_ranges (gsl_histogram2d * *h*, const   Function
        double *xrange*[], size_t *xsize*, const double *yrange*[], size_t
        ysize)

This function sets the ranges of the existing histogram *h* using the arrays *xrange* and *yrange* of size *xsize* and *ysize* respectively. The values of the histogram bins are reset to zero.

int gsl_histogram2d_set_ranges_uniform (gsl_histogram2d * *h*,   Function
        double *xmin*, double *xmax*, double *ymin*, double *ymax*)

This function sets the ranges of the existing histogram *h* to cover the ranges *xmin* to *xmax* and *ymin* to *ymax* uniformly. The values of the histogram bins are reset to zero.

void gsl_histogram2d_free (gsl_histogram2d * *h*)   Function

This function frees the 2D histogram *h* and all of the memory associated with it.

21.15 Copying 2D Histograms

int gsl_histogram2d_memcpy (gsl_histogram2d * *dest*, const   Function
        gsl_histogram2d * *src*)

This function copies the histogram *src* into the pre-existing histogram *dest*, making *dest* into an exact copy of *src*. The two histograms must be of the same size.

gsl_histogram2d * gsl_histogram2d_clone (const   Function
        gsl_histogram2d * *src*)

This function returns a pointer to a newly created histogram which is an exact copy of the histogram *src*.

21.16 Updating and accessing 2D histogram elements

You can access the bins of a two-dimensional histogram either by specifying a pair of (x, y) coordinates or by using the bin indices (i, j) directly. The functions for accessing the histogram through (x, y) coordinates use binary searches in the x and y directions to identify the bin which covers the appropriate range.

int gsl_histogram2d_increment (gsl_histogram2d * h, double x,   *Function*
        double y)

This function updates the histogram h by adding one (1.0) to the bin whose x and y ranges contain the coordinates (x,y).

If the point (x, y) lies inside the valid ranges of the histogram then the function returns zero to indicate success. If (x, y) lies outside the limits of the histogram then the function returns GSL_EDOM, and none of the bins are modified. The error handler is not called, since it is often necessary to compute histograms for a small range of a larger dataset, ignoring any coordinates outside the range of interest.

int gsl_histogram2d_accumulate (gsl_histogram2d * h, double   *Function*
        x, double y, double *weight*)

This function is similar to gsl_histogram2d_increment but increases the value of the appropriate bin in the histogram h by the floating-point number *weight*.

double gsl_histogram2d_get (const gsl_histogram2d * h, size_t   *Function*
        i, size_t j)

This function returns the contents of the (i,j)-th bin of the histogram h. If (i,j) lies outside the valid range of indices for the histogram then the error handler is called with an error code of GSL_EDOM and the function returns 0.

int gsl_histogram2d_get_xrange (const gsl_histogram2d * h,   *Function*
        size_t i, double * *xlower*, double * *xupper*)
int gsl_histogram2d_get_yrange (const gsl_histogram2d * h,   *Function*
        size_t j, double * *ylower*, double * *yupper*)

These functions find the upper and lower range limits of the *i*-th and *j*-th bins in the x and y directions of the histogram h. The range limits are stored in *xlower* and *xupper* or *ylower* and *yupper*. The lower limits are inclusive (i.e. events with these coordinates are included in the bin) and the upper limits are exclusive (i.e. events with the value of the upper limit are not included and fall in the neighboring higher bin, if it exists). The functions return 0 to indicate success. If *i* or *j* lies outside the valid range of indices for the histogram then the error handler is called with an error code of GSL_EDOM.

`double gsl_histogram2d_xmax (const gsl_histogram2d * h)`	Function
`double gsl_histogram2d_xmin (const gsl_histogram2d * h)`	Function
`size_t gsl_histogram2d_nx (const gsl_histogram2d * h)`	Function
`double gsl_histogram2d_ymax (const gsl_histogram2d * h)`	Function
`double gsl_histogram2d_ymin (const gsl_histogram2d * h)`	Function
`size_t gsl_histogram2d_ny (const gsl_histogram2d * h)`	Function

These functions return the maximum upper and minimum lower range limits and the number of bins for the x and y directions of the histogram h. They provide a way of determining these values without accessing the gsl_histogram2d struct directly.

`void gsl_histogram2d_reset (gsl_histogram2d * h)`   Function

This function resets all the bins of the histogram h to zero.

21.17 Searching 2D histogram ranges

The following functions are used by the access and update routines to locate the bin which corresponds to a given (x, y) coordinate.

`int gsl_histogram2d_find (const gsl_histogram2d * h, double x,`   Function
`        double y, size_t * i, size_t * j)`

This function finds and sets the indices i and j to the to the bin which covers the coordinates (x,y). The bin is located using a binary search. The search includes an optimization for histograms with uniform ranges, and will return the correct bin immediately in this case. If (x, y) is found then the function sets the indices (i,j) and returns GSL_SUCCESS. If (x, y) lies outside the valid range of the histogram then the function returns GSL_EDOM and the error handler is invoked.

21.18 2D Histogram Statistics

`double gsl_histogram2d_max_val (const gsl_histogram2d * h)`   Function

This function returns the maximum value contained in the histogram bins.

`void gsl_histogram2d_max_bin (const gsl_histogram2d * h,`   Function
`        size_t * i, size_t * j)`

This function finds the indices of the bin containing the maximum value in the histogram h and stores the result in (i,j). In the case where several bins contain the same maximum value the first bin found is returned.

`double gsl_histogram2d_min_val (const gsl_histogram2d * h)`   Function

This function returns the minimum value contained in the histogram bins.

`void gsl_histogram2d_min_bin (const gsl_histogram2d * h,`   Function
`        size_t * i, size_t * j)`

This function finds the indices of the bin containing the minimum value in the histogram h and stores the result in (i,j). In the case where several bins contain the same maximum value the first bin found is returned.

double gsl_histogram2d_xmean (const gsl_histogram2d * h)   Function
  This function returns the mean of the histogrammed x variable, where the
  histogram is regarded as a probability distribution. Negative bin values are
  ignored for the purposes of this calculation.

double gsl_histogram2d_ymean (const gsl_histogram2d * h)   Function
  This function returns the mean of the histogrammed y variable, where the
  histogram is regarded as a probability distribution. Negative bin values are
  ignored for the purposes of this calculation.

double gsl_histogram2d_xsigma (const gsl_histogram2d * h)   Function
  This function returns the standard deviation of the histogrammed x variable,
  where the histogram is regarded as a probability distribution. Negative bin
  values are ignored for the purposes of this calculation.

double gsl_histogram2d_ysigma (const gsl_histogram2d * h)   Function
  This function returns the standard deviation of the histogrammed y variable,
  where the histogram is regarded as a probability distribution. Negative bin
  values are ignored for the purposes of this calculation.

double gsl_histogram2d_cov (const gsl_histogram2d * h)   Function
  This function returns the covariance of the histogrammed x and y variables,
  where the histogram is regarded as a probability distribution. Negative bin
  values are ignored for the purposes of this calculation.

double gsl_histogram2d_sum (const gsl_histogram2d * h)   Function
  This function returns the sum of all bin values. Negative bin values are
  included in the sum.

21.19 2D Histogram Operations

int gsl_histogram2d_equal_bins_p (const gsl_histogram2d * h1,   Function
      const gsl_histogram2d * h2)
  This function returns 1 if all the individual bin ranges of the two histograms
  are identical, and 0 otherwise.

int gsl_histogram2d_add (gsl_histogram2d * h1, const   Function
      gsl_histogram2d * h2)
  This function adds the contents of the bins in histogram h2 to the corresponding bins of histogram h1, i.e. $h'_1(i,j) = h_1(i,j) + h_2(i,j)$. The two
  histograms must have identical bin ranges.

int gsl_histogram2d_sub (gsl_histogram2d * h1, const   Function
      gsl_histogram2d * h2)
  This function subtracts the contents of the bins in histogram h2 from the
  corresponding bins of histogram h1, i.e. $h'_1(i,j) = h_1(i,j) - h_2(i,j)$. The
  two histograms must have identical bin ranges.

int gsl_histogram2d_mul (gsl_histogram2d * *h1*, const   Function
        gsl_histogram2d * *h2*)
  This function multiplies the contents of the bins of histogram *h1* by the
  contents of the corresponding bins in histogram *h2*, i.e. $h'_1(i,j) = h_1(i,j) * h_2(i,j)$. The two histograms must have identical bin ranges.

int gsl_histogram2d_div (gsl_histogram2d * *h1*, const   Function
        gsl_histogram2d * *h2*)
  This function divides the contents of the bins of histogram *h2*, i.e. the contents
  of the corresponding bins in histogram *h2*, i.e. $h'_1(i,j) = h_1(i,j)/h_2(i,j)$.
  The two histograms must have identical bin ranges.

int gsl_histogram2d_scale (gsl_histogram2d * *h*, double *scale*)   Function
  This function multiplies the contents of the bins of histogram *h* by the
  constant *scale*, i.e. $h'_1(i,j) = h_1(i,j) * scale$.

int gsl_histogram2d_shift (gsl_histogram2d * *h*, double *offset*)   Function
  This function shifts the contents of the bins of histogram *h* by the constant
  offset, i.e. $h'_1(i,j) = h_1(i,j) + offset$.

21.20 Reading and writing 2D histograms

The library provides functions for reading and writing two dimensional histograms to a file as binary data or formatted text.

int gsl_histogram2d_fwrite (FILE * *stream*, const   Function
        gsl_histogram2d * *h*)
  This function writes the ranges and bins of the histogram *h* to the stream
  stream in binary format. The return value is 0 for success and GSL_EFAILED
  if there was a problem writing to the file. Since the data is written in the
  native binary format it may not be portable between different architectures.

int gsl_histogram2d_fread (FILE * *stream*, gsl_histogram2d * *h*)   Function
  This function reads into the histogram *h* from the stream *stream* in binary
  format. The histogram *h* must be preallocated with the correct size since
  the function uses the number of x and y bins in *h* to determine how many
  bytes to read. The return value is 0 for success and GSL_EFAILED if there was
  a problem reading from the file. The data is assumed to have been written
  in the native binary format on the same architecture.

int gsl_histogram2d_fprintf (FILE * *stream*, const   Function
        gsl_histogram2d * *h*, const char * *range_format*, const char *
        bin_format)
  This function writes the ranges and bins of the histogram *h* line-by-line to
  the stream *stream* using the format specifiers *range_format* and *bin_format*.
  These should be one of the %g, %e or %f formats for floating point numbers.
  The function returns 0 for success and GSL_EFAILED if there was a problem
  writing to the file. The histogram output is formatted in five columns, and
  the columns are separated by spaces, like this,

```
xrange[0]  xrange[1]  yrange[0]   yrange[1]  bin(0,0)
xrange[0]  xrange[1]  yrange[1]   yrange[2]  bin(0,1)
xrange[0]  xrange[1]  yrange[2]   yrange[3]  bin(0,2)
....
xrange[0]  xrange[1]  yrange[ny-1] yrange[ny] bin(0,ny-1)

xrange[1]  xrange[2]  yrange[0]   yrange[1]  bin(1,0)
xrange[1]  xrange[2]  yrange[1]   yrange[2]  bin(1,1)
xrange[1]  xrange[2]  yrange[1]   yrange[2]  bin(1,2)
....
xrange[1]  xrange[2]  yrange[ny-1] yrange[ny] bin(1,ny-1)


....


xrange[nx-1] xrange[nx]  yrange[0]   yrange[1]  bin(nx-1,0)
xrange[nx-1] xrange[nx]  yrange[1]   yrange[2]  bin(nx-1,1)
xrange[nx-1] xrange[nx]  yrange[1]   yrange[2]  bin(nx-1,2)
....
xrange[nx-1] xrange[nx]  yrange[ny-1] yrange[ny] bin(nx-1,ny-1)
```

Each line contains the lower and upper limits of the bin and the contents of the bin. Since the upper limits of the each bin are the lower limits of the neighboring bins there is duplication of these values but this allows the histogram to be manipulated with line-oriented tools.

int gsl_histogram2d_fscanf (FILE * *stream*, gsl_histogram2d *   Function
       h)
This function reads formatted data from the stream *stream* into the histogram *h*. The data is assumed to be in the five-column format used by gsl_histogram2d_fprintf. The histogram *h* must be preallocated with the correct lengths since the function uses the sizes of *h* to determine how many numbers to read. The function returns 0 for success and GSL_EFAILED if there was a problem reading from the file.

21.21 Resampling from 2D histograms

As in the one-dimensional case, a two-dimensional histogram made by counting events can be regarded as a measurement of a probability distribution. Allowing for statistical error, the height of each bin represents the probability of an event where (x,y) falls in the range of that bin. For a two-dimensional histogram the probability distribution takes the form $p(x, y)dxdy$ where,

$$p(x, y) = n_{ij}/(NA_{ij})$$

In this equation n_{ij} is the number of events in the bin which contains (x, y), A_{ij} is the area of the bin and N is the total number of events. The distribution of events within each bin is assumed to be uniform.

gsl_histogram2d_pdf   Data Type

> size_t nx, ny
>> This is the number of histogram bins used to approximate the probability distribution function in the x and y directions.
>
> double * xrange
>> The ranges of the bins in the x-direction are stored in an array of $nx + 1$ elements pointed to by xrange.
>
> double * yrange
>> The ranges of the bins in the y-direction are stored in an array of $ny + 1$ pointed to by yrange.
>
> double * sum
>> The cumulative probability for the bins is stored in an array of $nx*ny$ elements pointed to by sum.

The following functions allow you to create a gsl_histogram2d_pdf struct which represents a two dimensional probability distribution and generate random samples from it.

gsl_histogram2d_pdf * gsl_histogram2d_pdf_alloc (size_t nx,   Function
         size_t ny)
> This function allocates memory for a two-dimensional probability distribution of size nx-by-ny and returns a pointer to a newly initialized gsl_histogram2d_pdf struct. If insufficient memory is available a null pointer is returned and the error handler is invoked with an error code of GSL_ENOMEM.

int gsl_histogram2d_pdf_init (gsl_histogram2d_pdf * p, const   Function
         gsl_histogram2d * h)
> This function initializes the two-dimensional probability distribution calculated p from the histogram h. If any of the bins of h are negative then the error handler is invoked with an error code of GSL_EDOM because a probability distribution cannot contain negative values.

void gsl_histogram2d_pdf_free (gsl_histogram2d_pdf * p)   Function
> This function frees the two-dimensional probability distribution function p and all of the memory associated with it.

int gsl_histogram2d_pdf_sample (const gsl_histogram2d_pdf *   Function
         p, double r1, double r2, double * x, double * y)
> This function uses two uniform random numbers between zero and one, r1 and r2, to compute a single random sample from the two-dimensional probability distribution p.

21.22 Example programs for 2D histograms

This program demonstrates two features of two-dimensional histograms. First a 10-by-10 two-dimensional histogram is created with x and y running from 0 to 1. Then a few sample points are added to the histogram, at (0.3,0.3) with a height of 1, at (0.8,0.1) with a height of 5 and at (0.7,0.9) with a height of 0.5. This histogram with three events is used to generate a random sample of 1000 simulated events, which are printed out.

```
#include <stdio.h>
#include <gsl/gsl_rng.h>
#include <gsl/gsl_histogram2d.h>

int
main (void)
{
  const gsl_rng_type * T;
  gsl_rng * r;

  gsl_histogram2d * h = gsl_histogram2d_alloc (10, 10);

  gsl_histogram2d_set_ranges_uniform (h,
                                      0.0, 1.0,
                                      0.0, 1.0);

  gsl_histogram2d_accumulate (h, 0.3, 0.3, 1);
  gsl_histogram2d_accumulate (h, 0.8, 0.1, 5);
  gsl_histogram2d_accumulate (h, 0.7, 0.9, 0.5);

  gsl_rng_env_setup ();

  T = gsl_rng_default;
  r = gsl_rng_alloc (T);

  {
    int i;
    gsl_histogram2d_pdf * p
      = gsl_histogram2d_pdf_alloc (h->nx, h->ny);

    gsl_histogram2d_pdf_init (p, h);

    for (i = 0; i < 1000; i++) {
      double x, y;
      double u = gsl_rng_uniform (r);
      double v = gsl_rng_uniform (r);

      gsl_histogram2d_pdf_sample (p, u, v, &x, &y);

      printf ("%g %g\n", x, y);
```

```
    }

    gsl_histogram2d_pdf_free (p);
}

gsl_histogram2d_free (h);
gsl_rng_free (r);

return 0;
}
```

The following plot shows the distribution of the simulated events. Using a higher resolution grid we can see the original underlying histogram and also the statistical fluctuations caused by the events being uniformly distributed over the area of the original bins.

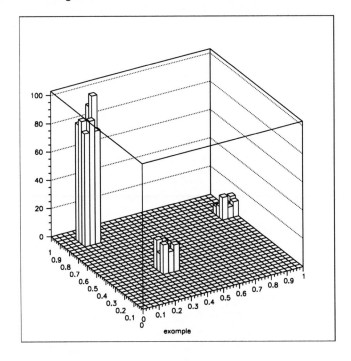

22 N-tuples

This chapter describes functions for creating and manipulating *ntuples*, sets of values associated with events. The ntuples are stored in files. Their values can be extracted in any combination and *booked* in a histogram using a selection function.

The values to be stored are held in a user-defined data structure, and an ntuple is created associating this data structure with a file. The values are then written to the file (normally inside a loop) using the ntuple functions described below.

A histogram can be created from ntuple data by providing a selection function and a value function. The selection function specifies whether an event should be included in the subset to be analyzed or not. The value function computes the entry to be added to the histogram for each event.

All the ntuple functions are defined in the header file 'gsl_ntuple.h'

22.1 The ntuple struct

Ntuples are manipulated using the `gsl_ntuple` struct. This struct contains information on the file where the ntuple data is stored, a pointer to the current ntuple data row and the size of the user-defined ntuple data struct.

```
typedef struct {
    FILE * file;
    void * ntuple_data;
    size_t size;
} gsl_ntuple;
```

22.2 Creating ntuples

gsl_ntuple * gsl_ntuple_create (char * *filename*, void *   *Function*
        ntuple_data, size_t *size*)

This function creates a new write-only ntuple file *filename* for ntuples of size *size* and returns a pointer to the newly created ntuple struct. Any existing file with the same name is truncated to zero length and overwritten. A pointer to memory for the current ntuple row *ntuple_data* must be supplied—this is used to copy ntuples in and out of the file.

22.3 Opening an existing ntuple file

gsl_ntuple * gsl_ntuple_open (char * *filename*, void *   *Function*
        ntuple_data, size_t *size*)

This function opens an existing ntuple file *filename* for reading and returns a pointer to a corresponding ntuple struct. The ntuples in the file must have size *size*. A pointer to memory for the current ntuple row *ntuple_data* must be supplied—this is used to copy ntuples in and out of the file.

22.4 Writing ntuples

int gsl_ntuple_write (gsl_ntuple * *ntuple*)   Function
  This function writes the current ntuple *ntuple->ntuple_data* of size *ntuple->size* to the corresponding file.

int gsl_ntuple_bookdata (gsl_ntuple * *ntuple*)   Function
  This function is a synonym for gsl_ntuple_write.

22.5 Reading ntuples

int gsl_ntuple_read (gsl_ntuple * *ntuple*)   Function
  This function reads the current row of the ntuple file for *ntuple* and stores the values in *ntuple->data*.

22.6 Closing an ntuple file

int gsl_ntuple_close (gsl_ntuple * *ntuple*)   Function
  This function closes the ntuple file *ntuple* and frees its associated allocated memory.

22.7 Histogramming ntuple values

Once an ntuple has been created its contents can be histogrammed in various ways using the function gsl_ntuple_project. Two user-defined functions must be provided, a function to select events and a function to compute scalar values. The selection function and the value function both accept the ntuple row as a first argument and other parameters as a second argument.

The *selection function* determines which ntuple rows are selected for histogramming. It is defined by the following struct,

```
typedef struct {
  int (* function) (void * ntuple_data, void * params);
  void * params;
} gsl_ntuple_select_fn;
```

The struct component *function* should return a non-zero value for each ntuple row that is to be included in the histogram.

The *value function* computes scalar values for those ntuple rows selected by the selection function,

```
typedef struct {
  double (* function) (void * ntuple_data, void * params);
  void * params;
} gsl_ntuple_value_fn;
```

In this case the struct component *function* should return the value to be added to the histogram for the ntuple row.

int gsl_ntuple_project (gsl_histogram * h, gsl_ntuple *   *Function*
       ntuple, gsl_ntuple_value_fn * *value_func*,
       gsl_ntuple_select_fn * *select_func*)

This function updates the histogram *h* from the ntuple *ntuple* using the
functions *value_func* and *select_func*. For each ntuple row where the selection function *select_func* is non-zero the corresponding value of that row is
computed using the function *value_func* and added to the histogram. Those
ntuple rows where *select_func* returns zero are ignored. New entries are
added to the histogram, so subsequent calls can be used to accumulate further data in the same histogram.

22.8 Examples

The following example programs demonstrate the use of ntuples in managing
a large dataset. The first program creates a set of 10,000 simulated "events",
each with 3 associated values (x, y, z). These are generated from a gaussian
distribution with unit variance, for demonstration purposes, and written to the
ntuple file 'test.dat'.

```
#include <gsl/gsl_ntuple.h>
#include <gsl/gsl_rng.h>
#include <gsl/gsl_randist.h>

struct data
{
  double x;
  double y;
  double z;
};

int
main (void)
{
  const gsl_rng_type * T;
  gsl_rng * r;

  struct data ntuple_row;
  int i;

  gsl_ntuple *ntuple
    = gsl_ntuple_create ("test.dat", &ntuple_row,
                         sizeof (ntuple_row));

  gsl_rng_env_setup ();

  T = gsl_rng_default;
  r = gsl_rng_alloc (T);

  for (i = 0; i < 10000; i++)
```

```
        {
          ntuple_row.x = gsl_ran_ugaussian (r);
          ntuple_row.y = gsl_ran_ugaussian (r);
          ntuple_row.z = gsl_ran_ugaussian (r);

          gsl_ntuple_write (ntuple);
        }

      gsl_ntuple_close (ntuple);
      gsl_rng_free (r);

      return 0;
    }
```

The next program analyses the ntuple data in the file 'test.dat'. The analysis procedure is to compute the squared-magnitude of each event, $E^2 = x^2 + y^2 + z^2$, and select only those which exceed a lower limit of 1.5. The selected events are then histogrammed using their E^2 values.

```
    #include <math.h>
    #include <gsl/gsl_ntuple.h>
    #include <gsl/gsl_histogram.h>

    struct data
    {
      double x;
      double y;
      double z;
    };

    int sel_func (void *ntuple_data, void *params);
    double val_func (void *ntuple_data, void *params);

    int
    main (void)
    {
      struct data ntuple_row;

      gsl_ntuple *ntuple
        = gsl_ntuple_open ("test.dat", &ntuple_row,
                           sizeof (ntuple_row));
      double lower = 1.5;

      gsl_ntuple_select_fn S;
      gsl_ntuple_value_fn V;

      gsl_histogram *h = gsl_histogram_alloc (100);
      gsl_histogram_set_ranges_uniform(h, 0.0, 10.0);

      S.function = &sel_func;
```

```
      S.params = &lower;

      V.function = &val_func;
      V.params = 0;

      gsl_ntuple_project (h, ntuple, &V, &S);
      gsl_histogram_fprintf (stdout, h, "%f", "%f");
      gsl_histogram_free (h);
      gsl_ntuple_close (ntuple);

      return 0;
    }

    int
    sel_func (void *ntuple_data, void *params)
    {
      struct data * data = (struct data *) ntuple_data;
      double x, y, z, E2, scale;
      scale = *(double *) params;

      x = data->x;
      y = data->y;
      z = data->z;

      E2 = x * x + y * y + z * z;

      return E2 > scale;
    }

    double
    val_func (void *ntuple_data, void *params)
    {
      struct data * data = (struct data *) ntuple_data;
      double x, y, z;

      x = data->x;
      y = data->y;
      z = data->z;

      return x * x + y * y + z * z;
    }
```

The following plot shows the distribution of the selected events. Note the cut-off at the lower bound.

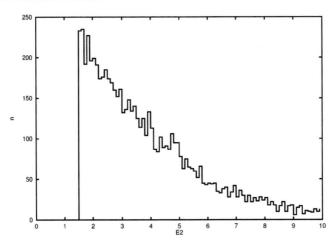

22.9 References and Further Reading

Further information on the use of ntuples can be found in the documentation for the CERN packages PAW and HBOOK (available online).

23 Monte Carlo Integration

This chapter describes routines for multidimensional Monte Carlo integration. These include the traditional Monte Carlo method and adaptive algorithms such as VEGAS and MISER which use importance sampling and stratified sampling techniques. Each algorithm computes an estimate of a multidimensional definite integral of the form,

$$I = \int_{x_l}^{x_u} dx \int_{y_l}^{y_u} dy \, ... f(x, y, ...)$$

over a hypercubic region $((x_l, x_u), (y_l, y_u), ...)$ using a fixed number of function calls. The routines also provide a statistical estimate of the error on the result. This error estimate should be taken as a guide rather than as a strict error bound—random sampling of the region may not uncover all the important features of the function, resulting in an underestimate of the error.

The functions are defined in separate header files for each routine, 'gsl_monte_plain.h', 'gsl_monte_miser.h' and 'gsl_monte_vegas.h'.

23.1 Interface

All of the Monte Carlo integration routines use the same general form of interface. There is an allocator to allocate memory for control variables and workspace, a routine to initialize those control variables, the integrator itself, and a function to free the space when done.

Each integration function requires a random number generator to be supplied, and returns an estimate of the integral and its standard deviation. The accuracy of the result is determined by the number of function calls specified by the user. If a known level of accuracy is required this can be achieved by calling the integrator several times and averaging the individual results until the desired accuracy is obtained.

Random sample points used within the Monte Carlo routines are always chosen strictly within the integration region, so that endpoint singularities are automatically avoided.

The function to be integrated has its own datatype, defined in the header file 'gsl_monte.h'.

gsl_monte_function   Data Type
    This data type defines a general function with parameters for Monte Carlo integration.

    double (* f) (double * x, size_t dim, void * params)
        this function should return the value $f(x, params)$ for the argument x and parameters params, where x is an array of size dim giving the coordinates of the point where the function is to be evaluated.

    size_t dim
        the number of dimensions for x.

void * params
:   a pointer to the parameters of the function.

Here is an example for a quadratic function in two dimensions,

$$f(x, y) = ax^2 + bxy + cy^2$$

with $a = 3$, $b = 2$, $c = 1$. The following code defines a gsl_monte_function F which you could pass to an integrator:

```
struct my_f_params { double a; double b; double c; };

double
my_f (double x[], size_t dim, void * p) {
   struct my_f_params * fp = (struct my_f_params *)p;

   if (dim != 2)
      {
         fprintf (stderr, "error: dim != 2");
         abort ();
      }

   return  fp->a * x[0] * x[0]
            + fp->b * x[0] * x[1]
              + fp->c * x[1] * x[1];
}

gsl_monte_function F;
struct my_f_params params = { 3.0, 2.0, 1.0 };

F.f = &my_f;
F.dim = 2;
F.params = &params;
```

The function $f(x)$ can be evaluated using the following macro,

```
#define GSL_MONTE_FN_EVAL(F,x)
   (*((F)->f))(x,(F)->dim,(F)->params)
```

23.2 PLAIN Monte Carlo

The plain Monte Carlo algorithm samples points randomly from the integration region to estimate the integral and its error. Using this algorithm the estimate of the integral $E(f; N)$ for N randomly distributed points x_i is given by,

$$E(f; N) = V\langle f \rangle = \frac{V}{N} \sum_{i}^{N} f(x_i)$$

where V is the volume of the integration region. The error on this estimate $\sigma(E; N)$ is calculated from the estimated variance of the mean,

$$\sigma^2(E; N) = \frac{V}{N} \sum_{i}^{N} (f(x_i) - \langle f \rangle)^2.$$

For large N this variance decreases asymptotically as $\text{Var}(f)/N$, where $\text{Var}(f)$ is the true variance of the function over the integration region. The error estimate itself should decrease as $\sigma(f)/\sqrt{N}$. The familiar law of errors decreasing as $1/\sqrt{N}$ applies—to reduce the error by a factor of 10 requires a 100-fold increase in the number of sample points.

The functions described in this section are declared in the header file 'gsl_monte_plain.h'.

gsl_monte_plain_state * gsl_monte_plain_alloc (size_t dim)   Function
  This function allocates and initializes a workspace for Monte Carlo integration in dim dimensions.

int gsl_monte_plain_init (gsl_monte_plain_state* s)   Function
  This function initializes a previously allocated integration state. This allows an existing workspace to be reused for different integrations.

int gsl_monte_plain_integrate (gsl_monte_function * f, double   Function
        * xl, double * xu, size_t dim, size_t calls, gsl_rng * r,
        gsl_monte_plain_state * s, double * result, double * abserr)
  This routines uses the plain Monte Carlo algorithm to integrate the function f over the dim-dimensional hypercubic region defined by the lower and upper limits in the arrays xl and xu, each of size dim. The integration uses a fixed number of function calls calls, and obtains random sampling points using the random number generator r. A previously allocated workspace s must be supplied. The result of the integration is returned in result, with an estimated absolute error abserr.

void gsl_monte_plain_free (gsl_monte_plain_state * s)   Function
  This function frees the memory associated with the integrator state s.

23.3 MISER

The MISER algorithm of Press and Farrar is based on recursive stratified sampling. This technique aims to reduce the overall integration error by concentrating integration points in the regions of highest variance.

The idea of stratified sampling begins with the observation that for two disjoint regions a and b with Monte Carlo estimates of the integral $E_a(f)$ and $E_b(f)$ and variances $\sigma_a^2(f)$ and $\sigma_b^2(f)$, the variance $\text{Var}(f)$ of the combined estimate $E(f) = \frac{1}{2}(E_a(f) + E_b(f))$ is given by,

$$\text{Var}(f) = \frac{\sigma_a^2(f)}{4N_a} + \frac{\sigma_b^2(f)}{4N_b}.$$

It can be shown that this variance is minimized by distributing the points such that,

$$\frac{N_a}{N_a + N_b} = \frac{\sigma_a}{\sigma_a + \sigma_b}.$$

Hence the smallest error estimate is obtained by allocating sample points in proportion to the standard deviation of the function in each sub-region.

The MISER algorithm proceeds by bisecting the integration region along one coordinate axis to give two sub-regions at each step. The direction is chosen by examining all d possible bisections and selecting the one which will minimize the combined variance of the two sub-regions. The variance in the sub-regions is estimated by sampling with a fraction of the total number of points available to the current step. The same procedure is then repeated recursively for each of the two half-spaces from the best bisection. The remaining sample points are allocated to the sub-regions using the formula for N_a and N_b. This recursive allocation of integration points continues down to a user-specified depth where each sub-region is integrated using a plain Monte Carlo estimate. These individual values and their error estimates are then combined upwards to give an overall result and an estimate of its error.

The functions described in this section are declared in the header file 'gsl_monte_miser.h'.

gsl_monte_miser_state * gsl_monte_miser_alloc (size_t *dim*)   Function
   This function allocates and initializes a workspace for Monte Carlo integration in *dim* dimensions. The workspace is used to maintain the state of the integration.

int gsl_monte_miser_init (gsl_monte_miser_state* *s*)   Function
   This function initializes a previously allocated integration state. This allows an existing workspace to be reused for different integrations.

int gsl_monte_miser_integrate (gsl_monte_function * f, double   Function
          * xl, double * xu, size_t dim, size_t calls, gsl_rng * r,
          gsl_monte_miser_state * s, double * result, double * abserr)
This routines uses the MISER Monte Carlo algorithm to integrate the function
f over the dim-dimensional hypercubic region defined by the lower and upper
limits in the arrays xl and xu, each of size dim. The integration uses a fixed
number of function calls calls, and obtains random sampling points using
the random number generator r. A previously allocated workspace s must
be supplied. The result of the integration is returned in result, with an
estimated absolute error abserr.

void gsl_monte_miser_free (gsl_monte_miser_state * s)   Function
This function frees the memory associated with the integrator state s.

    The MISER algorithm has several configurable parameters. The following
variables can be accessed through the gsl_monte_miser_state struct,

double estimate_frac   Variable
This parameter specifies the fraction of the currently available number of
function calls which are allocated to estimating the variance at each recursive
step. The default value is 0.1.

size_t min_calls   Variable
This parameter specifies the minimum number of function calls required for
each estimate of the variance. If the number of function calls allocated to
the estimate using estimate_frac falls below min_calls then min_calls are
used instead. This ensures that each estimate maintains a reasonable level
of accuracy. The default value of min_calls is 16 * dim.

size_t min_calls_per_bisection   Variable
This parameter specifies the minimum number of function calls required to
proceed with a bisection step. When a recursive step has fewer calls available
than min_calls_per_bisection it performs a plain Monte Carlo estimate of the
current sub-region and terminates its branch of the recursion. The default
value of this parameter is 32 * min_calls.

double alpha   Variable
This parameter controls how the estimated variances for the two sub-regions
of a bisection are combined when allocating points. With recursive sampling
the overall variance should scale better than $1/N$, since the values from the
sub-regions will be obtained using a procedure which explicitly minimizes
their variance. To accommodate this behavior the MISER algorithm allows
the total variance to depend on a scaling parameter α,

$$\text{Var}(f) = \frac{\sigma_a}{N_a^\alpha} + \frac{\sigma_b}{N_b^\alpha}.$$

The authors of the original paper describing MISER recommend the value
$\alpha = 2$ as a good choice, obtained from numerical experiments, and this is
used as the default value in this implementation.

`double dither`   Variable

This parameter introduces a random fractional variation of size *dither* into each bisection, which can be used to break the symmetry of integrands which are concentrated near the exact center of the hypercubic integration region. The default value of dither is zero, so no variation is introduced. If needed, a typical value of *dither* is 0.1.

23.4 VEGAS

The VEGAS algorithm of Lepage is based on importance sampling. It samples points from the probability distribution described by the function $|f|$, so that the points are concentrated in the regions that make the largest contribution to the integral.

In general, if the Monte Carlo integral of f is sampled with points distributed according to a probability distribution described by the function g, we obtain an estimate $E_g(f; N)$,

$$E_g(f; N) = E(f/g; N)$$

with a corresponding variance,

$$\text{Var}_g(f; N) = \text{Var}(f/g; N).$$

If the probability distribution is chosen as $g = |f|/I(|f|)$ then it can be shown that the variance $V_g(f; N)$ vanishes, and the error in the estimate will be zero. In practice it is not possible to sample from the exact distribution g for an arbitrary function, so importance sampling algorithms aim to produce efficient approximations to the desired distribution.

The VEGAS algorithm approximates the exact distribution by making a number of passes over the integration region while histogramming the function f. Each histogram is used to define a sampling distribution for the next pass. Asymptotically this procedure converges to the desired distribution. In order to avoid the number of histogram bins growing like K^d the probability distribution is approximated by a separable function: $g(x_1, x_2, \ldots) = g_1(x_1)g_2(x_2)\ldots$ so that the number of bins required is only Kd. This is equivalent to locating the peaks of the function from the projections of the integrand onto the coordinate axes. The efficiency of VEGAS depends on the validity of this assumption. It is most efficient when the peaks of the integrand are well-localized. If an integrand can be rewritten in a form which is approximately separable this will increase the efficiency of integration with VEGAS.

VEGAS incorporates a number of additional features, and combines both stratified sampling and importance sampling. The integration region is divided into a number of "boxes", with each box getting a fixed number of points (the goal is 2). Each box can then have a fractional number of bins, but if the ratio of bins-per-box is less than two, Vegas switches to a kind variance reduction (rather than importance sampling).

gsl_monte_vegas_state * gsl_monte_vegas_alloc (size_t *dim*)   Function
    This function allocates and initializes a workspace for Monte Carlo integra-
    tion in *dim* dimensions. The workspace is used to maintain the state of the
    integration.

int gsl_monte_vegas_init (gsl_monte_vegas_state* *s*)   Function
    This function initializes a previously allocated integration state. This allows
    an existing workspace to be reused for different integrations.

int gsl_monte_vegas_integrate (gsl_monte_function * *f*, double   Function
        * *xl*, double * *xu*, size_t *dim*, size_t *calls*, gsl_rng * *r*,
        gsl_monte_vegas_state * *s*, double * *result*, double * *abserr*)
    This routines uses the VEGAS Monte Carlo algorithm to integrate the func-
    tion *f* over the *dim*-dimensional hypercubic region defined by the lower and
    upper limits in the arrays *xl* and *xu*, each of size *dim*. The integration uses
    a fixed number of function calls *calls*, and obtains random sampling points
    using the random number generator *r*. A previously allocated workspace *s*
    must be supplied. The result of the integration is returned in *result*, with
    an estimated absolute error *abserr*. The result and its error estimate are
    based on a weighted average of independent samples. The chi-squared per
    degree of freedom for the weighted average is returned via the state struct
    component, *s->chisq*, and must be consistent with 1 for the weighted average
    to be reliable.

void gsl_monte_vegas_free (gsl_monte_vegas_state * *s*)   Function
    This function frees the memory associated with the integrator state *s*.

The VEGAS algorithm computes a number of independent estimates of the integral internally, according to the iterations parameter described below, and
returns their weighted average. Random sampling of the integrand can occasionally produce an estimate where the error is zero, particularly if the function
is constant in some regions. An estimate with zero error causes the weighted
average to break down and must be handled separately. In the original Fortran
implementations of VEGAS the error estimate is made non-zero by substituting
a small value (typically 1e-30). The implementation in GSL differs from this
and avoids the use of an arbitrary constant—it either assigns the value a weight
which is the average weight of the preceding estimates or discards it according
to the following procedure,

current estimate has zero error, weighted average has finite error
    The current estimate is assigned a weight which is the average weight of
    the preceding estimates.

current estimate has finite error, previous estimates had zero error
    The previous estimates are discarded and the weighted averaging procedure
    begins with the current estimate.

current estimate has zero error, previous estimates had zero error
    The estimates are averaged using the arithmetic mean, but no error is
    computed.

The VEGAS algorithm is highly configurable. The following variables can be accessed through the `gsl_monte_vegas_state` struct,

`double result`   Variable
`double sigma`   Variable

These parameters contain the raw value of the integral *result* and its error *sigma* from the last iteration of the algorithm.

`double chisq`   Variable

This parameter gives the chi-squared per degree of freedom for the weighted estimate of the integral. The value of *chisq* should be close to 1. A value of *chisq* which differs significantly from 1 indicates that the values from different iterations are inconsistent. In this case the weighted error will be under-estimated, and further iterations of the algorithm are needed to obtain reliable results.

`double alpha`   Variable

The parameter `alpha` controls the stiffness of the rebinning algorithm. It is typically set between one and two. A value of zero prevents rebinning of the grid. The default value is 1.5.

`size_t iterations`   Variable

The number of iterations to perform for each call to the routine. The default value is 5 iterations.

`int stage`   Variable

Setting this determines the *stage* of the calculation. Normally, `stage = 0` which begins with a new uniform grid and empty weighted average. Calling vegas with `stage = 1` retains the grid from the previous run but discards the weighted average, so that one can "tune" the grid using a relatively small number of points and then do a large run with `stage = 1` on the optimized grid. Setting `stage = 2` keeps the grid and the weighted average from the previous run, but may increase (or decrease) the number of histogram bins in the grid depending on the number of calls available. Choosing `stage = 3` enters at the main loop, so that nothing is changed, and is equivalent to performing additional iterations in a previous call.

`int mode`   Variable

The possible choices are `GSL_VEGAS_MODE_IMPORTANCE`, `GSL_VEGAS_MODE_STRATIFIED`, `GSL_VEGAS_MODE_IMPORTANCE_ONLY`. This determines whether VEGAS will use importance sampling or stratified sampling, or whether it can pick on its own. In low dimensions VEGAS uses strict stratified sampling (more precisely, stratified sampling is chosen if there are fewer than 2 bins per box).

int verbose   Variable
FILE * ostream   Variable
These parameters set the level of information printed by VEGAS. All information is written to the stream *ostream*. The default setting of *verbose* is -1, which turns off all output. A *verbose* value of 0 prints summary information about the weighted average and final result, while a value of 1 also displays the grid coordinates. A value of 2 prints information from the rebinning procedure for each iteration.

23.5 Examples

The example program below uses the Monte Carlo routines to estimate the value of the following 3-dimensional integral from the theory of random walks,

$$ I = \int_{-\pi}^{+\pi} \frac{dk_x}{2\pi} \int_{-\pi}^{+\pi} \frac{dk_y}{2\pi} \int_{-\pi}^{+\pi} \frac{dk_z}{2\pi} \frac{1}{(1 - \cos(k_x)\cos(k_y)\cos(k_z))}. $$

The analytic value of this integral can be shown to be $I = \Gamma(1/4)^4/(4\pi^3) = 1.393203929685676859...$. The integral gives the mean time spent at the origin by a random walk on a body-centered cubic lattice in three dimensions.

For simplicity we will compute the integral over the region $(0,0,0)$ to (π,π,π) and multiply by 8 to obtain the full result. The integral is slowly varying in the middle of the region but has integrable singularities at the corners $(0,0,0)$, $(0,\pi,\pi)$, $(\pi,0,\pi)$ and $(\pi,\pi,0)$. The Monte Carlo routines only select points which are strictly within the integration region and so no special measures are needed to avoid these singularities.

```
#include <stdlib.h>
#include <gsl/gsl_math.h>
#include <gsl/gsl_monte.h>
#include <gsl/gsl_monte_plain.h>
#include <gsl/gsl_monte_miser.h>
#include <gsl/gsl_monte_vegas.h>

/* Computation of the integral,

      I = int (dx dy dz)/(2pi)^3  1/(1-cos(x)cos(y)cos(z))

   over (-pi,-pi,-pi) to (+pi, +pi, +pi).  The exact answer
   is Gamma(1/4)^4/(4 pi^3).  This example is taken from
   C.Itzykson, J.M.Drouffe, "Statistical Field Theory -
   Volume 1", Section 1.1, p21, which cites the original
   paper M.L.Glasser, I.J.Zucker, Proc.Natl.Acad.Sci.USA 74
   1800 (1977) */

/* For simplicity we compute the integral over the region
   (0,0,0) -> (pi,pi,pi) and multiply by 8 */

double exact = 1.3932039296856768591842462603255;
```

```c
double
g (double *k, size_t dim, void *params)
{
  double A = 1.0 / (M_PI * M_PI * M_PI);
  return A / (1.0 - cos (k[0]) * cos (k[1]) * cos (k[2]));
}

void
display_results (char *title, double result, double error)
{
  printf ("%s ==================\n", title);
  printf ("result = % .6f\n", result);
  printf ("sigma  = % .6f\n", error);
  printf ("exact  = % .6f\n", exact);
  printf ("error  = % .6f = %.2g sigma\n", result - exact,
          fabs (result - exact) / error);
}

int
main (void)
{
  double res, err;

  double xl[3] = { 0, 0, 0 };
  double xu[3] = { M_PI, M_PI, M_PI };

  const gsl_rng_type *T;
  gsl_rng *r;

  gsl_monte_function G = { &g, 3, 0 };

  size_t calls = 500000;

  gsl_rng_env_setup ();

  T = gsl_rng_default;
  r = gsl_rng_alloc (T);

  {
    gsl_monte_plain_state *s = gsl_monte_plain_alloc (3);
    gsl_monte_plain_integrate (&G, xl, xu, 3, calls, r, s,
                               &res, &err);
    gsl_monte_plain_free (s);

    display_results ("plain", res, err);
  }
```

```
       {
         gsl_monte_miser_state *s = gsl_monte_miser_alloc (3);
         gsl_monte_miser_integrate (&G, xl, xu, 3, calls, r, s,
                                    &res, &err);
         gsl_monte_miser_free (s);

         display_results ("miser", res, err);
       }


       {
         gsl_monte_vegas_state *s = gsl_monte_vegas_alloc (3);

         gsl_monte_vegas_integrate (&G, xl, xu, 3, 10000, r, s,
                                    &res, &err);
         display_results ("vegas warm-up", res, err);

         printf ("converging...\n");

         do
           {
             gsl_monte_vegas_integrate (&G, xl, xu, 3, calls/5, r, s,
                                        &res, &err);
             printf ("result = % .6f sigma = % .6f "
                     "chisq/dof = %.1f\n", res, err, s->chisq);
           }
         while (fabs (s->chisq - 1.0) > 0.5);

         display_results ("vegas final", res, err);

         gsl_monte_vegas_free (s);
       }

       gsl_rng_free (r);

       return 0;
     }
```

With 500,000 function calls the plain Monte Carlo algorithm achieves a fractional error of 1%. The estimated error sigma is roughly consistent with the actual error—the computed result differs from the true result by about 1.4 standard deviations,

```
     plain ==================
     result =  1.412209
     sigma  =  0.013436
     exact  =  1.393204
     error  =  0.019005 = 1.4 sigma
```

The MISER algorithm reduces the error by a factor of four, and also correctly estimates the error,

```
miser ===================
result =  1.391322
sigma  =  0.003461
exact  =  1.393204
error  = -0.001882 = 0.54 sigma
```

In the case of the VEGAS algorithm the program uses an initial warm-up run of 10,000 function calls to prepare, or "warm up", the grid. This is followed by a main run with five iterations of 100,000 function calls. The chi-squared per degree of freedom for the five iterations are checked for consistency with 1, and the run is repeated if the results have not converged. In this case the estimates are consistent on the first pass.

```
vegas warm-up ===================
result =  1.392673
sigma  =  0.003410
exact  =  1.393204
error  = -0.000531 = 0.16 sigma
converging...
result =  1.393281 sigma =  0.000362 chisq/dof = 1.5
vegas final ===================
result =  1.393281
sigma  =  0.000362
exact  =  1.393204
error  =  0.000077 = 0.21 sigma
```

If the value of chisq had differed significantly from 1 it would indicate inconsistent results, with a correspondingly underestimated error. The final estimate from VEGAS (using a similar number of function calls) is significantly more accurate than the other two algorithms.

23.6 References and Further Reading

The MISER algorithm is described in the following article by Press and Farrar,

W.H. Press, G.R. Farrar, *Recursive Stratified Sampling for Multidimensional Monte Carlo Integration*, Computers in Physics, v4 (1990), pp190–195.

The VEGAS algorithm is described in the following papers,

G.P. Lepage, *A New Algorithm for Adaptive Multidimensional Integration*, Journal of Computational Physics 27, 192–203, (1978)

G.P. Lepage, *VEGAS: An Adaptive Multi-dimensional Integration Program*, Cornell preprint CLNS 80-447, March 1980

24 Simulated Annealing

Stochastic search techniques are used when the structure of a space is not well understood or is not smooth, so that techniques like Newton's method (which requires calculating Jacobian derivative matrices) cannot be used. In particular, these techniques are frequently used to solve combinatorial optimization problems, such as the traveling salesman problem.

The goal is to find a point in the space at which a real valued *energy function* (or *cost function*) is minimized. Simulated annealing is a minimization technique which has given good results in avoiding local minima; it is based on the idea of taking a random walk through the space at successively lower temperatures, where the probability of taking a step is given by a Boltzmann distribution.

The functions described in this chapter are declared in the header file 'gsl_siman.h'.

24.1 Simulated Annealing algorithm

The simulated annealing algorithm takes random walks through the problem space, looking for points with low energies; in these random walks, the probability of taking a step is determined by the Boltzmann distribution,

$$p = e^{-(E_{i+1} - E_i)/(kT)}$$

if $E_{i+1} > E_i$, and $p = 1$ when $E_{i+1} \leq E_i$.

In other words, a step will occur if the new energy is lower. If the new energy is higher, the transition can still occur, and its likelihood is proportional to the temperature T and inversely proportional to the energy difference $E_{i+1} - E_i$.

The temperature T is initially set to a high value, and a random walk is carried out at that temperature. Then the temperature is lowered very slightly according to a *cooling schedule*, for example: $T \to T/\mu_T$ where μ_T is slightly greater than 1.

The slight probability of taking a step that gives higher energy is what allows simulated annealing to frequently get out of local minima.

24.2 Simulated Annealing functions

void gsl_siman_solve (const gsl_rng * r, void * x0_p,   *Function*
        gsl_siman_Efunc_t *Ef*, gsl_siman_step_t *take_step*,
        gsl_siman_metric_t *distance*, gsl_siman_print_t
        print_position, gsl_siman_copy_t *copyfunc*,
        gsl_siman_copy_construct_t *copy_constructor*,
        gsl_siman_destroy_t *destructor*, size_t *element_size*,
        gsl_siman_params_t *params*)

This function performs a simulated annealing search through a given space. The space is specified by providing the functions *Ef* and *distance*. The simulated annealing steps are generated using the random number generator *r* and the function *take_step*.

The starting configuration of the system should be given by *x0_p*. The routine offers two modes for updating configurations, a fixed-size mode and a variable-size mode. In the fixed-size mode the configuration is stored as a single block of memory of size *element_size*. Copies of this configuration are created, copied and destroyed internally using the standard library functions malloc, memcpy and free. The function pointers *copyfunc*, *copy_constructor* and *destructor* should be null pointers in fixed-size mode. In the variable-size mode the functions *copyfunc*, *copy_constructor* and *destructor* are used to create, copy and destroy configurations internally. The variable *element_size* should be zero in the variable-size mode.

The *params* structure (described below) controls the run by providing the temperature schedule and other tunable parameters to the algorithm.

On exit the best result achieved during the search is placed in *x0_p. If the annealing process has been successful this should be a good approximation to the optimal point in the space.

If the function pointer *print_position* is not null, a debugging log will be printed to stdout with the following columns:

        #-iter #-evals temperature position energy best_energy

and the output of the function *print_position* itself. If *print_position* is null then no information is printed.

The simulated annealing routines require several user-specified functions to define the configuration space and energy function. The prototypes for these functions are given below.

gsl_siman_Efunc_t   *Data Type*

This function type should return the energy of a configuration *xp*.

        double (*gsl_siman_Efunc_t) (void *xp)

gsl_siman_step_t   *Data Type*

This function type should modify the configuration *xp* using a random step taken from the generator *r*, up to a maximum distance of *step_size*.

        void (*gsl_siman_step_t) (const gsl_rng *r, void *xp,
                                  double step_size)

gsl_siman_metric_t   Data Type
> This function type should return the distance between two configurations
> xp and yp.
>
>> double (*gsl_siman_metric_t) (void *xp, void *yp)

gsl_siman_print_t   Data Type
> This function type should print the contents of the configuration xp.
>
>> void (*gsl_siman_print_t) (void *xp)

gsl_siman_copy_t   Data Type
> This function type should copy the configuration source into dest.
>
>> void (*gsl_siman_copy_t) (void *source, void *dest)

gsl_siman_copy_construct_t   Data Type
> This function type should create a new copy of the configuration xp.
>
>> void * (*gsl_siman_copy_construct_t) (void *xp)

gsl_siman_destroy_t   Data Type
> This function type should destroy the configuration xp, freeing its memory.
>
>> void (*gsl_siman_destroy_t) (void *xp)

gsl_siman_params_t   Data Type
> These are the parameters that control a run of gsl_siman_solve. This
> structure contains all the information needed to control the search, beyond
> the energy function, the step function and the initial guess.
>
> int n_tries
>> The number of points to try for each step.
>
> int iters_fixed_T
>> The number of iterations at each temperature.
>
> double step_size
>> The maximum step size in the random walk.
>
> double k, t_initial, mu_t, t_min
>> The parameters of the Boltzmann distribution and cooling schedule.

24.3 Examples

The simulated annealing package is clumsy, and it has to be because it is written in C, for C callers, and tries to be polymorphic at the same time. But here we provide some examples which can be pasted into your application with little change and should make things easier.

24.3.1 Trivial example

The first example, in one dimensional cartesian space, sets up an energy function which is a damped sine wave; this has many local minima, but only one global minimum, somewhere between 1.0 and 1.5. The initial guess given is 15.5, which is several local minima away from the global minimum.

```
#include <math.h>
#include <stdlib.h>
#include <string.h>
#include <gsl/gsl_siman.h>

/* set up parameters for this simulated annealing run */

/* how many points do we try before stepping */
#define N_TRIES 200
/* how many iterations for each T? */
#define ITERS_FIXED_T 1000
/* max step size in random walk */
#define STEP_SIZE 1.0
/* Boltzmann constant */
#define K 1.0
/* initial temperature */
#define T_INITIAL 0.008
/* damping factor for temperature */
#define MU_T 1.003
#define T_MIN 2.0e-6

gsl_siman_params_t params
  = {N_TRIES, ITERS_FIXED_T, STEP_SIZE,
     K, T_INITIAL, MU_T, T_MIN};

/* now some functions to test in one dimension */
double E1(void *xp)
{
  double x = * ((double *) xp);
  return exp(-pow((x-1.0),2.0))*sin(8*x);
}

double M1(void *xp, void *yp)
{
  double x = *((double *) xp);
  double y = *((double *) yp);
```

```
    return fabs(x - y);
}

void S1(const gsl_rng * r, void *xp, double step_size)
{
  double old_x = *((double *) xp);
  double new_x;
  double u = gsl_rng_uniform(r);
  new_x = u * 2 * step_size - step_size + old_x;
  memcpy(xp, &new_x, sizeof(new_x));
}

void P1(void *xp)
{
  printf ("%12g", *((double *) xp));
}

int
main(int argc, char *argv[])
{
  const gsl_rng_type * T;
  gsl_rng * r;

  double x_initial = 15.5;

  gsl_rng_env_setup();

  T = gsl_rng_default;
  r = gsl_rng_alloc(T);

  gsl_siman_solve(r, &x_initial, E1, S1, M1, P1,
                  NULL, NULL, NULL,
                  sizeof(double), params);

  gsl_rng_free (r);
  return 0;
}
```

Here are a couple of plots that are generated by running `siman_test` in the following way:

```
$ ./siman_test | awk '!/^#/ {print $1, $4}'
  | graph -y 1.34 1.4 -W0 -X generation -Y position
  | plot -Tps > siman-test.eps
$ ./siman_test | awk '!/^#/ {print $1, $5}'
  | graph -y -0.88 -0.83 -W0 -X generation -Y energy
  | plot -Tps > siman-energy.eps
```

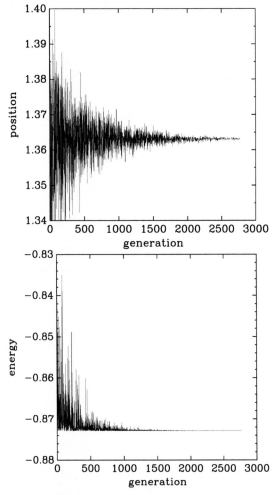

Example of a simulated annealing run: at higher temperatures (early in the plot) you see that the solution can fluctuate, but at lower temperatures it converges.

24.3.2 Traveling Salesman Problem

The TSP (*Traveling Salesman Problem*) is the classic combinatorial optimization problem. I have provided a very simple version of it, based on the coordinates of twelve cities in the southwestern United States. This should maybe be called the *Flying Salesman Problem*, since I am using the great-circle distance between cities, rather than the driving distance. Also: I assume the earth is a sphere, so I don't use geoid distances.

The `gsl_siman_solve` routine finds a route which is 3490.62 Kilometers long; this is confirmed by an exhaustive search of all possible routes with the same initial city.

The full code can be found in 'siman/siman_tsp.c', but I include here some plots generated in the following way:

```
$ ./siman_tsp > tsp.output
$ grep -v "^#" tsp.output
  | awk '{print $1, $NF}'
  | graph -y 3300 6500 -W0 -X generation -Y distance
    -L "TSP - 12 southwest cities"
  | plot -Tps > 12-cities.eps
$ grep initial_city_coord tsp.output
  | awk '{print $2, $3}'
  | graph -X "longitude (- means west)" -Y "latitude"
    -L "TSP - initial-order" -f 0.03 -S 1 0.1
  | plot -Tps > initial-route.eps
$ grep final_city_coord tsp.output
  | awk '{print $2, $3}'
  | graph -X "longitude (- means west)" -Y "latitude"
    -L "TSP - final-order" -f 0.03 -S 1 0.1
  | plot -Tps > final-route.eps
```

This is the output showing the initial order of the cities; longitude is negative, since it is west and I want the plot to look like a map.

```
# initial coordinates of cities (longitude and latitude)
###initial_city_coord: -105.95 35.68 Santa Fe
###initial_city_coord: -112.07 33.54 Phoenix
###initial_city_coord: -106.62 35.12 Albuquerque
###initial_city_coord: -103.2 34.41 Clovis
###initial_city_coord: -107.87 37.29 Durango
###initial_city_coord: -96.77 32.79 Dallas
###initial_city_coord: -105.92 35.77 Tesuque
###initial_city_coord: -107.84 35.15 Grants
###initial_city_coord: -106.28 35.89 Los Alamos
###initial_city_coord: -106.76 32.34 Las Cruces
###initial_city_coord: -108.58 37.35 Cortez
###initial_city_coord: -108.74 35.52 Gallup
###initial_city_coord: -105.95 35.68 Santa Fe
```

The optimal route turns out to be:

```
# final coordinates of cities (longitude and latitude)
###final_city_coord: -105.95 35.68 Santa Fe
###final_city_coord: -103.2 34.41 Clovis
###final_city_coord: -96.77 32.79 Dallas
###final_city_coord: -106.76 32.34 Las Cruces
###final_city_coord: -112.07 33.54 Phoenix
###final_city_coord: -108.74 35.52 Gallup
###final_city_coord: -108.58 37.35 Cortez
###final_city_coord: -107.87 37.29 Durango
###final_city_coord: -107.84 35.15 Grants
###final_city_coord: -106.62 35.12 Albuquerque
###final_city_coord: -106.28 35.89 Los Alamos
###final_city_coord: -105.92 35.77 Tesuque
###final_city_coord: -105.95 35.68 Santa Fe
```

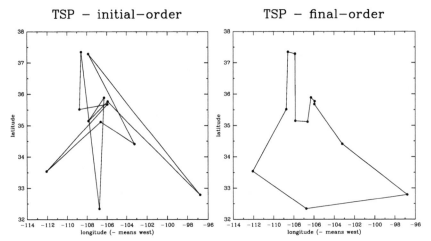

Initial and final (optimal) route for the 12 southwestern cities Flying Salesman Problem.

Here's a plot of the cost function (energy) versus generation (point in the calculation at which a new temperature is set) for this problem:

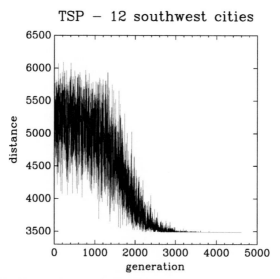

Example of a simulated annealing run for the 12 southwestern cities Flying Salesman Problem.

24.4 References and Further Reading

Further information is available in the following book,

Modern Heuristic Techniques for Combinatorial Problems, Colin R. Reeves (ed.), McGraw-Hill, 1995 (ISBN 0-07-709239-2).

25 Ordinary Differential Equations

This chapter describes functions for solving ordinary differential equation (ODE) initial value problems. The library provides a variety of low-level methods, such as Runge-Kutta and Bulirsch-Stoer routines, and higher-level components for adaptive step-size control. The components can be combined by the user to achieve the desired solution, with full access to any intermediate steps.

These functions are declared in the header file 'gsl_odeiv.h'.

25.1 Defining the ODE System

The routines solve the general n-dimensional first-order system,

$$\frac{dy_i(t)}{dt} = f_i(t, y_1(t), \ldots y_n(t))$$

for $i = 1, \ldots, n$. The stepping functions rely on the vector of derivatives f_i and the Jacobian matrix, $J_{ij} = \partial f_i(t, y(t))/\partial y_j$. A system of equations is defined using the gsl_odeiv_system datatype.

gsl_odeiv_system   Data Type
This data type defines a general ODE system with arbitrary parameters.

int (* function) (double t, const double y[], double dydt[], void *
params)
This function should store the vector elements $f_i(t, y, params)$ in the array *dydt*, for arguments (t,y) and parameters *params*. The function should return GSL_SUCCESS if the calculation was completed successfully. Any other return value indicates an error.

int (* jacobian) (double t, const double y[], double * dfdy, double
dfdt[], void * params);
This function should store the vector of derivative elements $\partial f_i(t, y, params)/\partial t$ in the array *dfdt* and the Jacobian matrix J_{ij} in the array *dfdy*, regarded as a row-ordered matrix J(i,j) = dfdy[i * dimension + j] where dimension is the dimension of the system. The function should return GSL_SUCCESS if the calculation was completed successfully. Any other return value indicates an error.

Some of the simpler solver algorithms do not make use of the Jacobian matrix, so it is not always strictly necessary to provide it (the jacobian element of the struct can be replaced by a null pointer for those algorithms). However, it is useful to provide the Jacobian to allow the solver algorithms to be interchanged—the best algorithms make use of the Jacobian.

size_t dimension;
This is the dimension of the system of equations.

void * params
This is a pointer to the arbitrary parameters of the system.

25.2 Stepping Functions

The lowest level components are the *stepping functions* which advance a solution from time t to $t + h$ for a fixed step-size h and estimate the resulting local error.

gsl_odeiv_step * gsl_odeiv_step_alloc (const   *Function*
        gsl_odeiv_step_type * *T*, size_t *dim*)
    This function returns a pointer to a newly allocated instance of a stepping function of type T for a system of *dim* dimensions.

int gsl_odeiv_step_reset (gsl_odeiv_step * *s*)   *Function*
    This function resets the stepping function *s*. It should be used whenever the next use of *s* will not be a continuation of a previous step.

void gsl_odeiv_step_free (gsl_odeiv_step * *s*)   *Function*
    This function frees all the memory associated with the stepping function *s*.

const char * gsl_odeiv_step_name (const gsl_odeiv_step * *s*)   *Function*
    This function returns a pointer to the name of the stepping function. For example,

        printf ("step method is '%s'\n",
                gsl_odeiv_step_name (s));

    would print something like step method is 'rkf45'.

unsigned int gsl_odeiv_step_order (const gsl_odeiv_step * *s*)   *Function*
    This function returns the order of the stepping function on the previous step. This order can vary if the stepping function itself is adaptive.

int gsl_odeiv_step_apply (gsl_odeiv_step * *s*, double *t*, double   *Function*
        h, double *y*[], double *yerr*[], const double *dydt_in*[],
        double *dydt_out*[], const gsl_odeiv_system * *dydt*)
    This function applies the stepping function *s* to the system of equations defined by *dydt*, using the step size *h* to advance the system from time *t* and state *y* to time *t+h*. The new state of the system is stored in *y* on output, with an estimate of the absolute error in each component stored in *yerr*. If the argument *dydt_in* is not null it should point an array containing the derivatives for the system at time *t* on input. This is optional as the derivatives will be computed internally if they are not provided, but allows the reuse of existing derivative information. On output the new derivatives of the system at time *t+h* will be stored in *dydt_out* if it is not null.

    If the user-supplied functions defined in the system *dydt* return a status other than GSL_SUCCESS the step will be aborted. In this case, the elements of *y* will be restored to their pre-step values and the error code from the user-supplied function will be returned. The step-size *h* will be set to the step-size which caused the error. If the function is called again with a smaller step-size, e.g. $h/10$, it should be possible to get closer to any singularity. To distinguish between error codes from the user-supplied functions and those from gsl_odeiv_step_apply itself, any user-defined return values should be distinct from the standard GSL error codes.

The following algorithms are available,

`gsl_odeiv_step_rk2`   Step Type
    Embedded Runge-Kutta (2, 3) method.

`gsl_odeiv_step_rk4`   Step Type
    4th order (classical) Runge-Kutta. The error estimate is obtained by halving
    the step-size. For more efficient estimate of the error, use the Runge-Kutta-
    Fehlberg method described below.

`gsl_odeiv_step_rkf45`   Step Type
    Embedded Runge-Kutta-Fehlberg (4, 5) method. This method is a good
    general-purpose integrator.

`gsl_odeiv_step_rkck`   Step Type
    Embedded Runge-Kutta Cash-Karp (4, 5) method.

`gsl_odeiv_step_rk8pd`   Step Type
    Embedded Runge-Kutta Prince-Dormand (8,9) method.

`gsl_odeiv_step_rk2imp`   Step Type
    Implicit 2nd order Runge-Kutta at Gaussian points.

`gsl_odeiv_step_rk4imp`   Step Type
    Implicit 4th order Runge-Kutta at Gaussian points.

`gsl_odeiv_step_bsimp`   Step Type
    Implicit Bulirsch-Stoer method of Bader and Deuflhard. This algorithm
    requires the Jacobian.

`gsl_odeiv_step_gear1`   Step Type
    M=1 implicit Gear method.

`gsl_odeiv_step_gear2`   Step Type
    M=2 implicit Gear method.

25.3 Adaptive Step-size Control

The control function examines the proposed change to the solution produced by a stepping function and attempts to determine the optimal step-size for a user-specified level of error.

gsl_odeiv_control * gsl_odeiv_control_standard_new (double   *Function*
        eps_abs, double eps_rel, double a_y, double a_dydt)

The standard control object is a four parameter heuristic based on absolute and relative errors eps_abs and eps_rel, and scaling factors a_y and a_dydt for the system state $y(t)$ and derivatives $y'(t)$ respectively.

The step-size adjustment procedure for this method begins by computing the desired error level D_i for each component,

$$D_i = \epsilon_{abs} + \epsilon_{rel} * (a_y |y_i| + a_{dydt} h |y_i'|)$$

and comparing it with the observed error $E_i = |yerr_i|$. If the observed error E exceeds the desired error level D by more than 10% for any component then the method reduces the step-size by an appropriate factor,

$$h_{new} = h_{old} * S * (E/D)^{-1/q}$$

where q is the consistency order of the method (e.g. $q = 4$ for 4(5) embedded RK), and S is a safety factor of 0.9. The ratio E/D is taken to be the maximum of the ratios E_i/D_i.

If the observed error E is less than 50% of the desired error level D for the maximum ratio E_i/D_i then the algorithm takes the opportunity to increase the step-size to bring the error in line with the desired level,

$$h_{new} = h_{old} * S * (E/D)^{-1/(q+1)}$$

This encompasses all the standard error scaling methods. To avoid uncontrolled changes in the stepsize, the overall scaling factor is limited to the range 1/5 to 5.

gsl_odeiv_control * gsl_odeiv_control_y_new (double eps_abs,   *Function*
        double eps_rel)

This function creates a new control object which will keep the local error on each step within an absolute error of eps_abs and relative error of eps_rel with respect to the solution $y_i(t)$. This is equivalent to the standard control object with a_y=1 and a_dydt=0.

gsl_odeiv_control * gsl_odeiv_control_yp_new (double eps_abs,   *Function*
        double eps_rel)

This function creates a new control object which will keep the local error on each step within an absolute error of eps_abs and relative error of eps_rel with respect to the derivatives of the solution $y_i'(t)$. This is equivalent to the standard control object with a_y=0 and a_dydt=1.

gsl_odeiv_control * gsl_odeiv_control_scaled_new (double   Function
        eps_abs, double eps_rel, double a_y, double a_dydt, const
        double scale_abs[], size_t dim)
This function creates a new control object which uses the same algorithm
as gsl_odeiv_control_standard_new but with an absolute error which is
scaled for each component by the array scale_abs. The formula for D_i for
this control object is,

$$D_i = \epsilon_{abs}s_i + \epsilon_{rel} * (a_y|y_i| + a_{dydt}h|y_i'|)$$

where s_i is the i-th component of the array scale_abs. The same error control
heuristic is used by the Matlab ODE suite.

gsl_odeiv_control * gsl_odeiv_control_alloc (const   Function
        gsl_odeiv_control_type * T)
This function returns a pointer to a newly allocated instance of a control
function of type T. This function is only needed for defining new types of control functions. For most purposes the standard control functions described
above should be sufficient.

int gsl_odeiv_control_init (gsl_odeiv_control * c, double   Function
        eps_abs, double eps_rel, double a_y, double a_dydt)
This function initializes the control function c with the parameters eps_abs
(absolute error), eps_rel (relative error), a_y (scaling factor for y) and a_dydt
(scaling factor for derivatives).

void gsl_odeiv_control_free (gsl_odeiv_control * c)   Function
This function frees all the memory associated with the control function c.

int gsl_odeiv_control_hadjust (gsl_odeiv_control * c,   Function
        gsl_odeiv_step * s, const double y[], const double yerr[],
        const double dydt[], double * h)
This function adjusts the step-size h using the control function c, and the
current values of y, yerr and dydt. The stepping function step is also needed
to determine the order of the method. If the error in the y-values yerr
is found to be too large then the step-size h is reduced and the function
returns GSL_ODEIV_HADJ_DEC. If the error is sufficiently small then h may
be increased and GSL_ODEIV_HADJ_INC is returned. The function returns
GSL_ODEIV_HADJ_NIL if the step-size is unchanged. The goal of the function
is to estimate the largest step-size which satisfies the user-specified accuracy
requirements for the current point.

const char * gsl_odeiv_control_name (const gsl_odeiv_control   Function
        * c)
This function returns a pointer to the name of the control function. For
example,

        printf ("control method is '%s'\n",
                gsl_odeiv_control_name (c));

would print something like control method is 'standard'

25.4 Evolution

The highest level of the system is the evolution function which combines the results of a stepping function and control function to reliably advance the solution forward over an interval (t_0, t_1). If the control function signals that the step-size should be decreased the evolution function backs out of the current step and tries the proposed smaller step-size. This process is continued until an acceptable step-size is found.

gsl_odeiv_evolve * gsl_odeiv_evolve_alloc (size_t *dim*)   Function
    This function returns a pointer to a newly allocated instance of an evolution function for a system of *dim* dimensions.

int gsl_odeiv_evolve_apply (gsl_odeiv_evolve * e,   Function
        gsl_odeiv_control * *con*, gsl_odeiv_step * *step*, const
        gsl_odeiv_system * *dydt*, double * t, double t1, double * h,
        double y[])
    This function advances the system (e, *dydt*) from time t and position y using the stepping function *step*. The new time and position are stored in t and y on output. The initial step-size is taken as h, but this will be modified using the control function c to achieve the appropriate error bound if necessary. The routine may make several calls to *step* in order to determine the optimum step-size. An estimate of the local error for the step can be obtained from the components of the array e->yerr[]. If the step-size has been changed the value of h will be modified on output. The maximum time t1 is guaranteed not to be exceeded by the time-step. On the final time-step the value of t will be set to t1 exactly.

    If the user-supplied functions defined in the system *dydt* return a status other than GSL_SUCCESS the step will be aborted. In this case, t and y will be restored to their pre-step values and the error code from the user-supplied function will be returned. To distinguish between error codes from the user-supplied functions and those from gsl_odeiv_evolve_apply itself, any user-defined return values should be distinct from the standard GSL error codes.

int gsl_odeiv_evolve_reset (gsl_odeiv_evolve * e)   Function
    This function resets the evolution function e. It should be used whenever the next use of e will not be a continuation of a previous step.

void gsl_odeiv_evolve_free (gsl_odeiv_evolve * e)   Function
    This function frees all the memory associated with the evolution function e.

Where a system has discontinuous changes in the derivatives at known times it is advisable to evolve the system between each discontinuity in sequence. For example, if a step-change in an external driving force occurs at times $t_a, t_b, t_c, \ldots$ then evolving over the ranges (t_0, t_a), (t_a, t_b), $\ldots$, (t_c, t_1) is more efficient than using the single range (t_0, t_1).

Evolving the system directly through a discontinuity with a strict tolerance may result in extremely small steps being taken at the edge of the discontinuity (e.g. down to the limit of machine precision). In this case it may be necessary to impose a minimum step size hmin suitable for the problem:

```
while (t < t1)
{
    gsl_odeiv_evolve_apply (e, c, s, &sys, &t, t1, &h, y);
    if (h < hmin) { h = hmin; } ;
}
```

The value of h returned by `gsl_odeiv_evolve_apply` is always a suggested value and can be modified whenever needed.

25.5 Examples

The following program solves the second-order nonlinear Van der Pol oscillator equation,

$$x''(t) + \mu x'(t)(x(t)^2 - 1) + x(t) = 0$$

This can be converted into a first order system suitable for use with the routines described in this chapter by introducing a separate variable for the velocity, $y = x'(t)$,

$$x' = y$$
$$y' = -x + \mu y(1 - x^2)$$

The program begins by defining functions for these derivatives and their Jacobian,

```
#include <stdio.h>
#include <gsl/gsl_errno.h>
#include <gsl/gsl_matrix.h>
#include <gsl/gsl_odeiv.h>

int
func (double t, const double y[], double f[],
      void *params)
{
    double mu = *(double *)params;
    f[0] = y[1];
    f[1] = -y[0] - mu*y[1]*(y[0]*y[0] - 1);
    return GSL_SUCCESS;
}

int
jac (double t, const double y[], double *dfdy,
     double dfdt[], void *params)
{
    double mu = *(double *)params;
    gsl_matrix_view dfdy_mat
      = gsl_matrix_view_array (dfdy, 2, 2);
    gsl_matrix * m = &dfdy_mat.matrix;
```

```c
  gsl_matrix_set (m, 0, 0, 0.0);
  gsl_matrix_set (m, 0, 1, 1.0);
  gsl_matrix_set (m, 1, 0, -2.0*mu*y[0]*y[1] - 1.0);
  gsl_matrix_set (m, 1, 1, -mu*(y[0]*y[0] - 1.0));
  dfdt[0] = 0.0;
  dfdt[1] = 0.0;
  return GSL_SUCCESS;
}

int
main (void)
{
  const gsl_odeiv_step_type * T
    = gsl_odeiv_step_rk8pd;

  gsl_odeiv_step * s
    = gsl_odeiv_step_alloc (T, 2);
  gsl_odeiv_control * c
    = gsl_odeiv_control_y_new (1e-6, 0.0);
  gsl_odeiv_evolve * e
    = gsl_odeiv_evolve_alloc (2);

  double mu = 10;
  gsl_odeiv_system sys = {func, jac, 2, &mu};

  double t = 0.0, t1 = 100.0;
  double h = 1e-6;
  double y[2] = { 1.0, 0.0 };

  while (t < t1)
    {
      int status = gsl_odeiv_evolve_apply (e, c, s,
                                           &sys,
                                           &t, t1,
                                           &h, y);

      if (status != GSL_SUCCESS)
          break;

      printf ("%.5e %.5e %.5e\n", t, y[0], y[1]);
    }

  gsl_odeiv_evolve_free (e);
  gsl_odeiv_control_free (c);
  gsl_odeiv_step_free (s);
  return 0;
}
```

For functions with multiple parameters, the appropriate information can be passed in through the *params* argument using a pointer to a struct.

The main loop of the program evolves the solution from $(y, y') = (1, 0)$ at $t = 0$ to $t = 100$. The step-size h is automatically adjusted by the controller to maintain an absolute accuracy of 10^{-6} in the function values y.

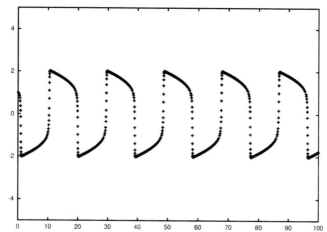

Numerical solution of the Van der Pol oscillator equation
using Prince-Dormand 8th order Runge-Kutta.

To obtain the values at user-specified positions, rather than those chosen automatically by the control function, the main loop can be modified to advance the solution from one chosen point to the next. For example, the following main loop prints the solution at the points $t_i = 0, 1, 2, \ldots, 100$,

```
for (i = 1; i <= 100; i++)
  {
    double ti = i * t1 / 100.0;

    while (t < ti)
      {
        gsl_odeiv_evolve_apply (e, c, s,
                                &sys,
                                &t, ti, &h,
                                y);
      }

    printf ("%.5e %.5e %.5e\n", t, y[0], y[1]);
  }
```

Note that arbitrary values of t_i can be used for each stage of the integration. The equally spaced points in this example are just used as an illustration.

It is also possible to work with a non-adaptive integrator, using only the stepping function itself. The following program uses the rk4 fourth-order Runge-Kutta stepping function with a fixed stepsize of 0.01,

```
int
main (void)
{
  const gsl_odeiv_step_type * T
    = gsl_odeiv_step_rk4;

  gsl_odeiv_step * s
    = gsl_odeiv_step_alloc (T, 2);

  double mu = 10;
  gsl_odeiv_system sys = {func, jac, 2, &mu};

  double t = 0.0, t1 = 100.0;
  double h = 1e-2;
  double y[2] = { 1.0, 0.0 }, y_err[2];
  double dydt_in[2], dydt_out[2];

  /* initialise dydt_in from system parameters */
  GSL_ODEIV_FN_EVAL(&sys, t, y, dydt_in);

  while (t < t1)
    {
      int status = gsl_odeiv_step_apply (s, t, h,
                                         y, y_err,
                                         dydt_in,
                                         dydt_out,
                                         &sys);

      if (status != GSL_SUCCESS)
          break;

      dydt_in[0] = dydt_out[0];
      dydt_in[1] = dydt_out[1];

      t += h;

      printf ("%.5e %.5e %.5e\n", t, y[0], y[1]);
    }

  gsl_odeiv_step_free (s);
  return 0;
}
```

The derivatives must be initialized for the starting point $t = 0$ before the first step is taken. Subsequent steps use the output derivatives *dydt_out* as inputs to the next step by copying their values into *dydt_in*.

25.6 References and Further Reading

Many of the basic Runge-Kutta formulas can be found in the Handbook of Mathematical Functions,

Abramowitz & Stegun (eds.), *Handbook of Mathematical Functions*, Section 25.5.

The implicit Bulirsch-Stoer algorithm bsimp is described in the following paper,

G. Bader and P. Deuflhard, "A Semi-Implicit Mid-Point Rule for Stiff Systems of Ordinary Differential Equations.", Numer. Math. 41, 373–398, 1983.

26 Interpolation

This chapter describes functions for performing interpolation. The library provides a variety of interpolation methods, including Cubic splines and Akima splines. The interpolation types are interchangeable, allowing different methods to be used without recompiling. Interpolations can be defined for both normal and periodic boundary conditions. Additional functions are available for computing derivatives and integrals of interpolating functions.

These interpolation methods produce curves that pass through each datapoint. To interpolate noisy data with a smoothing curve see Chapter 38 [Basis Splines], page 469.

The functions described in this section are declared in the header files 'gsl_interp.h' and 'gsl_spline.h'.

26.1 Introduction

Given a set of data points $(x_1, y_1) \ldots (x_n, y_n)$ the routines described in this section compute a continuous interpolating function $y(x)$ such that $y(x_i) = y_i$. The interpolation is piecewise smooth, and its behavior at the end-points is determined by the type of interpolation used.

26.2 Interpolation Functions

The interpolation function for a given dataset is stored in a gsl_interp object. These are created by the following functions.

gsl_interp * gsl_interp_alloc (const gsl_interp_type * T,   *Function*
      size_t size)
    This function returns a pointer to a newly allocated interpolation object of type T for size data-points.

int gsl_interp_init (gsl_interp * interp, const double xa[],   *Function*
      const double ya[], size_t size)
    This function initializes the interpolation object interp for the data (xa,ya) where xa and ya are arrays of size size. The interpolation object (gsl_interp) does not save the data arrays xa and ya and only stores the static state computed from the data. The xa data array is always assumed to be strictly ordered, with increasing x values; the behavior for other arrangements is not defined.

void gsl_interp_free (gsl_interp * interp)   *Function*
    This function frees the interpolation object interp.

26.3 Interpolation Types

The interpolation library provides five interpolation types:

gsl_interp_linear   *Interpolation Type*
> Linear interpolation. This interpolation method does not require any additional memory.

gsl_interp_polynomial   *Interpolation Type*
> Polynomial interpolation. This method should only be used for interpolating small numbers of points because polynomial interpolation introduces large oscillations, even for well-behaved datasets. The number of terms in the interpolating polynomial is equal to the number of points.

gsl_interp_cspline   *Interpolation Type*
> Cubic spline with natural boundary conditions. The resulting curve is piecewise cubic on each interval, with matching first and second derivatives at the supplied data-points. The second derivative is chosen to be zero at the first point and last point.

gsl_interp_cspline_periodic   *Interpolation Type*
> Cubic spline with periodic boundary conditions. The resulting curve is piecewise cubic on each interval, with matching first and second derivatives at the supplied data-points. The derivatives at the first and last points are also matched. Note that the last point in the data must have the same y-value as the first point, otherwise the resulting periodic interpolation will have a discontinuity at the boundary.

gsl_interp_akima   *Interpolation Type*
> Non-rounded Akima spline with natural boundary conditions. This method uses the non-rounded corner algorithm of Wodicka.

gsl_interp_akima_periodic   *Interpolation Type*
> Non-rounded Akima spline with periodic boundary conditions. This method uses the non-rounded corner algorithm of Wodicka.

The following related functions are available:

const char * gsl_interp_name (const gsl_interp * *interp*)   Function
> This function returns the name of the interpolation type used by *interp*. For example,

        printf ("interp uses '%s' interpolation.\n",
                gsl_interp_name (interp));

> would print something like,

        interp uses 'cspline' interpolation.

unsigned int gsl_interp_min_size (const gsl_interp * *interp*)   Function
> This function returns the minimum number of points required by the interpolation type of *interp*. For example, Akima spline interpolation requires a minimum of 5 points.

26.4 Index Look-up and Acceleration

The state of searches can be stored in a `gsl_interp_accel` object, which is a kind of iterator for interpolation lookups. It caches the previous value of an index lookup. When the subsequent interpolation point falls in the same interval its index value can be returned immediately.

size_t gsl_interp_bsearch (const double x_array[], double x,   Function
   size_t index_lo, size_t index_hi)
This function returns the index i of the array x_array such that x_array[i] <= x < x_array[i+1]. The index is searched for in the range [index_lo,index_hi]. An inline version of this function is used when HAVE_INLINE is defined.

gsl_interp_accel * gsl_interp_accel_alloc (void)   Function
This function returns a pointer to an accelerator object, which is a kind of iterator for interpolation lookups. It tracks the state of lookups, thus allowing for application of various acceleration strategies.

size_t gsl_interp_accel_find (gsl_interp_accel * a, const   Function
   double x_array[], size_t size, double x)
This function performs a lookup action on the data array x_array of size size, using the given accelerator a. This is how lookups are performed during evaluation of an interpolation. The function returns an index i such that x_array[i] <= x < x_array[i+1]. An inline version of this function is used when HAVE_INLINE is defined.

void gsl_interp_accel_free (gsl_interp_accel* acc)   Function
This function frees the accelerator object acc.

26.5 Evaluation of Interpolating Functions

double gsl_interp_eval (const gsl_interp * interp, const double   Function
   xa[], const double ya[], double x, gsl_interp_accel * acc)
int gsl_interp_eval_e (const gsl_interp * interp, const double   Function
   xa[], const double ya[], double x, gsl_interp_accel * acc,
   double * y)
These functions return the interpolated value of y for a given point x, using the interpolation object interp, data arrays xa and ya and the accelerator acc.

double gsl_interp_eval_deriv (const gsl_interp * interp, const   Function
   double xa[], const double ya[], double x, gsl_interp_accel
   * acc)
int gsl_interp_eval_deriv_e (const gsl_interp * interp, const   Function
   double xa[], const double ya[], double x, gsl_interp_accel
   * acc, double * d)
These functions return the derivative d of an interpolated function for a given point x, using the interpolation object interp, data arrays xa and ya and the accelerator acc.

double gsl_interp_eval_deriv2 (const gsl_interp * *interp*,   Function
            const double xa[], const double ya[], double *x*,
            gsl_interp_accel * *acc*)
int gsl_interp_eval_deriv2_e (const gsl_interp * *interp*, const   Function
            double xa[], const double ya[], double *x*, gsl_interp_accel
            * *acc*, double * *d2*)

These functions return the second derivative *d2* of an interpolated function
for a given point *x*, using the interpolation object *interp*, data arrays *xa* and
ya and the accelerator *acc*.

double gsl_interp_eval_integ (const gsl_interp * *interp*, const   Function
            double xa[], const double ya[], double *a*, double *b*,
            gsl_interp_accel * *acc*)
int gsl_interp_eval_integ_e (const gsl_interp * *interp*, const   Function
            double xa[], const double ya[], double *a*, double *b*,
            gsl_interp_accel * *acc*, double * *result*)

These functions return the numerical integral *result* of an interpolated function over the range [*a*, *b*], using the interpolation object *interp*, data arrays
xa and *ya* and the accelerator *acc*.

26.6 Higher-level Interface

The functions described in the previous sections required the user to supply
pointers to the *x* and *y* arrays on each call. The following functions are equivalent to the corresponding gsl_interp functions but maintain a copy of this
data in the gsl_spline object. This removes the need to pass both *xa* and *ya*
as arguments on each evaluation. These functions are defined in the header file
'gsl_spline.h'.

gsl_spline * gsl_spline_alloc (const gsl_interp_type * *T*,   Function
            size_t *size*)

int gsl_spline_init (gsl_spline * *spline*, const double xa[],   Function
            const double ya[], size_t *size*)

void gsl_spline_free (gsl_spline * *spline*)   Function

const char * gsl_spline_name (const gsl_spline * *spline*)   Function

unsigned int gsl_spline_min_size (const gsl_spline * *spline*)   Function

double gsl_spline_eval (const gsl_spline * *spline*, double *x*,   Function
            gsl_interp_accel * *acc*)
int gsl_spline_eval_e (const gsl_spline * *spline*, double *x*,   Function
            gsl_interp_accel * *acc*, double * *y*)

double gsl_spline_eval_deriv (const gsl_spline * *spline*,   Function
            double *x*, gsl_interp_accel * *acc*)
int gsl_spline_eval_deriv_e (const gsl_spline * *spline*, double   Function
            x, gsl_interp_accel * *acc*, double * *d*)

double gsl_spline_eval_deriv2 (const gsl_spline * *spline*, double *x*, gsl_interp_accel * *acc*)	Function
int gsl_spline_eval_deriv2_e (const gsl_spline * *spline*, double *x*, gsl_interp_accel * *acc*, double * *d2*)	Function
double gsl_spline_eval_integ (const gsl_spline * *spline*, double *a*, double *b*, gsl_interp_accel * *acc*)	Function
int gsl_spline_eval_integ_e (const gsl_spline * *spline*, double *a*, double *b*, gsl_interp_accel * *acc*, double * *result*)	Function

26.7 Examples

The following program demonstrates the use of the interpolation and spline functions. It computes a cubic spline interpolation of the 10-point dataset (x_i, y_i) where $x_i = i + \sin(i)/2$ and $y_i = i + \cos(i^2)$ for $i = 0 \ldots 9$.

```
#include <stdlib.h>
#include <stdio.h>
#include <math.h>
#include <gsl/gsl_errno.h>
#include <gsl/gsl_spline.h>

int
main (void)
{
  int i;
  double xi, yi, x[10], y[10];

  printf ("#m=0,S=2\n");

  for (i = 0; i < 10; i++)
    {
      x[i] = i + 0.5 * sin (i);
      y[i] = i + cos (i * i);
      printf ("%g %g\n", x[i], y[i]);
    }

  printf ("#m=1,S=0\n");

  {
    gsl_interp_accel *acc
      = gsl_interp_accel_alloc ();
    gsl_spline *spline
      = gsl_spline_alloc (gsl_interp_cspline, 10);

    gsl_spline_init (spline, x, y, 10);

    for (xi = x[0]; xi < x[9]; xi += 0.01)
      {
        yi = gsl_spline_eval (spline, xi, acc);
```

```
        printf ("%g %g\n", xi, yi);
      }
    gsl_spline_free (spline);
    gsl_interp_accel_free (acc);
  }
  return 0;
}
```

The output is designed to be used with the GNU plotutils graph program,

```
$ ./a.out > interp.dat
$ graph -T ps < interp.dat > interp.ps
```

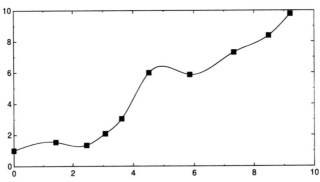

The result shows a smooth interpolation of the original points. The interpolation method can be changed simply by varying the first argument of gsl_spline_alloc.

The next program demonstrates a periodic cubic spline with 4 data points. Note that the first and last points must be supplied with the same y-value for a periodic spline.

```
#include <stdlib.h>
#include <stdio.h>
#include <math.h>
#include <gsl/gsl_errno.h>
#include <gsl/gsl_spline.h>

int
main (void)
{
  int N = 4;
  double x[4] = {0.00, 0.10,  0.27,   0.30};
  double y[4] = {0.15, 0.70, -0.10,   0.15};
            /* Note: y[0] == y[3] for periodic data */

  gsl_interp_accel *acc = gsl_interp_accel_alloc ();
  const gsl_interp_type *t = gsl_interp_cspline_periodic;
  gsl_spline *spline = gsl_spline_alloc (t, N);
```

```
      int i; double xi, yi;

      printf ("#m=0,S=5\n");
      for (i = 0; i < N; i++)
        {
          printf ("%g %g\n", x[i], y[i]);
        }

      printf ("#m=1,S=0\n");
      gsl_spline_init (spline, x, y, N);

      for (i = 0; i <= 100; i++)
        {
          xi = (1 - i / 100.0) * x[0] + (i / 100.0) * x[N-1];
          yi = gsl_spline_eval (spline, xi, acc);
          printf ("%g %g\n", xi, yi);
        }

      gsl_spline_free (spline);
      gsl_interp_accel_free (acc);
      return 0;
    }
```

The output can be plotted with GNU graph.

```
$ ./a.out > interp.dat
$ graph -T ps < interp.dat > interp.ps
```

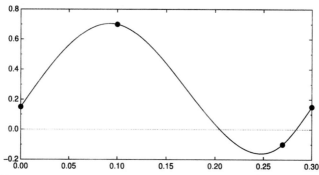

The result shows a periodic interpolation of the original points. The slope of the fitted curve is the same at the beginning and end of the data, and the second derivative is also.

26.8 References and Further Reading

Descriptions of the interpolation algorithms and further references can be found in the following books:

C.W. Ueberhuber, *Numerical Computation (Volume 1), Chapter 9 "Interpolation"*, Springer (1997), ISBN 3-540-62058-3.

D.M. Young, R.T. Gregory *A Survey of Numerical Mathematics (Volume 1), Chapter 6.8*, Dover (1988), ISBN 0-486-65691-8.

27 Numerical Differentiation

The functions described in this chapter compute numerical derivatives by finite differencing. An adaptive algorithm is used to find the best choice of finite difference and to estimate the error in the derivative. These functions are declared in the header file 'gsl_deriv.h'.

27.1 Functions

int gsl_deriv_central (const gsl_function * f, double x,   *Function*
   double h, double * result, double * abserr)

This function computes the numerical derivative of the function f at the point x using an adaptive central difference algorithm with a step-size of h. The derivative is returned in result and an estimate of its absolute error is returned in abserr.

The initial value of h is used to estimate an optimal step-size, based on the scaling of the truncation error and round-off error in the derivative calculation. The derivative is computed using a 5-point rule for equally spaced abscissae at $x - h$, $x - h/2$, x, $x + h/2$, $x + h$, with an error estimate taken from the difference between the 5-point rule and the corresponding 3-point rule $x - h$, x, $x + h$. Note that the value of the function at x does not contribute to the derivative calculation, so only 4-points are actually used.

int gsl_deriv_forward (const gsl_function * f, double x,   *Function*
   double h, double * result, double * abserr)

This function computes the numerical derivative of the function f at the point x using an adaptive forward difference algorithm with a step-size of h. The function is evaluated only at points greater than x, and never at x itself. The derivative is returned in result and an estimate of its absolute error is returned in abserr. This function should be used if $f(x)$ has a discontinuity at x, or is undefined for values less than x.

The initial value of h is used to estimate an optimal step-size, based on the scaling of the truncation error and round-off error in the derivative calculation. The derivative at x is computed using an "open" 4-point rule for equally spaced abscissae at $x + h/4$, $x + h/2$, $x + 3h/4$, $x + h$, with an error estimate taken from the difference between the 4-point rule and the corresponding 2-point rule $x + h/2$, $x + h$.

int gsl_deriv_backward (const gsl_function * f, double x,   *Function*
   double h, double * result, double * abserr)

This function computes the numerical derivative of the function f at the point x using an adaptive backward difference algorithm with a step-size of h. The function is evaluated only at points less than x, and never at x itself. The derivative is returned in result and an estimate of its absolute error is returned in abserr. This function should be used if $f(x)$ has a discontinuity at x, or is undefined for values greater than x.

This function is equivalent to calling `gsl_deriv_forward` with a negative step-size.

27.2 Examples

The following code estimates the derivative of the function $f(x) = x^{3/2}$ at $x = 2$ and at $x = 0$. The function $f(x)$ is undefined for $x < 0$ so the derivative at $x = 0$ is computed using `gsl_deriv_forward`.

```
#include <stdio.h>
#include <gsl/gsl_math.h>
#include <gsl/gsl_deriv.h>

double f (double x, void * params)
{
  return pow (x, 1.5);
}

int
main (void)
{
  gsl_function F;
  double result, abserr;

  F.function = &f;
  F.params = 0;

  printf ("f(x) = x^(3/2)\n");

  gsl_deriv_central (&F, 2.0, 1e-8, &result, &abserr);
  printf ("x = 2.0\n");
  printf ("f'(x) = %.10f +/- %.10f\n", result, abserr);
  printf ("exact = %.10f\n\n", 1.5 * sqrt(2.0));

  gsl_deriv_forward (&F, 0.0, 1e-8, &result, &abserr);
  printf ("x = 0.0\n");
  printf ("f'(x) = %.10f +/- %.10f\n", result, abserr);
  printf ("exact = %.10f\n", 0.0);

  return 0;
}
```

Here is the output of the program,

```
$ ./a.out
f(x) = x^(3/2)
x = 2.0
f'(x) = 2.1213203120 +/- 0.0000004064
exact = 2.1213203436

x = 0.0
```

```
f'(x) = 0.0000000160 +/- 0.0000000339
exact = 0.0000000000
```

27.3 References and Further Reading

The algorithms used by these functions are described in the following sources:

Abramowitz and Stegun, *Handbook of Mathematical Functions*, Section 25.3.4, and Table 25.5 (Coefficients for Differentiation).

S.D. Conte and Carl de Boor, *Elementary Numerical Analysis: An Algorithmic Approach*, McGraw-Hill, 1972.

28 Chebyshev Approximations

This chapter describes routines for computing Chebyshev approximations to univariate functions. A Chebyshev approximation is a truncation of the series $f(x) = \sum c_n T_n(x)$, where the Chebyshev polynomials $T_n(x) = \cos(n \arccos x)$ provide an orthogonal basis of polynomials on the interval $[-1, 1]$ with the weight function $1/\sqrt{1 - x^2}$. The first few Chebyshev polynomials are, $T_0(x) = 1$, $T_1(x) = x$, $T_2(x) = 2x^2 - 1$. For further information see Abramowitz & Stegun, Chapter 22.

The functions described in this chapter are declared in the header file 'gsl_chebyshev.h'.

28.1 Definitions

A Chebyshev series is stored using the following structure,

```
typedef struct
{
    double * c;      /* coefficients   c[0] .. c[order] */
    int order;       /* order of expansion              */
    double a;        /* lower interval point            */
    double b;        /* upper interval point            */
    ...
} gsl_cheb_series
```

The approximation is made over the range $[a, b]$ using order+1 terms, including the coefficient c[0]. The series is computed using the following convention,

$$f(x) = \frac{c_0}{2} + \sum_{n=1} c_n T_n(x)$$

which is needed when accessing the coefficients directly.

28.2 Creation and Calculation of Chebyshev Series

gsl_cheb_series * gsl_cheb_alloc (const size_t n)   *Function*
   This function allocates space for a Chebyshev series of order n and returns
   a pointer to a new gsl_cheb_series struct.

void gsl_cheb_free (gsl_cheb_series * cs)   *Function*
   This function frees a previously allocated Chebyshev series cs.

int gsl_cheb_init (gsl_cheb_series * cs, const gsl_function *   *Function*
            f, const double a, const double b)
   This function computes the Chebyshev approximation cs for the function f
   over the range (a, b) to the previously specified order. The computation of
   the Chebyshev approximation is an $O(n^2)$ process, and requires n function
   evaluations.

28.3 Auxiliary Functions

The following functions provide information about an existing Chebyshev series.

`size_t gsl_cheb_order (const gsl_cheb_series * cs)`   Function
  This function returns the order of Chebyshev series cs.

`size_t gsl_cheb_size (const gsl_cheb_series * cs)`   Function
`double * gsl_cheb_coeffs (const gsl_cheb_series * cs)`   Function
  These functions return the size of the Chebyshev coefficient array c[] and a pointer to its location in memory for the Chebyshev series cs.

28.4 Chebyshev Series Evaluation

`double gsl_cheb_eval (const gsl_cheb_series * cs, double x)`   Function
  This function evaluates the Chebyshev series cs at a given point x.

`int gsl_cheb_eval_err (const gsl_cheb_series * cs, const`   Function
         `double x, double * result, double * abserr)`
  This function computes the Chebyshev series cs at a given point x, estimating both the series result and its absolute error abserr. The error estimate is made from the first neglected term in the series.

`double gsl_cheb_eval_n (const gsl_cheb_series * cs, size_t`   Function
         `order, double x)`
  This function evaluates the Chebyshev series cs at a given point n, to (at most) the given order order.

`int gsl_cheb_eval_n_err (const gsl_cheb_series * cs, const`   Function
         `size_t order, const double x, double * result, double *`
         `abserr)`
  This function evaluates a Chebyshev series cs at a given point x, estimating both the series result and its absolute error abserr, to (at most) the given order order. The error estimate is made from the first neglected term in the series.

28.5 Derivatives and Integrals

The following functions allow a Chebyshev series to be differentiated or integrated, producing a new Chebyshev series. Note that the error estimate produced by evaluating the derivative series will be underestimated due to the contribution of higher order terms being neglected.

int gsl_cheb_calc_deriv (gsl_cheb_series * *deriv*, const   Function
        gsl_cheb_series * *cs*)

    This function computes the derivative of the series *cs*, storing the derivative coefficients in the previously allocated *deriv*. The two series *cs* and *deriv* must have been allocated with the same order.

int gsl_cheb_calc_integ (gsl_cheb_series * *integ*, const   Function
        gsl_cheb_series * *cs*)

    This function computes the integral of the series *cs*, storing the integral coefficients in the previously allocated *integ*. The two series *cs* and *integ* must have been allocated with the same order. The lower limit of the integration is taken to be the left hand end of the range a.

28.6 Examples

The following example program computes Chebyshev approximations to a step function. This is an extremely difficult approximation to make, due to the discontinuity, and was chosen as an example where approximation error is visible. For smooth functions the Chebyshev approximation converges extremely rapidly and errors would not be visible.

```
#include <stdio.h>
#include <gsl/gsl_math.h>
#include <gsl/gsl_chebyshev.h>

double
f (double x, void *p)
{
  if (x < 0.5)
    return 0.25;
  else
    return 0.75;
}

int
main (void)
{
  int i, n = 10000;

  gsl_cheb_series *cs = gsl_cheb_alloc (40);

  gsl_function F;
```

```
F.function = f;
F.params = 0;

gsl_cheb_init (cs, &F, 0.0, 1.0);

for (i = 0; i < n; i++)
  {
    double x = i / (double)n;
    double r10 = gsl_cheb_eval_n (cs, 10, x);
    double r40 = gsl_cheb_eval (cs, x);
    printf ("%g %g %g %g\n",
            x, GSL_FN_EVAL (&F, x), r10, r40);
  }

gsl_cheb_free (cs);

return 0;
}
```

The output from the program gives the original function, 10-th order approximation and 40-th order approximation, all sampled at intervals of 0.001 in x.

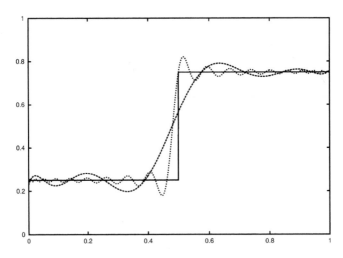

28.7 References and Further Reading

The following paper describes the use of Chebyshev series,

R. Broucke, "Ten Subroutines for the Manipulation of Chebyshev Series [C1] (Algorithm 446)". *Communications of the ACM* 16(4), 254–256 (1973)

29 Series Acceleration

The functions described in this chapter accelerate the convergence of a series using the Levin u-transform. This method takes a small number of terms from the start of a series and uses a systematic approximation to compute an extrapolated value and an estimate of its error. The u-transform works for both convergent and divergent series, including asymptotic series.

These functions are declared in the header file 'gsl_sum.h'.

29.1 Acceleration functions

The following functions compute the full Levin u-transform of a series with its error estimate. The error estimate is computed by propagating rounding errors from each term through to the final extrapolation.

These functions are intended for summing analytic series where each term is known to high accuracy, and the rounding errors are assumed to originate from finite precision. They are taken to be relative errors of order GSL_DBL_EPSILON for each term.

The calculation of the error in the extrapolated value is an $O(N^2)$ process, which is expensive in time and memory. A faster but less reliable method which estimates the error from the convergence of the extrapolated value is described in the next section. For the method described here a full table of intermediate values and derivatives through to $O(N)$ must be computed and stored, but this does give a reliable error estimate.

gsl_sum_levin_u_workspace * gsl_sum_levin_u_alloc (size_t n)   Function
This function allocates a workspace for a Levin u-transform of n terms. The size of the workspace is $O(2n^2 + 3n)$.

void gsl_sum_levin_u_free (gsl_sum_levin_u_workspace * w)   Function
This function frees the memory associated with the workspace w.

int gsl_sum_levin_u_accel (const double * array, size_t   Function
        array_size, gsl_sum_levin_u_workspace * w, double *
        sum_accel, double * abserr)
This function takes the terms of a series in array of size array_size and computes the extrapolated limit of the series using a Levin u-transform. Additional working space must be provided in w. The extrapolated sum is stored in sum_accel, with an estimate of the absolute error stored in abserr. The actual term-by-term sum is returned in w->sum_plain. The algorithm calculates the truncation error (the difference between two successive extrapolations) and round-off error (propagated from the individual terms) to choose an optimal number of terms for the extrapolation. All the terms of the series passed in through array should be non-zero.

29.2 Acceleration functions without error estimation

The functions described in this section compute the Levin u-transform of series and attempt to estimate the error from the "truncation error" in the extrapolation, the difference between the final two approximations. Using this method avoids the need to compute an intermediate table of derivatives because the error is estimated from the behavior of the extrapolated value itself. Consequently this algorithm is an $O(N)$ process and only requires $O(N)$ terms of storage. If the series converges sufficiently fast then this procedure can be acceptable. It is appropriate to use this method when there is a need to compute many extrapolations of series with similar convergence properties at high-speed. For example, when numerically integrating a function defined by a parameterized series where the parameter varies only slightly. A reliable error estimate should be computed first using the full algorithm described above in order to verify the consistency of the results.

gsl_sum_levin_utrunc_workspace * gsl_sum_levin_utrunc_alloc   *Function*
        (size_t n)
    This function allocates a workspace for a Levin u-transform of n terms, without error estimation. The size of the workspace is $O(3n)$.

void gsl_sum_levin_utrunc_free   *Function*
        (gsl_sum_levin_utrunc_workspace * w)
    This function frees the memory associated with the workspace w.

int gsl_sum_levin_utrunc_accel (const double * array, size_t   *Function*
        array_size, gsl_sum_levin_utrunc_workspace * w, double *
        sum_accel, double * abserr_trunc)
    This function takes the terms of a series in array of size array_size and computes the extrapolated limit of the series using a Levin u-transform. Additional working space must be provided in w. The extrapolated sum is stored in sum_accel. The actual term-by-term sum is returned in w->sum_plain. The algorithm terminates when the difference between two successive extrapolations reaches a minimum or is sufficiently small. The difference between these two values is used as estimate of the error and is stored in abserr_trunc. To improve the reliability of the algorithm the extrapolated values are replaced by moving averages when calculating the truncation error, smoothing out any fluctuations.

29.3 Examples

The following code calculates an estimate of $\zeta(2) = \pi^2/6$ using the series,

$$\zeta(2) = 1 + 1/2^2 + 1/3^2 + 1/4^2 + \ldots$$

After N terms the error in the sum is $O(1/N)$, making direct summation of the series converge slowly.

```c
#include <stdio.h>
#include <gsl/gsl_math.h>
#include <gsl/gsl_sum.h>

#define N 20

int
main (void)
{
  double t[N];
  double sum_accel, err;
  double sum = 0;
  int n;

  gsl_sum_levin_u_workspace * w
    = gsl_sum_levin_u_alloc (N);

  const double zeta_2 = M_PI * M_PI / 6.0;

  /* terms for zeta(2) = \sum_{n=1}^{\infty} 1/n^2 */

  for (n = 0; n < N; n++)
    {
      double np1 = n + 1.0;
      t[n] = 1.0 / (np1 * np1);
      sum += t[n];
    }

  gsl_sum_levin_u_accel (t, N, w, &sum_accel, &err);

  printf ("term-by-term sum = % .16f using %d terms\n",
          sum, N);

  printf ("term-by-term sum = % .16f using %d terms\n",
          w->sum_plain, w->terms_used);

  printf ("exact value     = % .16f\n", zeta_2);
  printf ("accelerated sum = % .16f using %d terms\n",
          sum_accel, w->terms_used);

  printf ("estimated error = % .16f\n", err);
```

```
        printf ("actual error    = % .16f\n",
                sum_accel - zeta_2);

        gsl_sum_levin_u_free (w);
        return 0;
    }
```

The output below shows that the Levin u-transform is able to obtain an estimate
of the sum to 1 part in 10^{10} using the first eleven terms of the series. The error
estimate returned by the function is also accurate, giving the correct number of
significant digits.

```
    $ ./a.out
      term-by-term sum =   1.5961632439130233 using 20 terms
      term-by-term sum =   1.5759958390005426 using 13 terms
      exact value      =   1.6449340668482264
      accelerated sum  =   1.6449340668166479 using 13 terms
      estimated error  =   0.0000000000508580
      actual error     =  -0.0000000000315785
```

Note that a direct summation of this series would require 10^{10} terms to achieve
the same precision as the accelerated sum does in 13 terms.

29.4 References and Further Reading

The algorithms used by these functions are described in the following papers,

> T. Fessler, W.F. Ford, D.A. Smith, HURRY: An acceleration algorithm for
> scalar sequences and series *ACM Transactions on Mathematical Software*,
> 9(3):346–354, 1983. and Algorithm 602 9(3):355–357, 1983.

The theory of the u-transform was presented by Levin,

> D. Levin, Development of Non-Linear Transformations for Improving Con-
> vergence of Sequences, *Intern. J. Computer Math.* B3:371–388, 1973.

A review paper on the Levin Transform is available online,

> Herbert H. H. Homeier, Scalar Levin-Type Sequence Transformations,
> http://arxiv.org/abs/math/0005209.

30 Wavelet Transforms

This chapter describes functions for performing Discrete Wavelet Transforms (DWTs). The library includes wavelets for real data in both one and two dimensions. The wavelet functions are declared in the header files 'gsl_wavelet.h' and 'gsl_wavelet2d.h'.

30.1 Definitions

The continuous wavelet transform and its inverse are defined by the relations,

$$w(s, \tau) = \int_{-\infty}^{\infty} f(t) * \psi_{s,\tau}^*(t) dt$$

and,

$$f(t) = \int_{0}^{\infty} ds \int_{-\infty}^{\infty} w(s, \tau) * \psi_{s,\tau}(t) d\tau$$

where the basis functions $\psi_{s,\tau}$ are obtained by scaling and translation from a single function, referred to as the *mother wavelet*.

The discrete version of the wavelet transform acts on equally-spaced samples, with fixed scaling and translation steps (s, τ). The frequency and time axes are sampled *dyadically* on scales of 2^j through a level parameter j. The resulting family of functions $\{\psi_{j,n}\}$ constitutes an orthonormal basis for square-integrable signals.

The discrete wavelet transform is an $O(N)$ algorithm, and is also referred to as the *fast wavelet transform*.

30.2 Initialization

The gsl_wavelet structure contains the filter coefficients defining the wavelet and any associated offset parameters.

gsl_wavelet * gsl_wavelet_alloc (const gsl_wavelet_type * T,   *Function*
        size_t k)

   This function allocates and initializes a wavelet object of type T. The parameter k selects the specific member of the wavelet family. A null pointer is returned if insufficient memory is available or if a unsupported member is selected.

The following wavelet types are implemented:

gsl_wavelet_daubechies   *Wavelet*
gsl_wavelet_daubechies_centered   *Wavelet*
   The is the Daubechies wavelet family of maximum phase with $k/2$ vanishing moments. The implemented wavelets are $k = 4, 6, \ldots, 20$, with k even.

gsl_wavelet_haar   Wavelet
gsl_wavelet_haar_centered   Wavelet
    This is the Haar wavelet. The only valid choice of k for the Haar wavelet is
    $k = 2$.

gsl_wavelet_bspline   Wavelet
gsl_wavelet_bspline_centered   Wavelet
    This is the biorthogonal B-spline wavelet family of order (i, j). The imple-
    mented values of $k = 100 * i + j$ are 103, 105, 202, 204, 206, 208, 301, 303,
    305 307, 309.

The centered forms of the wavelets align the coefficients of the various sub-bands
on edges. Thus the resulting visualization of the coefficients of the wavelet
transform in the phase plane is easier to understand.

const char * gsl_wavelet_name (const gsl_wavelet * w)   Function
    This function returns a pointer to the name of the wavelet family for w.

void gsl_wavelet_free (gsl_wavelet * w)   Function
    This function frees the wavelet object w.

    The gsl_wavelet_workspace structure contains scratch space of the same size
as the input data and is used to hold intermediate results during the transform.

gsl_wavelet_workspace * gsl_wavelet_workspace_alloc (size_t   Function
        n)
    This function allocates a workspace for the discrete wavelet transform. To
    perform a one-dimensional transform on n elements, a workspace of size n
    must be provided. For two-dimensional transforms of n-by-n matrices it is
    sufficient to allocate a workspace of size n, since the transform operates on
    individual rows and columns.

void gsl_wavelet_workspace_free (gsl_wavelet_workspace *   Function
        work)
    This function frees the allocated workspace work.

30.3 Transform Functions

    This sections describes the actual functions performing the discrete wavelet
transform. Note that the transforms use periodic boundary conditions. If the
signal is not periodic in the sample length then spurious coefficients will appear
at the beginning and end of each level of the transform.

30.3.1 Wavelet transforms in one dimension

int gsl_wavelet_transform (const gsl_wavelet * w, double *   Function
        data, size_t *stride*, size_t n, gsl_wavelet_direction *dir*,
        gsl_wavelet_workspace * *work*)
int gsl_wavelet_transform_forward (const gsl_wavelet * w,   Function
        double * *data*, size_t *stride*, size_t n,
        gsl_wavelet_workspace * *work*)
int gsl_wavelet_transform_inverse (const gsl_wavelet * w,   Function
        double * *data*, size_t *stride*, size_t n,
        gsl_wavelet_workspace * *work*)

These functions compute in-place forward and inverse discrete wavelet transforms of length n with stride *stride* on the array *data*. The length of the
transform n is restricted to powers of two. For the transform version of the
function the argument *dir* can be either forward $(+1)$ or backward (-1). A
workspace *work* of length n must be provided.

For the forward transform, the elements of the original array are replaced
by the discrete wavelet transform $f_i \rightarrow w_{j,k}$ in a packed triangular storage
layout, where j is the index of the level $j = 0 \ldots J - 1$ and k is the index of
the coefficient within each level, $k = 0 \ldots 2^j - 1$. The total number of levels
is $J = \log_2(n)$. The output data has the following form,

$$(s_{-1,0}, d_{0,0}, d_{1,0}, d_{1,1}, d_{2,0}, \cdots, d_{j,k}, \cdots, d_{J-1,2^{J-1}-1})$$

where the first element is the smoothing coefficient $s_{-1,0}$, followed by the
detail coefficients $d_{j,k}$ for each level j. The backward transform inverts these
coefficients to obtain the original data.

These functions return a status of GSL_SUCCESS upon successful completion.
GSL_EINVAL is returned if n is not an integer power of 2 or if insufficient
workspace is provided.

30.3.2 Wavelet transforms in two dimension

The library provides functions to perform two-dimensional discrete wavelet
transforms on square matrices. The matrix dimensions must be an integer
power of two. There are two possible orderings of the rows and columns in
the two-dimensional wavelet transform, referred to as the "standard" and "non-
standard" forms.

The "standard" transform performs a complete discrete wavelet transform on
the rows of the matrix, followed by a separate complete discrete wavelet transform on the columns of the resulting row-transformed matrix. This procedure
uses the same ordering as a two-dimensional fourier transform.

The "non-standard" transform is performed in interleaved passes on the rows
and columns of the matrix for each level of the transform. The first level of
the transform is applied to the matrix rows, and then to the matrix columns.
This procedure is then repeated across the rows and columns of the data for the
subsequent levels of the transform, until the full discrete wavelet transform is
complete. The non-standard form of the discrete wavelet transform is typically
used in image analysis.

The functions described in this section are declared in the header file 'gsl_wavelet2d.h'.

int gsl_wavelet2d_transform (const gsl_wavelet * w, double *   *Function*
      data, size_t tda, size_t size1, size_t size2,
      gsl_wavelet_direction dir, gsl_wavelet_workspace * work)

int gsl_wavelet2d_transform_forward (const gsl_wavelet * w,   *Function*
      double * data, size_t tda, size_t size1, size_t size2,
      gsl_wavelet_workspace * work)

int gsl_wavelet2d_transform_inverse (const gsl_wavelet * w,   *Function*
      double * data, size_t tda, size_t size1, size_t size2,
      gsl_wavelet_workspace * work)

These functions compute two-dimensional in-place forward and inverse discrete wavelet transforms in standard and non-standard forms on the array data stored in row-major form with dimensions size1 and size2 and physical row length tda. The dimensions must be equal (square matrix) and are restricted to powers of two. For the transform version of the function the argument dir can be either forward ($+1$) or backward (-1). A workspace work of the appropriate size must be provided. On exit, the appropriate elements of the array data are replaced by their two-dimensional wavelet transform.

The functions return a status of GSL_SUCCESS upon successful completion. GSL_EINVAL is returned if size1 and size2 are not equal and integer powers of 2, or if insufficient workspace is provided.

int gsl_wavelet2d_transform_matrix (const gsl_wavelet * w,   *Function*
      gsl_matrix * m, gsl_wavelet_direction dir,
      gsl_wavelet_workspace * work)

int gsl_wavelet2d_transform_matrix_forward (const   *Function*
      gsl_wavelet * w, gsl_matrix * m, gsl_wavelet_workspace *
      work)

int gsl_wavelet2d_transform_matrix_inverse (const   *Function*
      gsl_wavelet * w, gsl_matrix * m, gsl_wavelet_workspace *
      work)

These functions compute the two-dimensional in-place wavelet transform on a matrix a.

int gsl_wavelet2d_nstransform (const gsl_wavelet * w, double   *Function*
      * data, size_t tda, size_t size1, size_t size2,
      gsl_wavelet_direction dir, gsl_wavelet_workspace * work)

int gsl_wavelet2d_nstransform_forward (const gsl_wavelet * w,   *Function*
      double * data, size_t tda, size_t size1, size_t size2,
      gsl_wavelet_workspace * work)

int gsl_wavelet2d_nstransform_inverse (const gsl_wavelet * w,   *Function*
      double * data, size_t tda, size_t size1, size_t size2,
      gsl_wavelet_workspace * work)

These functions compute the two-dimensional wavelet transform in non-standard form.

```
int gsl_wavelet2d_nstransform_matrix (const gsl_wavelet * w,        Function
         gsl_matrix * m, gsl_wavelet_direction dir,
         gsl_wavelet_workspace * work)
int gsl_wavelet2d_nstransform_matrix_forward (const               Function
         gsl_wavelet * w, gsl_matrix * m, gsl_wavelet_workspace *
         work)
int gsl_wavelet2d_nstransform_matrix_inverse (const               Function
         gsl_wavelet * w, gsl_matrix * m, gsl_wavelet_workspace *
         work)
```

These functions compute the non-standard form of the two-dimensional in-place wavelet transform on a matrix a.

30.4 Examples

The following program demonstrates the use of the one-dimensional wavelet transform functions. It computes an approximation to an input signal (of length 256) using the 20 largest components of the wavelet transform, while setting the others to zero.

```
#include <stdio.h>
#include <math.h>
#include <gsl/gsl_sort.h>
#include <gsl/gsl_wavelet.h>

int
main (int argc, char **argv)
{
  int i, n = 256, nc = 20;
  double *data = malloc (n * sizeof (double));
  double *abscoeff = malloc (n * sizeof (double));
  size_t *p = malloc (n * sizeof (size_t));

  FILE * f;
  gsl_wavelet *w;
  gsl_wavelet_workspace *work;

  w = gsl_wavelet_alloc (gsl_wavelet_daubechies, 4);
  work = gsl_wavelet_workspace_alloc (n);

  f = fopen (argv[1], "r");
  for (i = 0; i < n; i++)
    fscanf (f, "%lg", &data[i]);
  fclose (f);

  gsl_wavelet_transform_forward (w, data, 1, n, work);

  for (i = 0; i < n; i++)
    abscoeff[i] = fabs (data[i]);
```

```
gsl_sort_index (p, abscoeff, 1, n);

for (i = 0; (i + nc) < n; i++)
  data[p[i]] = 0;

gsl_wavelet_transform_inverse (w, data, 1, n, work);

for (i = 0; i < n; i++)
  printf ("%g\n", data[i]);

gsl_wavelet_free (w);
gsl_wavelet_workspace_free (work);
free (data); free (abscoeff); free (p);
return 0;
}
```

The output can be used with the GNU plotutils graph program,

```
$ ./a.out ecg.dat > dwt.dat
$ graph -T ps -x 0 256 32 -h 0.3 -a dwt.dat > dwt.ps
```

The graphs below show an original and compressed version of a sample ECG recording from the MIT-BIH Arrhythmia Database, part of the PhysioNet archive of public-domain of medical datasets.

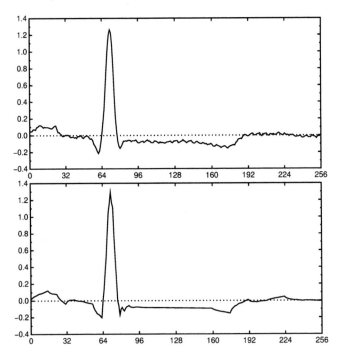

Original (upper) and wavelet-compressed (lower) ECG signals, using the 20 largest components of the Daubechies(4) discrete wavelet transform.

30.5 References and Further Reading

The mathematical background to wavelet transforms is covered in the original lectures by Daubechies,

Ingrid Daubechies. Ten Lectures on Wavelets. *CBMS-NSF Regional Conference Series in Applied Mathematics* (1992), SIAM, ISBN 0898712742.

An easy to read introduction to the subject with an emphasis on the application of the wavelet transform in various branches of science is,

Paul S. Addison. *The Illustrated Wavelet Transform Handbook.* Institute of Physics Publishing (2002), ISBN 0750306920.

For extensive coverage of signal analysis by wavelets, wavelet packets and local cosine bases see,

S. G. Mallat. *A wavelet tour of signal processing* (Second edition). Academic Press (1999), ISBN 012466606X.

The concept of multiresolution analysis underlying the wavelet transform is described in,

S. G. Mallat. Multiresolution Approximations and Wavelet Orthonormal Bases of $L^2(R)$. *Transactions of the American Mathematical Society*, 315(1), 1989, 69–87.

S. G. Mallat. A Theory for Multiresolution Signal Decomposition—The Wavelet Representation. *IEEE Transactions on Pattern Analysis and Machine Intelligence*, 11, 1989, 674–693.

The coefficients for the individual wavelet families implemented by the library can be found in the following papers,

I. Daubechies. Orthonormal Bases of Compactly Supported Wavelets. *Communications on Pure and Applied Mathematics*, 41 (1988) 909–996.

A. Cohen, I. Daubechies, and J.-C. Feauveau. Biorthogonal Bases of Compactly Supported Wavelets. *Communications on Pure and Applied Mathematics*, 45 (1992) 485–560.

The PhysioNet archive of physiological datasets can be found online at http://www.physionet.org/ and is described in the following paper,

Goldberger et al. PhysioBank, PhysioToolkit, and PhysioNet: Components of a New Research Resource for Complex Physiologic Signals. *Circulation* 101(23):e215-e220 2000.

31 Discrete Hankel Transforms

This chapter describes functions for performing Discrete Hankel Transforms (DHTs). The functions are declared in the header file 'gsl_dht.h'.

31.1 Definitions

The discrete Hankel transform acts on a vector of sampled data, where the samples are assumed to have been taken at points related to the zeroes of a Bessel function of fixed order; compare this to the case of the discrete Fourier transform, where samples are taken at points related to the zeroes of the sine or cosine function.

Specifically, let $f(t)$ be a function on the unit interval. Then the finite ν-Hankel transform of $f(t)$ is defined to be the set of numbers g_m given by,

$$g_m = \int_0^1 t\,dt\, J_\nu(j_{\nu,m}t) f(t),$$

so that,

$$f(t) = \sum_{m=1}^{\infty} \frac{2 J_\nu(j_{\nu,m}x)}{J_{\nu+1}(j_{\nu,m})^2} g_m.$$

Suppose that f is band-limited in the sense that $g_m = 0$ for $m > M$. Then we have the following fundamental sampling theorem.

$$g_m = \frac{2}{j_{\nu,M}^2} \sum_{k=1}^{M-1} f\left(\frac{j_{\nu,k}}{j_{\nu,M}}\right) \frac{J_\nu(j_{\nu,m}j_{\nu,k}/j_{\nu,M})}{J_{\nu+1}(j_{\nu,k})^2}.$$

It is this discrete expression which defines the discrete Hankel transform. The kernel in the summation above defines the matrix of the ν-Hankel transform of size $M - 1$. The coefficients of this matrix, being dependent on ν and M, must be precomputed and stored; the gsl_dht object encapsulates this data. The allocation function gsl_dht_alloc returns a gsl_dht object which must be properly initialized with gsl_dht_init before it can be used to perform transforms on data sample vectors, for fixed ν and M, using the gsl_dht_apply function. The implementation allows a scaling of the fundamental interval, for convenience, so that one can assume the function is defined on the interval $[0, X]$, rather than the unit interval.

Notice that by assumption $f(t)$ vanishes at the endpoints of the interval, consistent with the inversion formula and the sampling formula given above. Therefore, this transform corresponds to an orthogonal expansion in eigenfunctions of the Dirichlet problem for the Bessel differential equation.

31.2 Functions

`gsl_dht * gsl_dht_alloc (size_t size)`   Function
:   This function allocates a Discrete Hankel transform object of size *size*.

`int gsl_dht_init (gsl_dht * t, double nu, double xmax)`   Function
:   This function initializes the transform *t* for the given values of *nu* and *x*.

`gsl_dht * gsl_dht_new (size_t size, double nu, double xmax)`   Function
:   This function allocates a Discrete Hankel transform object of size *size* and initializes it for the given values of *nu* and *x*.

`void gsl_dht_free (gsl_dht * t)`   Function
:   This function frees the transform *t*.

`int gsl_dht_apply (const gsl_dht * t, double * f_in, double *`   Function
`        f_out)`
:   This function applies the transform *t* to the array *f_in* whose size is equal to the size of the transform. The result is stored in the array *f_out* which must be of the same length.

`double gsl_dht_x_sample (const gsl_dht * t, int n)`   Function
:   This function returns the value of the *n*-th sample point in the unit interval, $(j_{\nu,n+1}/j_{\nu,M})X$. These are the points where the function $f(t)$ is assumed to be sampled.

`double gsl_dht_k_sample (const gsl_dht * t, int n)`   Function
:   This function returns the value of the *n*-th sample point in "k-space", $j_{\nu,n+1}/X$.

31.3 References and Further Reading

The algorithms used by these functions are described in the following papers,

H. Fisk Johnson, Comp. Phys. Comm. 43, 181 (1987).

D. Lemoine, J. Chem. Phys. 101, 3936 (1994).

32 One dimensional Root-Finding

This chapter describes routines for finding roots of arbitrary one-dimensional functions. The library provides low level components for a variety of iterative solvers and convergence tests. These can be combined by the user to achieve the desired solution, with full access to the intermediate steps of the iteration. Each class of methods uses the same framework, so that you can switch between solvers at runtime without needing to recompile your program. Each instance of a solver keeps track of its own state, allowing the solvers to be used in multi-threaded programs.

The header file 'gsl_roots.h' contains prototypes for the root finding functions and related declarations.

32.1 Overview

One-dimensional root finding algorithms can be divided into two classes, *root bracketing* and *root polishing*. Algorithms which proceed by bracketing a root are guaranteed to converge. Bracketing algorithms begin with a bounded region known to contain a root. The size of this bounded region is reduced, iteratively, until it encloses the root to a desired tolerance. This provides a rigorous error estimate for the location of the root.

The technique of *root polishing* attempts to improve an initial guess to the root. These algorithms converge only if started "close enough" to a root, and sacrifice a rigorous error bound for speed. By approximating the behavior of a function in the vicinity of a root they attempt to find a higher order improvement of an initial guess. When the behavior of the function is compatible with the algorithm and a good initial guess is available a polishing algorithm can provide rapid convergence.

In GSL both types of algorithm are available in similar frameworks. The user provides a high-level driver for the algorithms, and the library provides the individual functions necessary for each of the steps. There are three main phases of the iteration. The steps are,

- initialize solver state, s, for algorithm T
- update s using the iteration T
- test s for convergence, and repeat iteration if necessary

The state for bracketing solvers is held in a gsl_root_fsolver struct. The updating procedure uses only function evaluations (not derivatives). The state for root polishing solvers is held in a gsl_root_fdfsolver struct. The updates require both the function and its derivative (hence the name fdf) to be supplied by the user.

32.2 Caveats

Note that root finding functions can only search for one root at a time. When there are several roots in the search area, the first root to be found will be returned; however it is difficult to predict which of the roots this will be. *In most cases, no error will be reported if you try to find a root in an area where there is more than one.*

Care must be taken when a function may have a multiple root (such as $f(x) = (x - x_0)^2$ or $f(x) = (x - x_0)^3$). It is not possible to use root-bracketing algorithms on even-multiplicity roots. For these algorithms the initial interval must contain a zero-crossing, where the function is negative at one end of the interval and positive at the other end. Roots with even-multiplicity do not cross zero, but only touch it instantaneously. Algorithms based on root bracketing will still work for odd-multiplicity roots (e.g. cubic, quintic, ...). Root polishing algorithms generally work with higher multiplicity roots, but at a reduced rate of convergence. In these cases the *Steffenson algorithm* can be used to accelerate the convergence of multiple roots.

While it is not absolutely required that f have a root within the search region, numerical root finding functions should not be used haphazardly to check for the *existence* of roots. There are better ways to do this. Because it is easy to create situations where numerical root finders can fail, it is a bad idea to throw a root finder at a function you do not know much about. In general it is best to examine the function visually by plotting before searching for a root.

32.3 Initializing the Solver

gsl_root_fsolver * gsl_root_fsolver_alloc (const   Function
        gsl_root_fsolver_type * T)

This function returns a pointer to a newly allocated instance of a solver of type T. For example, the following code creates an instance of a bisection solver,

```
const gsl_root_fsolver_type * T
  = gsl_root_fsolver_bisection;
gsl_root_fsolver * s
  = gsl_root_fsolver_alloc (T);
```

If there is insufficient memory to create the solver then the function returns a null pointer and the error handler is invoked with an error code of GSL_ENOMEM.

gsl_root_fdfsolver * gsl_root_fdfsolver_alloc (const   Function
        gsl_root_fdfsolver_type * T)

This function returns a pointer to a newly allocated instance of a derivative-based solver of type T. For example, the following code creates an instance of a Newton-Raphson solver,

```
const gsl_root_fdfsolver_type * T
  = gsl_root_fdfsolver_newton;
gsl_root_fdfsolver * s
  = gsl_root_fdfsolver_alloc (T);
```

If there is insufficient memory to create the solver then the function returns a null pointer and the error handler is invoked with an error code of GSL_ENOMEM.

int gsl_root_fsolver_set (gsl_root_fsolver * s, gsl_function *   Function
          f, double x_lower, double x_upper)
This function initializes, or reinitializes, an existing solver s to use the function f and the initial search interval [x_lower, x_upper].

int gsl_root_fdfsolver_set (gsl_root_fdfsolver * s,   Function
          gsl_function_fdf * fdf, double root)
This function initializes, or reinitializes, an existing solver s to use the function and derivative fdf and the initial guess root.

void gsl_root_fsolver_free (gsl_root_fsolver * s)   Function
void gsl_root_fdfsolver_free (gsl_root_fdfsolver * s)   Function
These functions free all the memory associated with the solver s.

const char * gsl_root_fsolver_name (const gsl_root_fsolver *   Function
          s)
const char * gsl_root_fdfsolver_name (const   Function
          gsl_root_fdfsolver * s)
These functions return a pointer to the name of the solver. For example,

```
printf ("s is a '%s' solver\n",
          gsl_root_fsolver_name (s));
```

would print something like s is a 'bisection' solver.

32.4 Providing the function to solve

You must provide a continuous function of one variable for the root finders to operate on, and, sometimes, its first derivative. In order to allow for general parameters the functions are defined by the following data types:

gsl_function   Data Type
This data type defines a general function with parameters.

> double (* function) (double x, void * params)
> this function should return the value $f(x, params)$ for argument x and parameters params

> void * params
> a pointer to the parameters of the function

Here is an example for the general quadratic function,

$$f(x) = ax^2 + bx + c$$

with $a = 3$, $b = 2$, $c = 1$. The following code defines a gsl_function F which you could pass to a root finder:

```
struct my_f_params { double a; double b; double c; };

double
my_f (double x, void * p) {
    struct my_f_params * params
      = (struct my_f_params *)p;
    double a = (params->a);
    double b = (params->b);
    double c = (params->c);

    return  (a * x + b) * x + c;
}

gsl_function F;
struct my_f_params params = { 3.0, 2.0, 1.0 };

F.function = &my_f;
F.params = &params;
```

The function $f(x)$ can be evaluated using the following macro,

```
#define GSL_FN_EVAL(F,x)
       (*((F)->function))(x,(F)->params)
```

gsl_function_fdf   Data Type
    This data type defines a general function with parameters and its first derivative.

    double (* f) (double x, void * params)
        this function should return the value of $f(x, params)$ for argument x and parameters params

    double (* df) (double x, void * params)
        this function should return the value of the derivative of f with respect to x, $f'(x, params)$, for argument x and parameters params

    void (* fdf) (double x, void * params, double * f, double * df)
        this function should set the values of the function f to $f(x, params)$ and its derivative df to $f'(x, params)$ for argument x and parameters params. This function provides an optimization of the separate functions for $f(x)$ and $f'(x)$—it is always faster to compute the function and its derivative at the same time.

    void * params
        a pointer to the parameters of the function

Here is an example where $f(x) = \exp(2x)$:

```
double
my_f (double x, void * params)
{
   return exp (2 * x);
}

double
my_df (double x, void * params)
{
   return 2 * exp (2 * x);
}

void
my_fdf (double x, void * params,
        double * f, double * df)
{
   double t = exp (2 * x);

   *f = t;
   *df = 2 * t;   /* uses existing value */
}

gsl_function_fdf FDF;

FDF.f = &my_f;
FDF.df = &my_df;
FDF.fdf = &my_fdf;
FDF.params = 0;
```

The function $f(x)$ can be evaluated using the following macro,

```
#define GSL_FN_FDF_EVAL_F(FDF,x)
        (*((FDF)->f))(x,(FDF)->params)
```

The derivative $f'(x)$ can be evaluated using the following macro,

```
#define GSL_FN_FDF_EVAL_DF(FDF,x)
        (*((FDF)->df))(x,(FDF)->params)
```

and both the function $y = f(x)$ and its derivative $dy = f'(x)$ can be evaluated at the same time using the following macro,

```
#define GSL_FN_FDF_EVAL_F_DF(FDF,x,y,dy)
        (*((FDF)->fdf))(x,(FDF)->params,(y),(dy))
```

The macro stores $f(x)$ in its y argument and $f'(x)$ in its dy argument—both of these should be pointers to double.

32.5 Search Bounds and Guesses

You provide either search bounds or an initial guess; this section explains how search bounds and guesses work and how function arguments control them.

A guess is simply an x value which is iterated until it is within the desired precision of a root. It takes the form of a double.

Search bounds are the endpoints of an interval which is iterated until the length of the interval is smaller than the requested precision. The interval is defined by two values, the lower limit and the upper limit. Whether the endpoints are intended to be included in the interval or not depends on the context in which the interval is used.

32.6 Iteration

The following functions drive the iteration of each algorithm. Each function performs one iteration to update the state of any solver of the corresponding type. The same functions work for all solvers so that different methods can be substituted at runtime without modifications to the code.

int gsl_root_fsolver_iterate (gsl_root_fsolver * s)   Function
int gsl_root_fdfsolver_iterate (gsl_root_fdfsolver * s)   Function
  These functions perform a single iteration of the solver s. If the iteration encounters an unexpected problem then an error code will be returned,

  GSL_EBADFUNC
        the iteration encountered a singular point where the function or its derivative evaluated to Inf or NaN.

  GSL_EZERODIV
        the derivative of the function vanished at the iteration point, preventing the algorithm from continuing without a division by zero.

The solver maintains a current best estimate of the root at all times. The bracketing solvers also keep track of the current best interval bounding the root. This information can be accessed with the following auxiliary functions,

double gsl_root_fsolver_root (const gsl_root_fsolver * s)   Function
double gsl_root_fdfsolver_root (const gsl_root_fdfsolver * s)   Function
  These functions return the current estimate of the root for the solver s.

double gsl_root_fsolver_x_lower (const gsl_root_fsolver * s)   Function
double gsl_root_fsolver_x_upper (const gsl_root_fsolver * s)   Function
  These functions return the current bracketing interval for the solver s.

32.7 Search Stopping Parameters

A root finding procedure should stop when one of the following conditions is true:

- A root has been found to within the user-specified precision.
- A user-specified maximum number of iterations has been reached.
- An error has occurred.

The handling of these conditions is under user control. The functions below allow the user to test the precision of the current result in several standard ways.

int gsl_root_test_interval (double x_lower, double x_upper,   *Function*
    double epsabs, double epsrel)
This function tests for the convergence of the interval $[x\_lower, x\_upper]$ with absolute error epsabs and relative error epsrel. The test returns GSL_SUCCESS if the following condition is achieved,

$$|a - b| < epsabs + epsrel \, \min(|a|, |b|)$$

when the interval $x = [a, b]$ does not include the origin. If the interval includes the origin then $\min(|a|, |b|)$ is replaced by zero (which is the minimum value of $|x|$ over the interval). This ensures that the relative error is accurately estimated for roots close to the origin.

This condition on the interval also implies that any estimate of the root r in the interval satisfies the same condition with respect to the true root r^*,

$$|r - r^*| < epsabs + epsrel \, r^*$$

assuming that the true root r^* is contained within the interval.

int gsl_root_test_delta (double x1, double x0, double epsabs,   *Function*
    double epsrel)
This function tests for the convergence of the sequence $\ldots, x0, x1$ with absolute error epsabs and relative error epsrel. The test returns GSL_SUCCESS if the following condition is achieved,

$$|x_1 - x_0| < epsabs + epsrel \, |x_1|$$

and returns GSL_CONTINUE otherwise.

int gsl_root_test_residual (double f, double epsabs)   *Function*
This function tests the residual value f against the absolute error bound epsabs. The test returns GSL_SUCCESS if the following condition is achieved,

$$|f| < epsabs$$

and returns GSL_CONTINUE otherwise. This criterion is suitable for situations where the precise location of the root, x, is unimportant provided a value can be found where the residual, $|f(x)|$, is small enough.

32.8 Root Bracketing Algorithms

The root bracketing algorithms described in this section require an initial interval which is guaranteed to contain a root—if a and b are the endpoints of the interval then $f(a)$ must differ in sign from $f(b)$. This ensures that the function crosses zero at least once in the interval. If a valid initial interval is used then these algorithm cannot fail, provided the function is well-behaved.

Note that a bracketing algorithm cannot find roots of even degree, since these do not cross the x-axis.

gsl_root_fsolver_bisection   Solver

The *bisection algorithm* is the simplest method of bracketing the roots of a function. It is the slowest algorithm provided by the library, with linear convergence.

On each iteration, the interval is bisected and the value of the function at the midpoint is calculated. The sign of this value is used to determine which half of the interval does not contain a root. That half is discarded to give a new, smaller interval containing the root. This procedure can be continued indefinitely until the interval is sufficiently small.

At any time the current estimate of the root is taken as the midpoint of the interval.

gsl_root_fsolver_falsepos   Solver

The *false position algorithm* is a method of finding roots based on linear interpolation. Its convergence is linear, but it is usually faster than bisection.

On each iteration a line is drawn between the endpoints $(a, f(a))$ and $(b, f(b))$ and the point where this line crosses the x-axis taken as a "midpoint". The value of the function at this point is calculated and its sign is used to determine which side of the interval does not contain a root. That side is discarded to give a new, smaller interval containing the root. This procedure can be continued indefinitely until the interval is sufficiently small.

The best estimate of the root is taken from the linear interpolation of the interval on the current iteration.

gsl_root_fsolver_brent   Solver

The *Brent-Dekker method* (referred to here as *Brent's method*) combines an interpolation strategy with the bisection algorithm. This produces a fast algorithm which is still robust.

On each iteration Brent's method approximates the function using an interpolating curve. On the first iteration this is a linear interpolation of the two endpoints. For subsequent iterations the algorithm uses an inverse quadratic fit to the last three points, for higher accuracy. The intercept of the interpolating curve with the x-axis is taken as a guess for the root. If it lies within the bounds of the current interval then the interpolating point is accepted, and used to generate a smaller interval. If the interpolating point is not accepted then the algorithm falls back to an ordinary bisection step.

The best estimate of the root is taken from the most recent interpolation or bisection.

32.9 Root Finding Algorithms using Derivatives

The root polishing algorithms described in this section require an initial guess for the location of the root. There is no absolute guarantee of convergence—the function must be suitable for this technique and the initial guess must be sufficiently close to the root for it to work. When these conditions are satisfied then convergence is quadratic.

These algorithms make use of both the function and its derivative.

`gsl_root_fdfsolver_newton`   Derivative Solver

Newton's Method is the standard root-polishing algorithm. The algorithm begins with an initial guess for the location of the root. On each iteration, a line tangent to the function f is drawn at that position. The point where this line crosses the x-axis becomes the new guess. The iteration is defined by the following sequence,

$$x_{i+1} = x_i - \frac{f(x_i)}{f'(x_i)}$$

Newton's method converges quadratically for single roots, and linearly for multiple roots.

`gsl_root_fdfsolver_secant`   Derivative Solver

The *secant method* is a simplified version of Newton's method which does not require the computation of the derivative on every step.

On its first iteration the algorithm begins with Newton's method, using the derivative to compute a first step,

$$x_1 = x_0 - \frac{f(x_0)}{f'(x_0)}$$

Subsequent iterations avoid the evaluation of the derivative by replacing it with a numerical estimate, the slope of the line through the previous two points,

$$x_{i+1} = x_i - \frac{f(x_i)}{f'_{est}} \text{ where } f'_{est} = \frac{f(x_i) - f(x_{i-1})}{x_i - x_{i-1}}$$

When the derivative does not change significantly in the vicinity of the root the secant method gives a useful saving. Asymptotically the secant method is faster than Newton's method whenever the cost of evaluating the derivative is more than 0.44 times the cost of evaluating the function itself. As with all methods of computing a numerical derivative the estimate can suffer from cancellation errors if the separation of the points becomes too small.

On single roots, the method has a convergence of order $(1 + \sqrt{5})/2$ (approximately 1.62). It converges linearly for multiple roots.

`gsl_root_fdfsolver_steffenson`   Derivative Solver

The *Steffenson Method* provides the fastest convergence of all the routines. It combines the basic Newton algorithm with an Aitken "delta-squared" acceleration. If the Newton iterates are x_i then the acceleration procedure generates a new sequence R_i,

$$R_i = x_i - \frac{(x_{i+1} - x_i)^2}{(x_{i+2} - 2x_{i+1} + x_i)}$$

which converges faster than the original sequence under reasonable conditions. The new sequence requires three terms before it can produce its first value so the method returns accelerated values on the second and subsequent iterations. On the first iteration it returns the ordinary Newton estimate. The Newton iterate is also returned if the denominator of the acceleration term ever becomes zero.

As with all acceleration procedures this method can become unstable if the function is not well-behaved.

32.10 Examples

For any root finding algorithm we need to prepare the function to be solved. For this example we will use the general quadratic equation described earlier. We first need a header file ('demo_fn.h') to define the function parameters,

```
struct quadratic_params
  {
    double a, b, c;
  };

double quadratic (double x, void *params);
double quadratic_deriv (double x, void *params);
void quadratic_fdf (double x, void *params,
                    double *y, double *dy);
```

We place the function definitions in a separate file ('demo_fn.c'),

```
double
quadratic (double x, void *params)
{
  struct quadratic_params *p
    = (struct quadratic_params *) params;

  double a = p->a;
  double b = p->b;
  double c = p->c;

  return (a * x + b) * x + c;
}

double
quadratic_deriv (double x, void *params)
```

```
{
  struct quadratic_params *p
    = (struct quadratic_params *) params;

  double a = p->a;
  double b = p->b;
  double c = p->c;

  return 2.0 * a * x + b;
}

void
quadratic_fdf (double x, void *params,
               double *y, double *dy)
{
  struct quadratic_params *p
    = (struct quadratic_params *) params;

  double a = p->a;
  double b = p->b;
  double c = p->c;

  *y = (a * x + b) * x + c;
  *dy = 2.0 * a * x + b;
}
```

The first program uses the function solver `gsl_root_fsolver_brent` for Brent's method and the general quadratic defined above to solve the following equation,

$$x^2 - 5 = 0$$

with solution $x = \sqrt{5} = 2.236068...$

```
#include <stdio.h>
#include <gsl/gsl_errno.h>
#include <gsl/gsl_math.h>
#include <gsl/gsl_roots.h>

#include "demo_fn.h"
#include "demo_fn.c"

int
main (void)
{
  int status;
  int iter = 0, max_iter = 100;
  const gsl_root_fsolver_type *T;
  gsl_root_fsolver *s;
  double r = 0, r_expected = sqrt (5.0);
  double x_lo = 0.0, x_hi = 5.0;
```

```
      gsl_function F;
      struct quadratic_params params = {1.0, 0.0, -5.0};

      F.function = &quadratic;
      F.params = &params;

      T = gsl_root_fsolver_brent;
      s = gsl_root_fsolver_alloc (T);
      gsl_root_fsolver_set (s, &F, x_lo, x_hi);

      printf ("using %s method\n",
              gsl_root_fsolver_name (s));

      printf ("%5s [%9s, %9s] %9s %10s %9s\n",
              "iter", "lower", "upper", "root",
              "err", "err(est)");

      do
        {
          iter++;
          status = gsl_root_fsolver_iterate (s);
          r = gsl_root_fsolver_root (s);
          x_lo = gsl_root_fsolver_x_lower (s);
          x_hi = gsl_root_fsolver_x_upper (s);
          status = gsl_root_test_interval (x_lo, x_hi,
                                           0, 0.001);

          if (status == GSL_SUCCESS)
            printf ("Converged:\n");

          printf ("%5d [%.7f, %.7f] %.7f %+.7f %.7f\n",
                  iter, x_lo, x_hi,
                  r, r - r_expected,
                  x_hi - x_lo);
        }
      while (status == GSL_CONTINUE && iter < max_iter);

      gsl_root_fsolver_free (s);

      return status;
    }
```

Here are the results of the iterations,

```
    $ ./a.out
    using brent method
      iter [    lower,     upper]     root         err  err(est)
         1 [1.0000000, 5.0000000] 1.0000000 -1.2360680 4.0000000
         2 [1.0000000, 3.0000000] 3.0000000 +0.7639320 2.0000000
         3 [2.0000000, 3.0000000] 2.0000000 -0.2360680 1.0000000
```

```
    4 [2.2000000, 3.0000000] 2.2000000 -0.0360680 0.8000000
    5 [2.2000000, 2.2366300] 2.2366300 +0.0005621 0.0366300
Converged:
    6 [2.2360634, 2.2366300] 2.2360634 -0.0000046 0.0005666
```

If the program is modified to use the bisection solver instead of Brent's method, by changing `gsl_root_fsolver_brent` to `gsl_root_fsolver_bisection` the slower convergence of the Bisection method can be observed,

```
$ ./a.out
using bisection method
  iter [    lower,     upper]      root         err  err(est)
    1 [0.0000000, 2.5000000] 1.2500000 -0.9860680 2.5000000
    2 [1.2500000, 2.5000000] 1.8750000 -0.3610680 1.2500000
    3 [1.8750000, 2.5000000] 2.1875000 -0.0485680 0.6250000
    4 [2.1875000, 2.5000000] 2.3437500 +0.1076820 0.3125000
    5 [2.1875000, 2.3437500] 2.2656250 +0.0295570 0.1562500
    6 [2.1875000, 2.2656250] 2.2265625 -0.0095055 0.0781250
    7 [2.2265625, 2.2656250] 2.2460938 +0.0100258 0.0390625
    8 [2.2265625, 2.2460938] 2.2363281 +0.0002601 0.0195312
    9 [2.2265625, 2.2363281] 2.2314453 -0.0046227 0.0097656
   10 [2.2314453, 2.2363281] 2.2338867 -0.0021813 0.0048828
   11 [2.2338867, 2.2363281] 2.2351074 -0.0009606 0.0024414
Converged:
   12 [2.2351074, 2.2363281] 2.2357178 -0.0003502 0.0012207
```

The next program solves the same function using a derivative solver instead.

```c
#include <stdio.h>
#include <gsl/gsl_errno.h>
#include <gsl/gsl_math.h>
#include <gsl/gsl_roots.h>

#include "demo_fn.h"
#include "demo_fn.c"

int
main (void)
{
  int status;
  int iter = 0, max_iter = 100;
  const gsl_root_fdfsolver_type *T;
  gsl_root_fdfsolver *s;
  double x0, x = 5.0, r_expected = sqrt (5.0);
  gsl_function_fdf FDF;
  struct quadratic_params params = {1.0, 0.0, -5.0};

  FDF.f = &quadratic;
  FDF.df = &quadratic_deriv;
  FDF.fdf = &quadratic_fdf;
  FDF.params = &params;
```

```
T = gsl_root_fdfsolver_newton;
s = gsl_root_fdfsolver_alloc (T);
gsl_root_fdfsolver_set (s, &FDF, x);

printf ("using %s method\n",
        gsl_root_fdfsolver_name (s));

printf ("%-5s %10s %10s %10s\n",
        "iter", "root", "err", "err(est)");
do
  {
    iter++;
    status = gsl_root_fdfsolver_iterate (s);
    x0 = x;
    x = gsl_root_fdfsolver_root (s);
    status = gsl_root_test_delta (x, x0, 0, 1e-3);

    if (status == GSL_SUCCESS)
      printf ("Converged:\n");

    printf ("%5d %10.7f %+10.7f %10.7f\n",
            iter, x, x - r_expected, x - x0);
  }
while (status == GSL_CONTINUE && iter < max_iter);

gsl_root_fdfsolver_free (s);
return status;
}
```

Here are the results for Newton's method,

```
$ ./a.out
using newton method
iter        root        err    err(est)
    1   3.0000000  +0.7639320 -2.0000000
    2   2.3333333  +0.0972654 -0.6666667
    3   2.2380952  +0.0020273 -0.0952381
Converged:
    4   2.2360689  +0.0000009 -0.0020263
```

Note that the error can be estimated more accurately by taking the difference between the current iterate and next iterate rather than the previous iterate. The other derivative solvers can be investigated by changing gsl_root_fdfsolver_newton to gsl_root_fdfsolver_secant or gsl_root_fdfsolver_steffenson.

32.11 References and Further Reading

For information on the Brent-Dekker algorithm see the following two papers,

R. P. Brent, "An algorithm with guaranteed convergence for finding a zero of a function", *Computer Journal*, 14 (1971) 422–425

J. C. P. Bus and T. J. Dekker, "Two Efficient Algorithms with Guaranteed Convergence for Finding a Zero of a Function", *ACM Transactions of Mathematical Software*, Vol. 1 No. 4 (1975) 330–345

33 One dimensional Minimization

This chapter describes routines for finding minima of arbitrary one-dimensional functions. The library provides low level components for a variety of iterative minimizers and convergence tests. These can be combined by the user to achieve the desired solution, with full access to the intermediate steps of the algorithms. Each class of methods uses the same framework, so that you can switch between minimizers at runtime without needing to recompile your program. Each instance of a minimizer keeps track of its own state, allowing the minimizers to be used in multi-threaded programs.

The header file 'gsl_min.h' contains prototypes for the minimization functions and related declarations. To use the minimization algorithms to find the maximum of a function simply invert its sign.

33.1 Overview

The minimization algorithms begin with a bounded region known to contain a minimum. The region is described by a lower bound a and an upper bound b, with an estimate of the location of the minimum x.

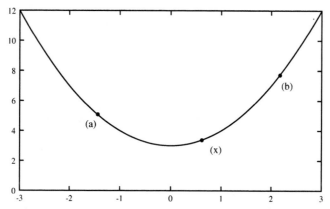

The value of the function at x must be less than the value of the function at the ends of the interval,

$$f(a) > f(x) < f(b)$$

This condition guarantees that a minimum is contained somewhere within the interval. On each iteration a new point x' is selected using one of the available algorithms. If the new point is a better estimate of the minimum, i.e. where $f(x') < f(x)$, then the current estimate of the minimum x is updated. The new point also allows the size of the bounded interval to be reduced, by choosing the most compact set of points which satisfies the constraint $f(a) > f(x) < f(b)$. The interval is reduced until it encloses the true minimum to a desired tolerance.

This provides a best estimate of the location of the minimum and a rigorous error estimate.

Several bracketing algorithms are available within a single framework. The user provides a high-level driver for the algorithm, and the library provides the individual functions necessary for each of the steps. There are three main phases of the iteration. The steps are,

- initialize minimizer state, s, for algorithm T
- update s using the iteration T
- test s for convergence, and repeat iteration if necessary

The state for the minimizers is held in a `gsl_min_fminimizer` struct. The updating procedure uses only function evaluations (not derivatives).

33.2 Caveats

Note that minimization functions can only search for one minimum at a time. When there are several minima in the search area, the first minimum to be found will be returned; however it is difficult to predict which of the minima this will be. *In most cases, no error will be reported if you try to find a minimum in an area where there is more than one.*

With all minimization algorithms it can be difficult to determine the location of the minimum to full numerical precision. The behavior of the function in the region of the minimum x^* can be approximated by a Taylor expansion,

$$y = f(x^*) + \frac{1}{2} f''(x^*)(x - x^*)^2$$

and the second term of this expansion can be lost when added to the first term at finite precision. This magnifies the error in locating x^*, making it proportional to $\sqrt{\epsilon}$ (where ϵ is the relative accuracy of the floating point numbers). For functions with higher order minima, such as x^4, the magnification of the error is correspondingly worse. The best that can be achieved is to converge to the limit of numerical accuracy in the function values, rather than the location of the minimum itself.

33.3 Initializing the Minimizer

`gsl_min_fminimizer * gsl_min_fminimizer_alloc (const`   *Function*
          `gsl_min_fminimizer_type * T)`

This function returns a pointer to a newly allocated instance of a minimizer of type T. For example, the following code creates an instance of a golden section minimizer,

```
const gsl_min_fminimizer_type * T
  = gsl_min_fminimizer_goldensection;
gsl_min_fminimizer * s
  = gsl_min_fminimizer_alloc (T);
```

If there is insufficient memory to create the minimizer then the function returns a null pointer and the error handler is invoked with an error code of `GSL_ENOMEM`.

int gsl_min_fminimizer_set (gsl_min_fminimizer * s,   *Function*
        gsl_function * f, double x_minimum, double x_lower, double
        x_upper)

This function sets, or resets, an existing minimizer s to use the function
f and the initial search interval [x_lower, x_upper], with a guess for the
location of the minimum x_minimum.

If the interval given does not contain a minimum, then the function returns
an error code of GSL_EINVAL.

int gsl_min_fminimizer_set_with_values (gsl_min_fminimizer *   *Function*
        s, gsl_function * f, double x_minimum, double f_minimum,
        double x_lower, double f_lower, double x_upper, double
        f_upper)

This function is equivalent to gsl_min_fminimizer_set but uses the values
f_minimum, f_lower and f_upper instead of computing f(x_minimum), f(x_
lower) and f(x_upper).

void gsl_min_fminimizer_free (gsl_min_fminimizer * s)   *Function*
This function frees all the memory associated with the minimizer s.

const char * gsl_min_fminimizer_name (const   *Function*
        gsl_min_fminimizer * s)

This function returns a pointer to the name of the minimizer. For example,

    printf ("s is a '%s' minimizer\n",
          gsl_min_fminimizer_name (s));

would print something like s is a 'brent' minimizer.

33.4 Providing the function to minimize

You must provide a continuous function of one variable for the minimizers to
operate on. In order to allow for general parameters the functions are defined
by a gsl_function data type (see Section 32.4 [Providing the function to solve],
page 391).

33.5 Iteration

The following functions drive the iteration of each algorithm. Each function
performs one iteration to update the state of any minimizer of the corresponding
type. The same functions work for all minimizers so that different methods can
be substituted at runtime without modifications to the code.

int gsl_min_fminimizer_iterate (gsl_min_fminimizer * s)   *Function*
This function performs a single iteration of the minimizer s. If the iteration
encounters an unexpected problem then an error code will be returned,

   GSL_EBADFUNC
       the iteration encountered a singular point where the function evaluated
       to Inf or NaN.

GSL_FAILURE
>   the algorithm could not improve the current best approximation or
>   bounding interval.

The minimizer maintains a current best estimate of the position of the minimum at all times, and the current interval bounding the minimum. This information can be accessed with the following auxiliary functions,

double gsl_min_fminimizer_x_minimum (const   Function
        gsl_min_fminimizer * s)
>   This function returns the current estimate of the position of the minimum
>   for the minimizer s.

double gsl_min_fminimizer_x_upper (const gsl_min_fminimizer *   Function
        s)
double gsl_min_fminimizer_x_lower (const gsl_min_fminimizer *   Function
        s)
>   These functions return the current upper and lower bound of the interval
>   for the minimizer s.

double gsl_min_fminimizer_f_minimum (const   Function
        gsl_min_fminimizer * s)
double gsl_min_fminimizer_f_upper (const gsl_min_fminimizer *   Function
        s)
double gsl_min_fminimizer_f_lower (const gsl_min_fminimizer *   Function
        s)
>   These functions return the value of the function at the current estimate of
>   the minimum and at the upper and lower bounds of the interval for the
>   minimizer s.

33.6 Stopping Parameters

A minimization procedure should stop when one of the following conditions
is true:

- A minimum has been found to within the user-specified precision.

- A user-specified maximum number of iterations has been reached.

- An error has occurred.

The handling of these conditions is under user control. The function below
allows the user to test the precision of the current result.

int gsl_min_test_interval (double x_lower, double x_upper,   Function
        double epsabs, double epsrel)
>   This function tests for the convergence of the interval $[x\_lower, x\_upper]$
>   with absolute error epsabs and relative error epsrel. The test returns GSL_
>   SUCCESS if the following condition is achieved,

$$|a - b| < epsabs + epsrel \min(|a|, |b|)$$

when the interval $x = [a, b]$ does not include the origin. If the interval includes the origin then $\min(|a|, |b|)$ is replaced by zero (which is the minimum value of $|x|$ over the interval). This ensures that the relative error is accurately estimated for minima close to the origin.

This condition on the interval also implies that any estimate of the minimum x_m in the interval satisfies the same condition with respect to the true minimum x_m^*,

$$|x_m - x_m^*| < epsabs + epsrel \, x_m^*$$

assuming that the true minimum x_m^* is contained within the interval.

33.7 Minimization Algorithms

The minimization algorithms described in this section require an initial interval which is guaranteed to contain a minimum—if a and b are the endpoints of the interval and x is an estimate of the minimum then $f(a) > f(x) < f(b)$. This ensures that the function has at least one minimum somewhere in the interval. If a valid initial interval is used then these algorithm cannot fail, provided the function is well-behaved.

`gsl_min_fminimizer_goldensection`   Minimizer

The *golden section algorithm* is the simplest method of bracketing the minimum of a function. It is the slowest algorithm provided by the library, with linear convergence.

On each iteration, the algorithm first compares the subintervals from the endpoints to the current minimum. The larger subinterval is divided in a golden section (using the famous ratio $(3 - \sqrt{5})/2 = 0.3189660\ldots$) and the value of the function at this new point is calculated. The new value is used with the constraint $f(a') > f(x') < f(b')$ to a select new interval containing the minimum, by discarding the least useful point. This procedure can be continued indefinitely until the interval is sufficiently small. Choosing the golden section as the bisection ratio can be shown to provide the fastest convergence for this type of algorithm.

`gsl_min_fminimizer_brent`   Minimizer

The *Brent minimization algorithm* combines a parabolic interpolation with the golden section algorithm. This produces a fast algorithm which is still robust.

The outline of the algorithm can be summarized as follows: on each iteration Brent's method approximates the function using an interpolating parabola through three existing points. The minimum of the parabola is taken as a guess for the minimum. If it lies within the bounds of the current interval then the interpolating point is accepted, and used to generate a smaller interval. If the interpolating point is not accepted then the algorithm falls back to an ordinary golden section step. The full details of Brent's method include some additional checks to improve convergence.

33.8 Examples

The following program uses the Brent algorithm to find the minimum of the function $f(x) = \cos(x) + 1$, which occurs at $x = \pi$. The starting interval is $(0, 6)$, with an initial guess for the minimum of 2.

```
#include <stdio.h>
#include <gsl/gsl_errno.h>
#include <gsl/gsl_math.h>
#include <gsl/gsl_min.h>

double fn1 (double x, void * params)
{
  return cos(x) + 1.0;
}

int
main (void)
{
  int status;
  int iter = 0, max_iter = 100;
  const gsl_min_fminimizer_type *T;
  gsl_min_fminimizer *s;
  double m = 2.0, m_expected = M_PI;
  double a = 0.0, b = 6.0;
  gsl_function F;

  F.function = &fn1;
  F.params = 0;

  T = gsl_min_fminimizer_brent;
  s = gsl_min_fminimizer_alloc (T);
  gsl_min_fminimizer_set (s, &F, m, a, b);

  printf ("using %s method\n",
          gsl_min_fminimizer_name (s));

  printf ("%5s [%9s, %9s] %9s %10s %9s\n",
          "iter", "lower", "upper", "min",
          "err", "err(est)");

  printf ("%5d [%.7f, %.7f] %.7f %+.7f %.7f\n",
          iter, a, b,
          m, m - m_expected, b - a);

  do
    {
      iter++;
      status = gsl_min_fminimizer_iterate (s);
```

```
        m = gsl_min_fminimizer_x_minimum (s);
        a = gsl_min_fminimizer_x_lower (s);
        b = gsl_min_fminimizer_x_upper (s);

        status
          = gsl_min_test_interval (a, b, 0.001, 0.0);

        if (status == GSL_SUCCESS)
          printf ("Converged:\n");

        printf ("%5d [%.7f, %.7f] "
                "%.7f %+.7f %.7f\n",
                iter, a, b,
                m, m - m_expected, b - a);
      }
    while (status == GSL_CONTINUE && iter < max_iter);

    gsl_min_fminimizer_free (s);

    return status;
  }
```

Here are the results of the minimization procedure.

```
$ ./a.out
     0 [0.0000000, 6.0000000] 2.0000000 -1.1415927 6.0000000
     1 [2.0000000, 6.0000000] 3.2758640 +0.1342713 4.0000000
     2 [2.0000000, 3.2831929] 3.2758640 +0.1342713 1.2831929
     3 [2.8689068, 3.2831929] 3.2758640 +0.1342713 0.4142862
     4 [2.8689068, 3.2831929] 3.2758640 +0.1342713 0.4142862
     5 [2.8689068, 3.2758640] 3.1460585 +0.0044658 0.4069572
     6 [3.1346075, 3.2758640] 3.1460585 +0.0044658 0.1412565
     7 [3.1346075, 3.1874620] 3.1460585 +0.0044658 0.0528545
     8 [3.1346075, 3.1460585] 3.1460585 +0.0044658 0.0114510
     9 [3.1346075, 3.1460585] 3.1424060 +0.0008133 0.0114510
    10 [3.1346075, 3.1424060] 3.1415885 -0.0000041 0.0077985
Converged:
    11 [3.1415885, 3.1424060] 3.1415927 -0.0000000 0.0008175
```

33.9 References and Further Reading

Further information on Brent's algorithm is available in the following book,

Richard Brent, *Algorithms for minimization without derivatives*, Prentice-Hall (1973), republished by Dover in paperback (2002), ISBN 0-486-41998-3.

34 Multidimensional Root-Finding

This chapter describes functions for multidimensional root-finding (solving nonlinear systems with n equations in n unknowns). The library provides low level components for a variety of iterative solvers and convergence tests. These can be combined by the user to achieve the desired solution, with full access to the intermediate steps of the iteration. Each class of methods uses the same framework, so that you can switch between solvers at runtime without needing to recompile your program. Each instance of a solver keeps track of its own state, allowing the solvers to be used in multi-threaded programs. The solvers are based on the original Fortran library MINPACK.

The header file 'gsl_multiroots.h' contains prototypes for the multidimensional root finding functions and related declarations.

34.1 Overview

The problem of multidimensional root finding requires the simultaneous solution of n equations, f_i, in n variables, x_i,

$$f_i(x_1, \ldots, x_n) = 0 \qquad \text{for } i = 1 \ldots n.$$

In general there are no bracketing methods available for n dimensional systems, and no way of knowing whether any solutions exist. All algorithms proceed from an initial guess using a variant of the Newton iteration,

$$x \to x' = x - J^{-1} f(x)$$

where x, f are vector quantities and J is the Jacobian matrix $J_{ij} = \partial f_i / \partial x_j$. Additional strategies can be used to enlarge the region of convergence. These include requiring a decrease in the norm $|f|$ on each step proposed by Newton's method, or taking steepest-descent steps in the direction of the negative gradient of $|f|$.

Several root-finding algorithms are available within a single framework. The user provides a high-level driver for the algorithms, and the library provides the individual functions necessary for each of the steps. There are three main phases of the iteration. The steps are,

- initialize solver state, s, for algorithm T
- update s using the iteration T
- test s for convergence, and repeat iteration if necessary

The evaluation of the Jacobian matrix can be problematic, either because programming the derivatives is intractable or because computation of the n^2 terms of the matrix becomes too expensive. For these reasons the algorithms provided by the library are divided into two classes according to whether the derivatives are available or not.

The state for solvers with an analytic Jacobian matrix is held in a gsl_multiroot_fdfsolver struct. The updating procedure requires both the function and its derivatives to be supplied by the user.

The state for solvers which do not use an analytic Jacobian matrix is held in a `gsl_multiroot_fsolver` struct. The updating procedure uses only function evaluations (not derivatives). The algorithms estimate the matrix J or J^{-1} by approximate methods.

34.2 Initializing the Solver

The following functions initialize a multidimensional solver, either with or without derivatives. The solver itself depends only on the dimension of the problem and the algorithm and can be reused for different problems.

`gsl_multiroot_fsolver * gsl_multiroot_fsolver_alloc (const`   *Function*
    `gsl_multiroot_fsolver_type * T, size_t n)`
  This function returns a pointer to a newly allocated instance of a solver of type T for a system of n dimensions. For example, the following code creates an instance of a hybrid solver, to solve a 3-dimensional system of equations.

    const gsl_multiroot_fsolver_type * T
      = gsl_multiroot_fsolver_hybrid;
    gsl_multiroot_fsolver * s
      = gsl_multiroot_fsolver_alloc (T, 3);

  If there is insufficient memory to create the solver then the function returns a null pointer and the error handler is invoked with an error code of GSL_ENOMEM.

`gsl_multiroot_fdfsolver * gsl_multiroot_fdfsolver_alloc`   *Function*
    `(const gsl_multiroot_fdfsolver_type * T, size_t n)`
  This function returns a pointer to a newly allocated instance of a derivative solver of type T for a system of n dimensions. For example, the following code creates an instance of a Newton-Raphson solver, for a 2-dimensional system of equations.

    const gsl_multiroot_fdfsolver_type * T
      = gsl_multiroot_fdfsolver_newton;
    gsl_multiroot_fdfsolver * s =
      gsl_multiroot_fdfsolver_alloc (T, 2);

  If there is insufficient memory to create the solver then the function returns a null pointer and the error handler is invoked with an error code of GSL_ENOMEM.

`int gsl_multiroot_fsolver_set (gsl_multiroot_fsolver * s,`   *Function*
    `gsl_multiroot_function * f, const gsl_vector * x)`
`int gsl_multiroot_fdfsolver_set (gsl_multiroot_fdfsolver * s,`   *Function*
    `gsl_multiroot_function_fdf * fdf, const gsl_vector * x)`
  These functions set, or reset, an existing solver s to use the function f or function and derivative fdf, and the initial guess x. Note that the initial position is copied from x, this argument is not modified by subsequent iterations.

```
void gsl_multiroot_fsolver_free (gsl_multiroot_fsolver * s)        Function
void gsl_multiroot_fdfsolver_free (gsl_multiroot_fdfsolver *       Function
     s)
```
These functions free all the memory associated with the solver s.

```
const char * gsl_multiroot_fsolver_name (const                     Function
     gsl_multiroot_fsolver * s)
const char * gsl_multiroot_fdfsolver_name (const                   Function
     gsl_multiroot_fdfsolver * s)
```
These functions return a pointer to the name of the solver. For example,

```
printf ("s is a '%s' solver\n",
        gsl_multiroot_fdfsolver_name (s));
```
would print something like s is a 'newton' solver.

34.3 Providing the function to solve

You must provide n functions of n variables for the root finders to operate on. In order to allow for general parameters the functions are defined by the following data types:

```
gsl_multiroot_function                                             Data Type
```
This data type defines a general system of functions with parameters.

```
int (* f) (const gsl_vector * x, void * params, gsl_vector * f)
```
this function should store the vector result $f(x, params)$ in f for argument x and parameters params, returning an appropriate error code if the function cannot be computed.

```
size_t n
```
the dimension of the system, i.e. the number of components of the vectors x and f.

```
void * params
```
a pointer to the parameters of the function.

Here is an example using Powell's test function,

$$f_1(x) = Ax_0 x_1 - 1, f_2(x) = \exp(-x_0) + \exp(-x_1) - (1 + 1/A)$$

with $A = 10^4$. The following code defines a gsl_multiroot_function system F which you could pass to a solver:

```
struct powell_params { double A; };

int
powell (gsl_vector * x, void * p, gsl_vector * f) {
    struct powell_params * params
      = *(struct powell_params *)p;
    const double A = (params->A);
    const double x0 = gsl_vector_get(x,0);
    const double x1 = gsl_vector_get(x,1);
```

```
    gsl_vector_set (f, 0, A * x0 * x1 - 1);
    gsl_vector_set (f, 1, (exp(-x0) + exp(-x1)
                           - (1.0 + 1.0/A)));
    return GSL_SUCCESS
}


gsl_multiroot_function F;
struct powell_params params = { 10000.0 };

F.f = &powell;
F.n = 2;
F.params = &params;
```

gsl_multiroot_function_fdf   Data Type

This data type defines a general system of functions with parameters and
the corresponding Jacobian matrix of derivatives,

int (* f) (const gsl_vector * x, void * params, gsl_vector * f)
> this function should store the vector result $f(x, params)$ in f for argu-
> ment x and parameters params, returning an appropriate error code if
> the function cannot be computed.

int (* df) (const gsl_vector * x, void * params, gsl_matrix * J)
> this function should store the n-by-n Jacobian matrix result $J_{ij} = \partial f_i(x, params)/\partial x_j$ in J for argument x and parameters params, re-
> turning an appropriate error code if the function cannot be computed.

int (* fdf) (const gsl_vector * x, void * params, gsl_vector * f,
gsl_matrix * J)
> This function should set the values of the f and J as above, for argu-
> ments x and parameters params. This function provides an optimiza-
> tion of the separate functions for $f(x)$ and $J(x)$—it is always faster to
> compute the function and its derivative at the same time.

size_t n
> the dimension of the system, i.e. the number of components of the
> vectors x and f.

void * params
> a pointer to the parameters of the function.

The example of Powell's test function defined above can be extended to include
analytic derivatives using the following code,

```
    int
    powell_df (gsl_vector * x, void * p, gsl_matrix * J)
    {
        struct powell_params * params
          = *(struct powell_params *)p;
        const double A = (params->A);
        const double x0 = gsl_vector_get(x,0);
        const double x1 = gsl_vector_get(x,1);
```

```
        gsl_matrix_set (J, 0, 0, A * x1);
        gsl_matrix_set (J, 0, 1, A * x0);
        gsl_matrix_set (J, 1, 0, -exp(-x0));
        gsl_matrix_set (J, 1, 1, -exp(-x1));
        return GSL_SUCCESS
    }


    int
    powell_fdf (gsl_vector * x, void * p,
                gsl_matrix * f, gsl_matrix * J) {
        struct powell_params * params
          = *(struct powell_params *)p;
        const double A = (params->A);
        const double x0 = gsl_vector_get(x,0);
        const double x1 = gsl_vector_get(x,1);

        const double u0 = exp(-x0);
        const double u1 = exp(-x1);

        gsl_vector_set (f, 0, A * x0 * x1 - 1);
        gsl_vector_set (f, 1, u0 + u1 - (1 + 1/A));

        gsl_matrix_set (J, 0, 0, A * x1);
        gsl_matrix_set (J, 0, 1, A * x0);
        gsl_matrix_set (J, 1, 0, -u0);
        gsl_matrix_set (J, 1, 1, -u1);
        return GSL_SUCCESS
    }


    gsl_multiroot_function_fdf FDF;

    FDF.f = &powell_f;
    FDF.df = &powell_df;
    FDF.fdf = &powell_fdf;
    FDF.n = 2;
    FDF.params = 0;
```

Note that the function powell_fdf is able to reuse existing terms from the
function when calculating the Jacobian, thus saving time.

34.4 Iteration

The following functions drive the iteration of each algorithm. Each function performs one iteration to update the state of any solver of the corresponding type. The same functions work for all solvers so that different methods can be substituted at runtime without modifications to the code.

int gsl_multiroot_fsolver_iterate (gsl_multiroot_fsolver * s)   Function
int gsl_multiroot_fdfsolver_iterate (gsl_multiroot_fdfsolver   Function
      * s)

These functions perform a single iteration of the solver s. If the iteration encounters an unexpected problem then an error code will be returned,

GSL_EBADFUNC

the iteration encountered a singular point where the function or its derivative evaluated to Inf or NaN.

GSL_ENOPROG

the iteration is not making any progress, preventing the algorithm from continuing.

The solver maintains a current best estimate of the root s->x and its function value s->f at all times. This information can be accessed with the following auxiliary functions,

gsl_vector * gsl_multiroot_fsolver_root (const   Function
      gsl_multiroot_fsolver * s)
gsl_vector * gsl_multiroot_fdfsolver_root (const   Function
      gsl_multiroot_fdfsolver * s)

These functions return the current estimate of the root for the solver s, given by s->x.

gsl_vector * gsl_multiroot_fsolver_f (const   Function
      gsl_multiroot_fsolver * s)
gsl_vector * gsl_multiroot_fdfsolver_f (const   Function
      gsl_multiroot_fdfsolver * s)

These functions return the function value $f(x)$ at the current estimate of the root for the solver s, given by s->f.

gsl_vector * gsl_multiroot_fsolver_dx (const   Function
      gsl_multiroot_fsolver * s)
gsl_vector * gsl_multiroot_fdfsolver_dx (const   Function
      gsl_multiroot_fdfsolver * s)

These functions return the last step dx taken by the solver s, given by s->dx.

34.5 Search Stopping Parameters

A root finding procedure should stop when one of the following conditions is true:

- A multidimensional root has been found to within the user-specified precision.

- A user-specified maximum number of iterations has been reached.

- An error has occurred.

The handling of these conditions is under user control. The functions below allow the user to test the precision of the current result in several standard ways.

int gsl_multiroot_test_delta (const gsl_vector * dx, const   Function
        gsl_vector * x, double $epsabs$, double $epsrel$)
This function tests for the convergence of the sequence by comparing the last step dx with the absolute error $epsabs$ and relative error $epsrel$ to the current position x. The test returns GSL_SUCCESS if the following condition is achieved,

$$|dx_i| < epsabs + epsrel\,|x_i|$$

for each component of x and returns GSL_CONTINUE otherwise.

int gsl_multiroot_test_residual (const gsl_vector * f, double   Function
        $epsabs$)
This function tests the residual value f against the absolute error bound $epsabs$. The test returns GSL_SUCCESS if the following condition is achieved,

$$\sum_i |f_i| < epsabs$$

and returns GSL_CONTINUE otherwise. This criterion is suitable for situations where the precise location of the root, x, is unimportant provided a value can be found where the residual is small enough.

34.6 Algorithms using Derivatives

The root finding algorithms described in this section make use of both the function and its derivative. They require an initial guess for the location of the root, but there is no absolute guarantee of convergence—the function must be suitable for this technique and the initial guess must be sufficiently close to the root for it to work. When the conditions are satisfied then convergence is quadratic.

`gsl_multiroot_fdfsolver_hybridsj`   Derivative Solver

This is a modified version of Powell's Hybrid method as implemented in the HYBRJ algorithm in MINPACK. Minpack was written by Jorge J. Moré, Burton S. Garbow and Kenneth E. Hillstrom. The Hybrid algorithm retains the fast convergence of Newton's method but will also reduce the residual when Newton's method is unreliable.

The algorithm uses a generalized trust region to keep each step under control. In order to be accepted a proposed new position x' must satisfy the condition $|D(x' - x)| < \delta$, where D is a diagonal scaling matrix and δ is the size of the trust region. The components of D are computed internally, using the column norms of the Jacobian to estimate the sensitivity of the residual to each component of x. This improves the behavior of the algorithm for badly scaled functions.

On each iteration the algorithm first determines the standard Newton step by solving the system $Jdx = -f$. If this step falls inside the trust region it is used as a trial step in the next stage. If not, the algorithm uses the linear combination of the Newton and gradient directions which is predicted to minimize the norm of the function while staying inside the trust region,

$$dx = -\alpha J^{-1} f(x) - \beta \nabla |f(x)|^2.$$

This combination of Newton and gradient directions is referred to as a *dogleg step*.

The proposed step is now tested by evaluating the function at the resulting point, x'. If the step reduces the norm of the function sufficiently then it is accepted and size of the trust region is increased. If the proposed step fails to improve the solution then the size of the trust region is decreased and another trial step is computed.

The speed of the algorithm is increased by computing the changes to the Jacobian approximately, using a rank-1 update. If two successive attempts fail to reduce the residual then the full Jacobian is recomputed. The algorithm also monitors the progress of the solution and returns an error if several steps fail to make any improvement,

`GSL_ENOPROG`

the iteration is not making any progress, preventing the algorithm from continuing.

`GSL_ENOPROGJ`

re-evaluations of the Jacobian indicate that the iteration is not making any progress, preventing the algorithm from continuing.

gsl_multiroot_fdfsolver_hybridj   Derivative Solver
  This algorithm is an unscaled version of hybridsj. The steps are controlled
  by a spherical trust region $|x' - x| < \delta$, instead of a generalized region. This
  can be useful if the generalized region estimated by hybridsj is inappropriate.

gsl_multiroot_fdfsolver_newton   Derivative Solver
  Newton's Method is the standard root-polishing algorithm. The algorithm
  begins with an initial guess for the location of the solution. On each iteration
  a linear approximation to the function F is used to estimate the step which
  will zero all the components of the residual. The iteration is defined by the
  following sequence,

  $$x \to x' = x - J^{-1} f(x)$$

  where the Jacobian matrix J is computed from the derivative functions
  provided by f. The step dx is obtained by solving the linear system,

  $$J\, dx = -f(x)$$

  using LU decomposition.

gsl_multiroot_fdfsolver_gnewton   Derivative Solver
  This is a modified version of Newton's method which attempts to improve
  global convergence by requiring every step to reduce the Euclidean norm of
  the residual, $|f(x)|$. If the Newton step leads to an increase in the norm
  then a reduced step of relative size,

  $$t = (\sqrt{1 + 6r} - 1)/(3r)$$

  is proposed, with r being the ratio of norms $|f(x')|^2/|f(x)|^2$. This procedure
  is repeated until a suitable step size is found.

34.7 Algorithms without Derivatives

  The algorithms described in this section do not require any derivative information to be supplied by the user. Any derivatives needed are approximated
  by finite differences. Note that if the finite-differencing step size chosen by
  these routines is inappropriate, an explicit user-supplied numerical derivative
  can always be used with the algorithms described in the previous section.

gsl_multiroot_fsolver_hybrids   Solver
  This is a version of the Hybrid algorithm which replaces calls to the Jacobian
  function by its finite difference approximation. The finite difference approximation is computed using gsl_multiroots_fdjac with a relative step size
  of GSL_SQRT_DBL_EPSILON. Note that this step size will not be suitable for
  all problems.

gsl_multiroot_fsolver_hybrid   .   Solver
  This is a finite difference version of the Hybrid algorithm without internal
  scaling.

`gsl_multiroot_fsolver_dnewton`   Solver

The *discrete Newton algorithm* is the simplest method of solving a multidimensional system. It uses the Newton iteration

$$x \to x - J^{-1} f(x)$$

where the Jacobian matrix J is approximated by taking finite differences of the function f. The approximation scheme used by this implementation is,

$$J_{ij} = (f_i(x + \delta_j) - f_i(x))/\delta_j$$

where δ_j is a step of size $\sqrt{\epsilon}|x_j|$ with ϵ being the machine precision ($\epsilon \approx 2.22 \times 10^{-16}$). The order of convergence of Newton's algorithm is quadratic, but the finite differences require n^2 function evaluations on each iteration. The algorithm may become unstable if the finite differences are not a good approximation to the true derivatives.

`gsl_multiroot_fsolver_broyden`   Solver

The *Broyden algorithm* is a version of the discrete Newton algorithm which attempts to avoids the expensive update of the Jacobian matrix on each iteration. The changes to the Jacobian are also approximated, using a rank-1 update,

$$J^{-1} \to J^{-1} - (J^{-1}df - dx)dx^T J^{-1}/dx^T J^{-1} df$$

where the vectors dx and df are the changes in x and f. On the first iteration the inverse Jacobian is estimated using finite differences, as in the discrete Newton algorithm.

This approximation gives a fast update but is unreliable if the changes are not small, and the estimate of the inverse Jacobian becomes worse as time passes. The algorithm has a tendency to become unstable unless it starts close to the root. The Jacobian is refreshed if this instability is detected (consult the source for details).

This algorithm is included only for demonstration purposes, and is not recommended for serious use.

34.8 Examples

The multidimensional solvers are used in a similar way to the one-dimensional root finding algorithms. This first example demonstrates the `hybrids` scaled-hybrid algorithm, which does not require derivatives. The program solves the Rosenbrock system of equations,

$$f_1(x,y) = a(1 - x), \quad f_2(x,y) = b(y - x^2)$$

with $a = 1, b = 10$. The solution of this system lies at $(x, y) = (1, 1)$ in a narrow valley.

The first stage of the program is to define the system of equations,

```
#include <stdlib.h>
#include <stdio.h>
#include <gsl/gsl_vector.h>
#include <gsl/gsl_multiroots.h>

struct rparams
  {
    double a;
    double b;
  };

int
rosenbrock_f (const gsl_vector * x, void *params,
              gsl_vector * f)
{
  double a = ((struct rparams *) params)->a;
  double b = ((struct rparams *) params)->b;

  const double x0 = gsl_vector_get (x, 0);
  const double x1 = gsl_vector_get (x, 1);

  const double y0 = a * (1 - x0);
  const double y1 = b * (x1 - x0 * x0);

  gsl_vector_set (f, 0, y0);
  gsl_vector_set (f, 1, y1);

  return GSL_SUCCESS;
}
```

The main program begins by creating the function object f, with the arguments (x,y) and parameters (a,b). The solver s is initialized to use this function, with the hybrids method.

```
int
main (void)
{
  const gsl_multiroot_fsolver_type *T;
  gsl_multiroot_fsolver *s;

  int status;
  size_t i, iter = 0;

  const size_t n = 2;
  struct rparams p = {1.0, 10.0};
  gsl_multiroot_function f = {&rosenbrock_f, n, &p};

  double x_init[2] = {-10.0, -5.0};
  gsl_vector *x = gsl_vector_alloc (n);
```

```
gsl_vector_set (x, 0, x_init[0]);
gsl_vector_set (x, 1, x_init[1]);

T = gsl_multiroot_fsolver_hybrids;
s = gsl_multiroot_fsolver_alloc (T, 2);
gsl_multiroot_fsolver_set (s, &f, x);

print_state (iter, s);

do
  {
    iter++;
    status = gsl_multiroot_fsolver_iterate (s);

    print_state (iter, s);

    if (status)    /* check if solver is stuck */
      break;

    status =
      gsl_multiroot_test_residual (s->f, 1e-7);
  }
while (status == GSL_CONTINUE && iter < 1000);

printf ("status = %s\n", gsl_strerror (status));

gsl_multiroot_fsolver_free (s);
gsl_vector_free (x);
return 0;
}
```

Note that it is important to check the return status of each solver step, in case
the algorithm becomes stuck. If an error condition is detected, indicating that
the algorithm cannot proceed, then the error can be reported to the user, a new
starting point chosen or a different algorithm used.

The intermediate state of the solution is displayed by the following function.
The solver state contains the vector s->x which is the current position, and the
vector s->f with corresponding function values.

```
int
print_state (size_t iter, gsl_multiroot_fsolver * s)
{
  printf ("iter = %3u x = % .3f % .3f "
          "f(x) = % .3e % .3e\n",
          iter,
          gsl_vector_get (s->x, 0),
          gsl_vector_get (s->x, 1),
          gsl_vector_get (s->f, 0),
          gsl_vector_get (s->f, 1));
}
```

Here are the results of running the program. The algorithm is started at $(-10, -5)$ far from the solution. Since the solution is hidden in a narrow valley the earliest steps follow the gradient of the function downhill, in an attempt to reduce the large value of the residual. Once the root has been approximately located, on iteration 8, the Newton behavior takes over and convergence is very rapid.

```
iter =  0 x = -10.000   -5.000   f(x) = 1.100e+01 -1.050e+03
iter =  1 x = -10.000   -5.000   f(x) = 1.100e+01 -1.050e+03
iter =  2 x =  -3.976   24.827   f(x) = 4.976e+00  9.020e+01
iter =  3 x =  -3.976   24.827   f(x) = 4.976e+00  9.020e+01
iter =  4 x =  -3.976   24.827   f(x) = 4.976e+00  9.020e+01
iter =  5 x =  -1.274   -5.680   f(x) = 2.274e+00 -7.302e+01
iter =  6 x =  -1.274   -5.680   f(x) = 2.274e+00 -7.302e+01
iter =  7 x =   0.249    0.298   f(x) = 7.511e-01  2.359e+00
iter =  8 x =   0.249    0.298   f(x) = 7.511e-01  2.359e+00
iter =  9 x =   1.000    0.878   f(x) = 1.268e-10 -1.218e+00
iter = 10 x =   1.000    0.989   f(x) = 1.124e-11 -1.080e-01
iter = 11 x =   1.000    1.000   f(x) = 0.000e+00  0.000e+00
status = success
```

Note that the algorithm does not update the location on every iteration. Some iterations are used to adjust the trust-region parameter, after trying a step which was found to be divergent, or to recompute the Jacobian, when poor convergence behavior is detected.

The next example program adds derivative information, in order to accelerate the solution. There are two derivative functions rosenbrock_df and rosenbrock_fdf. The latter computes both the function and its derivative simultaneously. This allows the optimization of any common terms. For simplicity we substitute calls to the separate f and df functions at this point in the code below.

```
int
rosenbrock_df (const gsl_vector * x, void *params,
               gsl_matrix * J)
{
  const double a = ((struct rparams *) params)->a;
  const double b = ((struct rparams *) params)->b;

  const double x0 = gsl_vector_get (x, 0);

  const double df00 = -a;
  const double df01 = 0;
  const double df10 = -2 * b  * x0;
  const double df11 = b;

  gsl_matrix_set (J, 0, 0, df00);
  gsl_matrix_set (J, 0, 1, df01);
  gsl_matrix_set (J, 1, 0, df10);
  gsl_matrix_set (J, 1, 1, df11);
```

```
    return GSL_SUCCESS;
}

int
rosenbrock_fdf (const gsl_vector * x, void *params,
                gsl_vector * f, gsl_matrix * J)
{
  rosenbrock_f (x, params, f);
  rosenbrock_df (x, params, J);

  return GSL_SUCCESS;
}
```

The main program now makes calls to the corresponding fdfsolver versions of the functions,

```
int
main (void)
{
  const gsl_multiroot_fdfsolver_type *T;
  gsl_multiroot_fdfsolver *s;

  int status;
  size_t i, iter = 0;

  const size_t n = 2;
  struct rparams p = {1.0, 10.0};
  gsl_multiroot_function_fdf f = {&rosenbrock_f,
                                  &rosenbrock_df,
                                  &rosenbrock_fdf,
                                  n, &p};

  double x_init[2] = {-10.0, -5.0};
  gsl_vector *x = gsl_vector_alloc (n);

  gsl_vector_set (x, 0, x_init[0]);
  gsl_vector_set (x, 1, x_init[1]);

  T = gsl_multiroot_fdfsolver_gnewton;
  s = gsl_multiroot_fdfsolver_alloc (T, n);
  gsl_multiroot_fdfsolver_set (s, &f, x);

  print_state (iter, s);

  do
    {
      iter++;

      status = gsl_multiroot_fdfsolver_iterate (s);
```

```
    print_state (iter, s);

    if (status)
      break;

    status = gsl_multiroot_test_residual (s->f, 1e-7);
  }
while (status == GSL_CONTINUE && iter < 1000);

printf ("status = %s\n", gsl_strerror (status));

gsl_multiroot_fdfsolver_free (s);
gsl_vector_free (x);
return 0;
}
```

The addition of derivative information to the hybrids solver does not make any significant difference to its behavior, since it able to approximate the Jacobian numerically with sufficient accuracy. To illustrate the behavior of a different derivative solver we switch to gnewton. This is a traditional Newton solver with the constraint that it scales back its step if the full step would lead "uphill". Here is the output for the gnewton algorithm,

```
iter = 0 x = -10.000  -5.000 f(x) =  1.100e+01 -1.050e+03
iter = 1 x =  -4.231 -65.317 f(x) =  5.231e+00 -8.321e+02
iter = 2 x =   1.000 -26.358 f(x) = -8.882e-16 -2.736e+02
iter = 3 x =   1.000   1.000 f(x) = -2.220e-16 -4.441e-15
status = success
```

The convergence is much more rapid, but takes a wide excursion out to the point $(-4.23, -65.3)$. This could cause the algorithm to go astray in a realistic application. The hybrid algorithm follows the downhill path to the solution more reliably.

34.9 References and Further Reading

The original version of the Hybrid method is described in the following articles by Powell,

M.J.D. Powell, "A Hybrid Method for Nonlinear Equations" (Chap 6, p 87–114) and "A Fortran Subroutine for Solving systems of Nonlinear Algebraic Equations" (Chap 7, p 115–161), in *Numerical Methods for Nonlinear Algebraic Equations*, P. Rabinowitz, editor. Gordon and Breach, 1970.

The following papers are also relevant to the algorithms described in this section,

J.J. Moré, M.Y. Cosnard, "Numerical Solution of Nonlinear Equations", *ACM Transactions on Mathematical Software*, Vol 5, No 1, (1979), p 64–85

C.G. Broyden, "A Class of Methods for Solving Nonlinear Simultaneous Equations", *Mathematics of Computation*, Vol 19 (1965), p 577–593

J.J. Moré, B.S. Garbow, K.E. Hillstrom, "Testing Unconstrained Optimization Software", ACM Transactions on Mathematical Software, Vol 7, No 1 (1981), p 17–41

35 Multidimensional Minimization

This chapter describes routines for finding minima of arbitrary multidimensional functions. The library provides low level components for a variety of iterative minimizers and convergence tests. These can be combined by the user to achieve the desired solution, while providing full access to the intermediate steps of the algorithms. Each class of methods uses the same framework, so that you can switch between minimizers at runtime without needing to recompile your program. Each instance of a minimizer keeps track of its own state, allowing the minimizers to be used in multi-threaded programs. The minimization algorithms can be used to maximize a function by inverting its sign.

The header file 'gsl_multimin.h' contains prototypes for the minimization functions and related declarations.

35.1 Overview

The problem of multidimensional minimization requires finding a point x such that the scalar function,

$$f(x_1, \ldots, x_n)$$

takes a value which is lower than at any neighboring point. For smooth functions the gradient $g = \nabla f$ vanishes at the minimum. In general there are no bracketing methods available for the minimization of n-dimensional functions. The algorithms proceed from an initial guess using a search algorithm which attempts to move in a downhill direction.

Algorithms making use of the gradient of the function perform a one-dimensional line minimisation along this direction until the lowest point is found to a suitable tolerance. The search direction is then updated with local information from the function and its derivatives, and the whole process repeated until the true n-dimensional minimum is found.

Algorithms which do not require the gradient of the function use different strategies. For example, the Nelder-Mead Simplex algorithm maintains $n + 1$ trial parameter vectors as the vertices of a n-dimensional simplex. On each iteration it tries to improve the worst vertex of the simplex by geometrical transformations. The iterations are continued until the overall size of the simplex has decreased sufficiently.

Both types of algorithms use a standard framework. The user provides a high-level driver for the algorithms, and the library provides the individual functions necessary for each of the steps. There are three main phases of the iteration. The steps are,

- initialize minimizer state, s, for algorithm T
- update s using the iteration T
- test s for convergence, and repeat iteration if necessary

Each iteration step consists either of an improvement to the line-minimisation in the current direction or an update to the search direction itself. The state for the minimizers is held in a `gsl_multimin_fdfminimizer` struct or a `gsl_multimin_fminimizer` struct.

35.2 Caveats

Note that the minimization algorithms can only search for one local minimum at a time. When there are several local minima in the search area, the first minimum to be found will be returned; however it is difficult to predict which of the minima this will be. In most cases, no error will be reported if you try to find a local minimum in an area where there is more than one.

It is also important to note that the minimization algorithms find local minima; there is no way to determine whether a minimum is a global minimum of the function in question.

35.3 Initializing the Multidimensional Minimizer

The following function initializes a multidimensional minimizer. The minimizer itself depends only on the dimension of the problem and the algorithm and can be reused for different problems.

`gsl_multimin_fdfminimizer * gsl_multimin_fdfminimizer_alloc`   Function
      `(const gsl_multimin_fdfminimizer_type * T, size_t n)`
`gsl_multimin_fminimizer * gsl_multimin_fminimizer_alloc`   Function
      `(const gsl_multimin_fminimizer_type * T, size_t n)`
  This function returns a pointer to a newly allocated instance of a minimizer of type T for an n-dimension function. If there is insufficient memory to create the minimizer then the function returns a null pointer and the error handler is invoked with an error code of `GSL_ENOMEM`.

`int gsl_multimin_fdfminimizer_set (gsl_multimin_fdfminimizer`   Function
      `* s, gsl_multimin_function_fdf * fdf, const gsl_vector * x,`
      `double step_size, double tol)`
  This function initializes the minimizer s to minimize the function fdf starting from the initial point x. The size of the first trial step is given by $step\_size$. The accuracy of the line minimization is specified by tol. The precise meaning of this parameter depends on the method used. Typically the line minimization is considered successful if the gradient of the function g is orthogonal to the current search direction p to a relative accuracy of tol, where $p \cdot g < tol|p||g|$. A tol value of 0.1 is suitable for most purposes, since line minimization only needs to be carried out approximately. Note that setting tol to zero will force the use of "exact" line-searches, which are extremely expensive.

`int gsl_multimin_fminimizer_set (gsl_multimin_fminimizer * s,`   Function
      `gsl_multimin_function * f, const gsl_vector * x, const`
      `gsl_vector * step_size)`
  This function initializes the minimizer s to minimize the function f, starting from the initial point x. The size of the initial trial steps is given in vector

step_size. The precise meaning of this parameter depends on the method used.

void **gsl_multimin_fdfminimizer_free**   *Function*
        (gsl_multimin_fdfminimizer * s)
void **gsl_multimin_fminimizer_free** (gsl_multimin_fminimizer *   *Function*
        s)
        This function frees all the memory associated with the minimizer s.

const char * **gsl_multimin_fdfminimizer_name** (const   *Function*
        gsl_multimin_fdfminimizer * s)
const char * **gsl_multimin_fminimizer_name** (const   *Function*
        gsl_multimin_fminimizer * s)
        This function returns a pointer to the name of the minimizer. For example,

        printf ("s is a '%s' minimizer\n",
                gsl_multimin_fdfminimizer_name (s));

would print something like s is a 'conjugate_pr' minimizer.

35.4 Providing a function to minimize

You must provide a parametric function of n variables for the minimizers to operate on. You may also need to provide a routine which calculates the gradient of the function and a third routine which calculates both the function value and the gradient together. In order to allow for general parameters the functions are defined by the following data types:

gsl_multimin_function_fdf   *Data Type*
        This data type defines a general function of n variables with parameters and the corresponding gradient vector of derivatives,

        double (* f) (const gsl_vector * x, void * *params*)
                this function should return the result $f(x, params)$ for argument x and parameters *params*. If the function cannot be computed, an error value of GSL_NAN should be returned.

        void (* df) (const gsl_vector * x, void * *params*, gsl_vector * g)
                this function should store the n-dimensional gradient result $g_i = \partial f(x, params)/\partial x_i$ in the vector g for argument x and parameters *params*, returning an appropriate error code if the function cannot be computed.

        void (* fdf) (const gsl_vector * x, void * *params*, double * f,
        gsl_vector * g)
                This function should set the values of the f and g as above, for arguments x and parameters *params*. This function provides an optimization of the separate functions for $f(x)$ and $g(x)$—it is always faster to compute the function and its derivative at the same time.

        size_t n
                the dimension of the system, i.e. the number of components of the vectors x.

void * params
    a pointer to the parameters of the function.

gsl_multimin_function   Data Type
    This data type defines a general function of n variables with parameters,

double (* f) (const gsl_vector * x, void * params)
    this function should return the result $f(x, params)$ for argument x and
    parameters params. If the function cannot be computed, an error value
    of GSL_NAN should be returned.

size_t n
    the dimension of the system, i.e. the number of components of the
    vectors x.

void * params
    a pointer to the parameters of the function.

The following example function defines a simple two-dimensional paraboloid
with five parameters,

```
/* Paraboloid centered on (p[0],p[1]), with
   scale factors (p[2],p[3]) and minimum p[4] */

double
my_f (const gsl_vector *v, void *params)
{
  double x, y;
  double *p = (double *)params;

  x = gsl_vector_get(v, 0);
  y = gsl_vector_get(v, 1);

  return p[2] * (x - p[0]) * (x - p[0]) +
         p[3] * (y - p[1]) * (y - p[1]) + p[4];
}

/* The gradient of f, df = (df/dx, df/dy). */
void
my_df (const gsl_vector *v, void *params,
       gsl_vector *df)
{
  double x, y;
  double *p = (double *)params;

  x = gsl_vector_get(v, 0);
  y = gsl_vector_get(v, 1);

  gsl_vector_set(df, 0, 2.0 * p[2] * (x - p[0]));
  gsl_vector_set(df, 1, 2.0 * p[3] * (y - p[1]));
}
```

```
/* Compute both f and df together. */
void
my_fdf (const gsl_vector *x, void *params,
        double *f, gsl_vector *df)
{
  *f = my_f(x, params);
  my_df(x, params, df);
}
```

The function can be initialized using the following code,

```
gsl_multimin_function_fdf my_func;
```

```
/* Paraboloid center at (1,2), scale factors (10, 20),
   minimum value 30 */
double p[5] = { 1.0, 2.0, 10.0, 20.0, 30.0 };
```

```
my_func.n = 2;  /* number of function components */
my_func.f = &my_f;
my_func.df = &my_df;
my_func.fdf = &my_fdf;
my_func.params = (void *)p;
```

35.5 Iteration

The following function drives the iteration of each algorithm. The function performs one iteration to update the state of the minimizer. The same function works for all minimizers so that different methods can be substituted at runtime without modifications to the code.

int gsl_multimin_fdfminimizer_iterate   Function
        (gsl_multimin_fdfminimizer * s)
int gsl_multimin_fminimizer_iterate (gsl_multimin_fminimizer   Function
        * s)

These functions perform a single iteration of the minimizer s. If the iteration encounters an unexpected problem then an error code will be returned.

The minimizer maintains a current best estimate of the minimum at all times. This information can be accessed with the following auxiliary functions,

gsl_vector * gsl_multimin_fdfminimizer_x (const   Function
          gsl_multimin_fdfminimizer * s)
gsl_vector * gsl_multimin_fminimizer_x (const   Function
          gsl_multimin_fminimizer * s)
double gsl_multimin_fdfminimizer_minimum (const   Function
          gsl_multimin_fdfminimizer * s)
double gsl_multimin_fminimizer_minimum (const   Function
          gsl_multimin_fminimizer * s)
gsl_vector * gsl_multimin_fdfminimizer_gradient (const   Function
          gsl_multimin_fdfminimizer * s)
double gsl_multimin_fminimizer_size (const   Function
          gsl_multimin_fminimizer * s)

These functions return the current best estimate of the location of the minimum, the value of the function at that point, its gradient, and minimizer
specific characteristic size for the minimizer s.

int gsl_multimin_fdfminimizer_restart   Function
          (gsl_multimin_fdfminimizer * s)
This function resets the minimizer s to use the current point as a new starting
point.

35.6 Stopping Criteria

A minimization procedure should stop when one of the following conditions
is true:

- A minimum has been found to within the user-specified precision.
- A user-specified maximum number of iterations has been reached.
- An error has occurred.

The handling of these conditions is under user control. The functions below
allow the user to test the precision of the current result.

int gsl_multimin_test_gradient (const gsl_vector * g, double   Function
          epsabs)
This function tests the norm of the gradient g against the absolute tolerance
epsabs. The gradient of a multidimensional function goes to zero at a minimum. The test returns GSL_SUCCESS if the following condition is achieved,

$$|g| < epsabs$$

and returns GSL_CONTINUE otherwise. A suitable choice of epsabs can be
made from the desired accuracy in the function for small variations in x.
The relationship between these quantities is given by $\delta f = g\, \delta x$.

int gsl_multimin_test_size (const double size, double epsabs)   Function
This function tests the minimizer specific characteristic size (if applicable
to the used minimizer) against absolute tolerance epsabs. The test returns
GSL_SUCCESS if the size is smaller than tolerance, otherwise GSL_CONTINUE
is returned.

35.7 Algorithms with Derivatives

There are several minimization methods available. The best choice of algorithm depends on the problem. The algorithms described in this section use the value of the function and its gradient at each evaluation point.

`gsl_multimin_fdfminimizer_conjugate_fr`   Minimizer

This is the Fletcher-Reeves conjugate gradient algorithm. The conjugate gradient algorithm proceeds as a succession of line minimizations. The sequence of search directions is used to build up an approximation to the curvature of the function in the neighborhood of the minimum.

An initial search direction p is chosen using the gradient, and line minimization is carried out in that direction. The accuracy of the line minimization is specified by the parameter *tol*. The minimum along this line occurs when the function gradient g and the search direction p are orthogonal. The line minimization terminates when $p \cdot g < tol \|p\| \|g\|$. The search direction is updated using the Fletcher-Reeves formula $p' = g' - \beta g$ where $\beta = -|g'|^2/|g|^2$, and the line minimization is then repeated for the new search direction.

`gsl_multimin_fdfminimizer_conjugate_pr`   Minimizer

This is the Polak-Ribiere conjugate gradient algorithm. It is similar to the Fletcher-Reeves method, differing only in the choice of the coefficient β. Both methods work well when the evaluation point is close enough to the minimum of the objective function that it is well approximated by a quadratic hypersurface.

`gsl_multimin_fdfminimizer_vector_bfgs2`   Minimizer
`gsl_multimin_fdfminimizer_vector_bfgs`   Minimizer

These methods use the vector Broyden-Fletcher-Goldfarb-Shanno (BFGS) algorithm. This is a quasi-Newton method which builds up an approximation to the second derivatives of the function f using the difference between successive gradient vectors. By combining the first and second derivatives the algorithm is able to take Newton-type steps towards the function minimum, assuming quadratic behavior in that region.

The `bfgs2` version of this minimizer is the most efficient version available, and is a faithful implementation of the line minimization scheme described in Fletcher's *Practical Methods of Optimization*, Algorithms 2.6.2 and 2.6.4. It supercedes the original `bfgs` routine and requires substantially fewer function and gradient evaluations. The user-supplied tolerance *tol* corresponds to the parameter σ used by Fletcher. A value of 0.1 is recommended for typical use (larger values correspond to less accurate line searches).

`gsl_multimin_fdfminimizer_steepest_descent`   Minimizer

The steepest descent algorithm follows the downhill gradient of the function at each step. When a downhill step is successful the step-size is increased by a factor of two. If the downhill step leads to a higher function value then the algorithm backtracks and the step size is decreased using the parameter *tol*. A suitable value of *tol* for most applications is 0.1. The steepest descent method is inefficient and is included only for demonstration purposes.

35.8 Algorithms without Derivatives

The algorithms described in this section use only the value of the function at each evaluation point.

gsl_multimin_fminimizer_nmsimplex2   Minimizer
gsl_multimin_fminimizer_nmsimplex   Minimizer

These methods use the Simplex algorithm of Nelder and Mead. They construct n vectors p_i from the starting vector x and the vector $step\_size$ as follows:

$$p_0 = (x_0, x_1, \cdots, x_n)$$
$$p_1 = (x_0 + step\_size_0, x_1, \cdots, x_n)$$
$$p_2 = (x_0, x_1 + step\_size_1, \cdots, x_n)$$
$$\cdots = \cdots$$
$$p_n = (x_0, x_1, \cdots, x_n + step\_size_n)$$

These vectors form the $n+1$ vertices of a simplex in n dimensions. On each iteration the algorithm tries to improve the parameter vector p_i corresponding to the highest function value by simple geometrical transformations. These are reflection, reflection followed by expansion, contraction and multiple contraction. Using these transformations the simplex moves through the parameter space towards the minimum, where it contracts itself.

After each iteration, the best vertex is returned. Note, that due to the nature of the algorithm not every step improves the current best parameter vector. Usually several iterations are required.

The minimizer-specific characteristic size is calculated as the average distance from the geometrical center of the simplex to all its vertices. This size can be used as a stopping criteria, as the simplex contracts itself near the minimum. The size is returned by the function gsl_multimin_fminimizer_size.

The nmsimplex2 version of this minimiser is a new $O(N)$ implementation of the earlier $O(N^2)$ nmsimplex minimiser. It calculates the size of simplex as the RMS distance of each vertex from the center rather than the mean distance, which has the advantage of allowing a linear update.

35.9 Examples

This example program finds the minimum of the paraboloid function defined earlier. The location of the minimum is offset from the origin in x and y, and the function value at the minimum is non-zero. The main program is given below, it requires the example function given earlier in this chapter.

```
int
main (void)
{
  size_t iter = 0;
  int status;

  const gsl_multimin_fdfminimizer_type *T;
  gsl_multimin_fdfminimizer *s;

  /* Position of the minimum (1,2), scale factors
     10,20, height 30. */
  double par[5] = { 1.0, 2.0, 10.0, 20.0, 30.0 };

  gsl_vector *x;
  gsl_multimin_function_fdf my_func;

  my_func.n = 2;
  my_func.f = &my_f;
  my_func.df = &my_df;
  my_func.fdf = &my_fdf;
  my_func.params = &par;

  /* Starting point, x = (5,7) */
  x = gsl_vector_alloc (2);
  gsl_vector_set (x, 0, 5.0);
  gsl_vector_set (x, 1, 7.0);

  T = gsl_multimin_fdfminimizer_conjugate_fr;
  s = gsl_multimin_fdfminimizer_alloc (T, 2);

  gsl_multimin_fdfminimizer_set (s, &my_func, x, 0.01, 1e-4);

  do
    {
      iter++;
      status = gsl_multimin_fdfminimizer_iterate (s);

      if (status)
        break;

      status = gsl_multimin_test_gradient (s->gradient, 1e-3);
```

```
          if (status == GSL_SUCCESS)
            printf ("Minimum found at:\n");

          printf ("%5d %.5f %.5f %10.5f\n", iter,
                   gsl_vector_get (s->x, 0),
                   gsl_vector_get (s->x, 1),
                   s->f);

      }
    while (status == GSL_CONTINUE && iter < 100);

    gsl_multimin_fdfminimizer_free (s);
    gsl_vector_free (x);

    return 0;
  }
```

The initial step-size is chosen as 0.01, a conservative estimate in this case, and the line minimization parameter is set at 0.0001. The program terminates when the norm of the gradient has been reduced below 0.001. The output of the program is shown below,

```
              x       y         f
     1 4.99629 6.99072  687.84780
     2 4.98886 6.97215  683.55456
     3 4.97400 6.93501  675.01278
     4 4.94429 6.86073  658.10798
     5 4.88487 6.71217  625.01340
     6 4.76602 6.41506  561.68440
     7 4.52833 5.82083  446.46694
     8 4.05295 4.63238  261.79422
     9 3.10219 2.25548   75.49762
    10 2.85185 1.62963   67.03704
    11 2.19088 1.76182   45.31640
    12 0.86892 2.02622   30.18555
   Minimum found at:
    13 1.00000 2.00000   30.00000
```

Note that the algorithm gradually increases the step size as it successfully moves downhill, as can be seen by plotting the successive points.

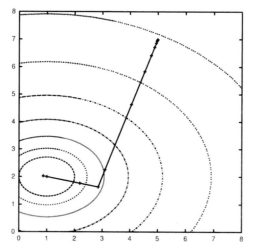

The conjugate gradient algorithm finds the minimum on its second direction because the function is purely quadratic. Additional iterations would be needed for a more complicated function.

Here is another example using the Nelder-Mead Simplex algorithm to minimize the same example object function, as above.

```
int
main(void)
{
  double par[5] = {1.0, 2.0, 10.0, 20.0, 30.0};

  const gsl_multimin_fminimizer_type *T =
    gsl_multimin_fminimizer_nmsimplex2;
  gsl_multimin_fminimizer *s = NULL;
  gsl_vector *ss, *x;
  gsl_multimin_function minex_func;

  size_t iter = 0;
  int status;
  double size;

  /* Starting point */
  x = gsl_vector_alloc (2);
  gsl_vector_set (x, 0, 5.0);
  gsl_vector_set (x, 1, 7.0);

  /* Set initial step sizes to 1 */
  ss = gsl_vector_alloc (2);
  gsl_vector_set_all (ss, 1.0);

  /* Initialize method and iterate */
  minex_func.n = 2;
```

```
minex_func.f = &my_f;
minex_func.params = (void *)&par;

s = gsl_multimin_fminimizer_alloc (T, 2);
gsl_multimin_fminimizer_set (s, &minex_func, x, ss);

do
  {
    iter++;
    status = gsl_multimin_fminimizer_iterate(s);

    if (status)
      break;

    size = gsl_multimin_fminimizer_size (s);
    status = gsl_multimin_test_size (size, 1e-2);

    if (status == GSL_SUCCESS)
      {
        printf ("converged to minimum at\n");
      }

    printf ("%5d %10.3e %10.3e f() = %7.3f size = %.3f\n",
            iter,
            gsl_vector_get (s->x, 0),
            gsl_vector_get (s->x, 1),
            s->fval, size);
  }
while (status == GSL_CONTINUE && iter < 100);

gsl_vector_free(x);
gsl_vector_free(ss);
gsl_multimin_fminimizer_free (s);

return status;
}
```

The minimum search stops when the Simplex size drops to 0.01. The output is shown below.

```
 1  6.500e+00  5.000e+00 f() = 512.500 size = 1.130
 2  5.250e+00  4.000e+00 f() = 290.625 size = 1.409
 3  5.250e+00  4.000e+00 f() = 290.625 size = 1.409
 4  5.500e+00  1.000e+00 f() = 252.500 size = 1.409
 5  2.625e+00  3.500e+00 f() = 101.406 size = 1.847
 6  2.625e+00  3.500e+00 f() = 101.406 size = 1.847
 7  0.000e+00  3.000e+00 f() =  60.000 size = 1.847
 8  2.094e+00  1.875e+00 f() =  42.275 size = 1.321
 9  2.578e-01  1.906e+00 f() =  35.684 size = 1.069
10  5.879e-01  2.445e+00 f() =  35.664 size = 0.841
```

```
11  1.258e+00  2.025e+00  f() =   30.680 size = 0.476
12  1.258e+00  2.025e+00  f() =   30.680 size = 0.367
13  1.093e+00  1.849e+00  f() =   30.539 size = 0.300
14  8.830e-01  2.004e+00  f() =   30.137 size = 0.172
15  8.830e-01  2.004e+00  f() =   30.137 size = 0.126
16  9.582e-01  2.060e+00  f() =   30.090 size = 0.106
17  1.022e+00  2.004e+00  f() =   30.005 size = 0.063
18  1.022e+00  2.004e+00  f() =   30.005 size = 0.043
19  1.022e+00  2.004e+00  f() =   30.005 size = 0.043
20  1.022e+00  2.004e+00  f() =   30.005 size = 0.027
21  1.022e+00  2.004e+00  f() =   30.005 size = 0.022
22  9.920e-01  1.997e+00  f() =   30.001 size = 0.016
23  9.920e-01  1.997e+00  f() =   30.001 size = 0.013
converged to minimum at
24  9.920e-01  1.997e+00  f() =   30.001 size = 0.008
```

The simplex size first increases, while the simplex moves towards the minimum. After a while the size begins to decrease as the simplex contracts around the minimum.

35.10 References and Further Reading

The conjugate gradient and BFGS methods are described in detail in the following book,

R. Fletcher, *Practical Methods of Optimization (Second Edition)* Wiley (1987), ISBN 0471915475.

A brief description of multidimensional minimization algorithms and more recent references can be found in,

C.W. Ueberhuber, *Numerical Computation (Volume 2)*, Chapter 14, Section 4.4 "Minimization Methods", p. 325–335, Springer (1997), ISBN 3-540-62057-5.

The simplex algorithm is described in the following paper,

J.A. Nelder and R. Mead, *A simplex method for function minimization*, Computer Journal vol. 7 (1965), 308–315.

36 Least-Squares Fitting

This chapter describes routines for performing least squares fits to experimental data using linear combinations of functions. The data may be weighted or unweighted, i.e. with known or unknown errors. For weighted data the functions compute the best fit parameters and their associated covariance matrix. For unweighted data the covariance matrix is estimated from the scatter of the points, giving a variance-covariance matrix.

The functions are divided into separate versions for simple one- or two-parameter regression and multiple-parameter fits. The functions are declared in the header file 'gsl_fit.h'.

36.1 Overview

Least-squares fits are found by minimizing χ^2 (chi-squared), the weighted sum of squared residuals over n experimental datapoints (x_i, y_i) for the model $Y(c, x)$,

$$\chi^2 = \sum_i w_i (y_i - Y(c, x_i))^2$$

The p parameters of the model are $c = \{c_0, c_1, \ldots\}$. The weight factors w_i are given by $w_i = 1/\sigma_i^2$, where σ_i is the experimental error on the data-point y_i. The errors are assumed to be gaussian and uncorrelated. For unweighted data the chi-squared sum is computed without any weight factors.

The fitting routines return the best-fit parameters c and their $p \times p$ covariance matrix. The covariance matrix measures the statistical errors on the best-fit parameters resulting from the errors on the data, σ_i, and is defined as $C_{ab} = \langle \delta c_a \delta c_b \rangle$ where $\langle \rangle$ denotes an average over the gaussian error distributions of the underlying datapoints.

The covariance matrix is calculated by error propagation from the data errors σ_i. The change in a fitted parameter δc_a caused by a small change in the data δy_i is given by

$$\delta c_a = \sum_i \frac{\partial c_a}{\partial y_i} \delta y_i$$

allowing the covariance matrix to be written in terms of the errors on the data,

$$C_{ab} = \sum_{i,j} \frac{\partial c_a}{\partial y_i} \frac{\partial c_b}{\partial y_j} \langle \delta y_i \delta y_j \rangle$$

For uncorrelated data the fluctuations of the underlying datapoints satisfy $\langle \delta y_i \delta y_j \rangle = \sigma_i^2 \delta_{ij}$, giving a corresponding parameter covariance matrix of

$$C_{ab} = \sum_i \frac{1}{w_i} \frac{\partial c_a}{\partial y_i} \frac{\partial c_b}{\partial y_i}$$

When computing the covariance matrix for unweighted data, i.e. data with unknown errors, the weight factors w_i in this sum are replaced by the single estimate $w = 1/\sigma^2$, where σ^2 is the computed variance of the residuals about the best-fit model, $\sigma^2 = \sum(y_i - Y(c, x_i))^2/(n - p)$. This is referred to as the *variance-covariance matrix*.

The standard deviations of the best-fit parameters are given by the square root of the corresponding diagonal elements of the covariance matrix, $\sigma_{c_a} = \sqrt{C_{aa}}$. The correlation coefficient of the fit parameters c_a and c_b is given by $\rho_{ab} = C_{ab}/\sqrt{C_{aa}C_{bb}}$.

36.2 Linear regression

The functions described in this section can be used to perform least-squares fits to a straight line model, $Y(c, x) = c_0 + c_1 x$.

int gsl_fit_linear (const double * x, const size_t xstride,   Function
         const double * y, const size_t ystride, size_t n, double *
         c0, double * c1, double * cov00, double * cov01, double *
         cov11, double * sumsq)

This function computes the best-fit linear regression coefficients (c0,c1) of the model $Y = c_0 + c_1 X$ for the dataset (x, y), two vectors of length n with strides *xstride* and *ystride*. The errors on y are assumed unknown so the variance-covariance matrix for the parameters (c0, c1) is estimated from the scatter of the points around the best-fit line and returned via the parameters (cov00, cov01, cov11). The sum of squares of the residuals from the best-fit line is returned in *sumsq*. Note: the correlation coefficient of the data can be computed using gsl_stats_correlation (see Section 20.6 [Correlation], page 291), it does not depend on the fit.

int gsl_fit_wlinear (const double * x, const size_t xstride,   Function
         const double * w, const size_t wstride, const double * y,
         const size_t ystride, size_t n, double * c0, double * c1,
         double * cov00, double * cov01, double * cov11, double *
         chisq)

This function computes the best-fit linear regression coefficients (c0,c1) of the model $Y = c_0 + c_1 X$ for the weighted dataset (x, y), two vectors of length n with strides *xstride* and *ystride*. The vector w, of length n and stride *wstride*, specifies the weight of each datapoint. The weight is the reciprocal of the variance for each datapoint in y.

The covariance matrix for the parameters (c0, c1) is computed using the weights and returned via the parameters (cov00, cov01, cov11). The weighted sum of squares of the residuals from the best-fit line, χ^2, is returned in *chisq*.

int gsl_fit_linear_est (double x, double $c0$, double $c1$, double   Function
         $cov00$, double $cov01$, double $cov11$, double * y, double *
         y_err)

This function uses the best-fit linear regression coefficients $c0$, $c1$ and their covariance $cov00$, $cov01$, $cov11$ to compute the fitted function y and its standard deviation y_err for the model $Y = c_0 + c_1 X$ at the point x.

36.3 Linear fitting without a constant term

The functions described in this section can be used to perform least-squares fits to a straight line model without a constant term, $Y = c_1 X$.

int gsl_fit_mul (const double * x, const size_t $xstride$, const   Function
         double * y, const size_t $ystride$, size_t n, double * $c1$,
         double * $cov11$, double * $sumsq$)

This function computes the best-fit linear regression coefficient $c1$ of the model $Y = c_1 X$ for the datasets (x, y), two vectors of length n with strides $xstride$ and $ystride$. The errors on y are assumed unknown so the variance of the parameter $c1$ is estimated from the scatter of the points around the best-fit line and returned via the parameter $cov11$. The sum of squares of the residuals from the best-fit line is returned in $sumsq$.

int gsl_fit_wmul (const double * x, const size_t $xstride$, const   Function
         double * w, const size_t $wstride$, const double * y, const
         size_t $ystride$, size_t n, double * $c1$, double * $cov11$, double
         * $sumsq$)

This function computes the best-fit linear regression coefficient $c1$ of the model $Y = c_1 X$ for the weighted datasets (x, y), two vectors of length n with strides $xstride$ and $ystride$. The vector w, of length n and stride $wstride$, specifies the weight of each datapoint. The weight is the reciprocal of the variance for each datapoint in y.

The variance of the parameter $c1$ is computed using the weights and returned via the parameter $cov11$. The weighted sum of squares of the residuals from the best-fit line, χ^2, is returned in $chisq$.

int gsl_fit_mul_est (double x, double $c1$, double $cov11$, double   Function
         * y, double * y_err)

This function uses the best-fit linear regression coefficient $c1$ and its covariance $cov11$ to compute the fitted function y and its standard deviation y_err for the model $Y = c_1 X$ at the point x.

36.4 Multi-parameter fitting

The functions described in this section perform least-squares fits to a general linear model, $y = Xc$ where y is a vector of n observations, X is an n by p matrix of predictor variables, and the elements of the vector c are the p unknown best-fit parameters which are to be estimated. The chi-squared value is given by $\chi^2 = \sum_i w_i(y_i - \sum_j X_{ij}c_j)^2$.

This formulation can be used for fits to any number of functions and/or variables by preparing the n-by-p matrix X appropriately. For example, to fit to a p-th order polynomial in x, use the following matrix,

$$X_{ij} = x_i^j$$

where the index i runs over the observations and the index j runs from 0 to $p - 1$.

To fit to a set of p sinusoidal functions with fixed frequencies $\omega_1, \omega_2, \ldots, \omega_p$, use,

$$X_{ij} = \sin(\omega_j x_i)$$

To fit to p independent variables $x_1, x_2, \ldots, x_p$, use,

$$X_{ij} = x_j(i)$$

where $x_j(i)$ is the i-th value of the predictor variable x_j.

The functions described in this section are declared in the header file 'gsl_multifit.h'.

The solution of the general linear least-squares system requires an additional working space for intermediate results, such as the singular value decomposition of the matrix X.

gsl_multifit_linear_workspace * gsl_multifit_linear_alloc   Function
          (size_t n, size_t p)
    This function allocates a workspace for fitting a model to n observations using p parameters.

void gsl_multifit_linear_free (gsl_multifit_linear_workspace   Function
          * work)
    This function frees the memory associated with the workspace w.

int gsl_multifit_linear (const gsl_matrix * X, const   Function
          gsl_vector * y, gsl_vector * c, gsl_matrix * cov, double *
          chisq, gsl_multifit_linear_workspace * work)
int gsl_multifit_linear_svd (const gsl_matrix * X, const   Function
          gsl_vector * y, double tol, size_t * rank, gsl_vector * c,
          gsl_matrix * cov, double * chisq,
          gsl_multifit_linear_workspace * work)
    These functions compute the best-fit parameters c of the model $y = Xc$ for the observations y and the matrix of predictor variables X. The variance-covariance matrix of the model parameters cov is estimated from the scatter of the observations about the best-fit. The sum of squares of the residuals

from the best-fit, χ^2, is returned in *chisq*. If the coefficient of determination is desired, it can be computed from the expression $R^2 = 1 - \chi^2/TSS$, where the total sum of squares (TSS) of the observations y may be computed from `gsl_stats_tss`.

The best-fit is found by singular value decomposition of the matrix X using the preallocated workspace provided in *work*. The modified Golub-Reinsch SVD algorithm is used, with column scaling to improve the accuracy of the singular values. Any components which have zero singular value (to machine precision) are discarded from the fit. In the second form of the function the components are discarded if the ratio of singular values s_i/s_0 falls below the user-specified tolerance *tol*, and the effective rank is returned in *rank*.

int gsl_multifit_wlinear (const gsl_matrix * X, const   *Function*
         gsl_vector * w, const gsl_vector * y, gsl_vector * c,
         gsl_matrix * cov, double * chisq,
         gsl_multifit_linear_workspace * work)

int gsl_multifit_wlinear_svd (const gsl_matrix * X, const   *Function*
         gsl_vector * w, const gsl_vector * y, double tol, size_t *
         rank, gsl_vector * c, gsl_matrix * cov, double * chisq,
         gsl_multifit_linear_workspace * work)

This function computes the best-fit parameters c of the weighted model $y = Xc$ for the observations y with weights w and the matrix of predictor variables X. The covariance matrix of the model parameters *cov* is computed with the given weights. The weighted sum of squares of the residuals from the best-fit, χ^2, is returned in *chisq*. If the coefficient of determination is desired, it can be computed from the expression $R^2 = 1 - \chi^2/WTSS$, where the weighted total sum of squares (WTSS) of the observations y may be computed from `gsl_stats_wtss`.

The best-fit is found by singular value decomposition of the matrix X using the preallocated workspace provided in *work*. Any components which have zero singular value (to machine precision) are discarded from the fit. In the second form of the function the components are discarded if the ratio of singular values s_i/s_0 falls below the user-specified tolerance *tol*, and the effective rank is returned in *rank*.

int gsl_multifit_linear_est (const gsl_vector * x, const   *Function*
         gsl_vector * c, const gsl_matrix * cov, double * y, double *
         y_err)

This function uses the best-fit multilinear regression coefficients c and their covariance matrix *cov* to compute the fitted function value y and its standard deviation *y_err* for the model $y = x.c$ at the point x.

int gsl_multifit_linear_residuals (const gsl_matrix * X,   *Function*
         const gsl_vector * y, const gsl_vector * c, gsl_vector * r)

This function computes the vector of residuals $r = y - Xc$ for the observations y, coefficients c and matrix of predictor variables X.

36.5 Examples

The following program computes a least squares straight-line fit to a simple dataset, and outputs the best-fit line and its associated one standard-deviation error bars.

```
#include <stdio.h>
#include <gsl/gsl_fit.h>

int
main (void)
{
  int i, n = 4;
  double x[4] = { 1970, 1980, 1990, 2000 };
  double y[4] = {   12,   11,   14,   13 };
  double w[4] = {  0.1,  0.2,  0.3,  0.4 };

  double c0, c1, cov00, cov01, cov11, chisq;

  gsl_fit_wlinear (x, 1, w, 1, y, 1, n,
                   &c0, &c1, &cov00, &cov01, &cov11,
                   &chisq);

  printf ("# best fit: Y = %g + %g X\n", c0, c1);
  printf ("# covariance matrix:\n");
  printf ("# [ %g, %g\n#   %g, %g]\n",
          cov00, cov01, cov01, cov11);
  printf ("# chisq = %g\n", chisq);

  for (i = 0; i < n; i++)
    printf ("data: %g %g %g\n",
                   x[i], y[i], 1/sqrt(w[i]));

  printf ("\n");

  for (i = -30; i < 130; i++)
    {
      double xf = x[0] + (i/100.0) * (x[n-1] - x[0]);
      double yf, yf_err;

      gsl_fit_linear_est (xf,
                          c0, c1,
                          cov00, cov01, cov11,
                          &yf, &yf_err);

      printf ("fit: %g %g\n", xf, yf);
      printf ("hi : %g %g\n", xf, yf + yf_err);
      printf ("lo : %g %g\n", xf, yf - yf_err);
    }
```

```
    return 0;
  }
```

The following commands extract the data from the output of the program and display it using the GNU plotutils graph utility,

```
$ ./demo > tmp
$ more tmp
# best fit: Y = -106.6 + 0.06 X
# covariance matrix:
# [ 39602, -19.9
#    -19.9, 0.01]
# chisq = 0.8

$ for n in data fit hi lo ;
  do
     grep "^$n" tmp | cut -d: -f2 > $n ;
  done
$ graph -T X -X x -Y y -y 0 20 -m 0 -S 2 -Ie data
     -S 0 -I a -m 1 fit -m 2 hi -m 2 lo
```

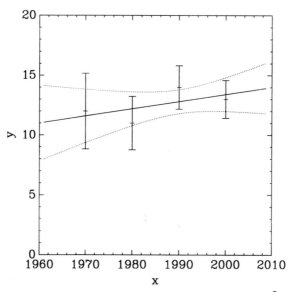

The next program performs a quadratic fit $y = c_0 + c_1 x + c_2 x^2$ to a weighted dataset using the generalised linear fitting function gsl_multifit_wlinear. The model matrix X for a quadratic fit is given by,

$$ X = \begin{pmatrix} 1 & x_0 & x_0^2 \\ 1 & x_1 & x_1^2 \\ 1 & x_2 & x_2^2 \\ \dots & \dots & \dots \end{pmatrix} $$

where the column of ones corresponds to the constant term c_0. The two remaining columns corresponds to the terms $c_1 x$ and $c_2 x^2$.

The program reads n lines of data in the format (x, y, err) where err is the error (standard deviation) in the value y.

```
#include <stdio.h>
#include <gsl/gsl_multifit.h>

int
main (int argc, char **argv)
{
  int i, n;
  double xi, yi, ei, chisq;
  gsl_matrix *X, *cov;
  gsl_vector *y, *w, *c;

  if (argc != 2)
    {
      fprintf (stderr,"usage: fit n < data\n");
      exit (-1);
    }

  n = atoi (argv[1]);

  X = gsl_matrix_alloc (n, 3);
  y = gsl_vector_alloc (n);
  w = gsl_vector_alloc (n);

  c = gsl_vector_alloc (3);
  cov = gsl_matrix_alloc (3, 3);

  for (i = 0; i < n; i++)
    {
      int count = fscanf (stdin, "%lg %lg %lg",
                          &xi, &yi, &ei);

      if (count != 3)
        {
          fprintf (stderr, "error reading file\n");
          exit (-1);
        }

      printf ("%g %g +/- %g\n", xi, yi, ei);

      gsl_matrix_set (X, i, 0, 1.0);
      gsl_matrix_set (X, i, 1, xi);
      gsl_matrix_set (X, i, 2, xi*xi);

      gsl_vector_set (y, i, yi);
```

```
      gsl_vector_set (w, i, 1.0/(ei*ei));
    }

  {
    gsl_multifit_linear_workspace * work
      = gsl_multifit_linear_alloc (n, 3);
    gsl_multifit_wlinear (X, w, y, c, cov,
                            &chisq, work);
    gsl_multifit_linear_free (work);
  }

#define C(i) (gsl_vector_get(c,(i)))
#define COV(i,j) (gsl_matrix_get(cov,(i),(j)))

  {
    printf ("# best fit: Y = %g + %g X + %g X^2\n",
            C(0), C(1), C(2));

    printf ("# covariance matrix:\n");
    printf ("[ %+.5e, %+.5e, %+.5e  \n",
              COV(0,0), COV(0,1), COV(0,2));
    printf ("  %+.5e, %+.5e, %+.5e  \n",
              COV(1,0), COV(1,1), COV(1,2));
    printf ("  %+.5e, %+.5e, %+.5e ]\n",
              COV(2,0), COV(2,1), COV(2,2));
    printf ("# chisq = %g\n", chisq);
  }

  gsl_matrix_free (X);
  gsl_vector_free (y);
  gsl_vector_free (w);
  gsl_vector_free (c);
  gsl_matrix_free (cov);

  return 0;
}
```

A suitable set of data for fitting can be generated using the following program. It outputs a set of points with gaussian errors from the curve $y = e^x$ in the region $0 < x < 2$.

```
#include <stdio.h>
#include <math.h>
#include <gsl/gsl_randist.h>

int
main (void)
{
  double x;
  const gsl_rng_type * T;
```

```
    gsl_rng * r;

    gsl_rng_env_setup ();

    T = gsl_rng_default;
    r = gsl_rng_alloc (T);

    for (x = 0.1; x < 2; x+= 0.1)
      {
        double y0 = exp (x);
        double sigma = 0.1 * y0;
        double dy = gsl_ran_gaussian (r, sigma);

        printf ("%g %g %g\n", x, y0 + dy, sigma);
      }

    gsl_rng_free(r);

    return 0;
}
```

The data can be prepared by running the resulting executable program,

```
$ ./generate > exp.dat
$ more exp.dat
0.1 0.97935 0.110517
0.2 1.3359 0.12214
0.3 1.52573 0.134986
0.4 1.60318 0.149182
0.5 1.81731 0.164872
0.6 1.92475 0.182212
....
```

To fit the data use the previous program, with the number of data points given as the first argument. In this case there are 19 data points.

```
$ ./fit 19 < exp.dat
0.1 0.97935 +/- 0.110517
0.2 1.3359 +/- 0.12214
...
# best fit: Y = 1.02318 + 0.956201 X + 0.876796 X^2
# covariance matrix:
[ +1.25612e-02, -3.64387e-02, +1.94389e-02
  -3.64387e-02, +1.42339e-01, -8.48761e-02
  +1.94389e-02, -8.48761e-02, +5.60243e-02 ]
# chisq = 23.0987
```

The parameters of the quadratic fit match the coefficients of the expansion of e^x, taking into account the errors on the parameters and the $O(x^3)$ difference between the exponential and quadratic functions for the larger values of x. The errors on the parameters are given by the square-root of the corresponding diagonal elements of the covariance matrix. The chi-squared per degree of freedom is 1.4, indicating a reasonable fit to the data.

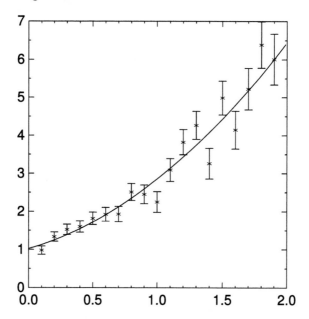

36.6 References and Further Reading

A summary of formulas and techniques for least squares fitting can be found in the "Statistics" chapter of the Annual Review of Particle Physics prepared by the Particle Data Group,

> *Review of Particle Properties*, R.M. Barnett et al., Physical Review D54, 1 (1996) http://pdg.lbl.gov/

The Review of Particle Physics is available online at the website given above.

The tests used to prepare these routines are based on the NIST Statistical Reference Datasets. The datasets and their documentation are available from NIST at the following website,

> http://www.nist.gov/itl/div898/strd/index.html.

37 Nonlinear Least-Squares Fitting

This chapter describes functions for multidimensional nonlinear least-squares fitting. The library provides low level components for a variety of iterative solvers and convergence tests. These can be combined by the user to achieve the desired solution, with full access to the intermediate steps of the iteration. Each class of methods uses the same framework, so that you can switch between solvers at runtime without needing to recompile your program. Each instance of a solver keeps track of its own state, allowing the solvers to be used in multi-threaded programs.

The header file 'gsl_multifit_nlin.h' contains prototypes for the multidimensional nonlinear fitting functions and related declarations.

37.1 Overview

The problem of multidimensional nonlinear least-squares fitting requires the minimization of the squared residuals of n functions, f_i, in p parameters, x_i,

$$\Phi(x) = \frac{1}{2}||F(x)||^2 = \frac{1}{2}\sum_{i=1}^{n} f_i(x_1,\ldots,x_p)^2$$

All algorithms proceed from an initial guess using the linearization,

$$\psi(p) = ||F(x+p)|| \approx ||F(x) + Jp||$$

where x is the initial point, p is the proposed step and J is the Jacobian matrix $J_{ij} = \partial f_i/\partial x_j$. Additional strategies are used to enlarge the region of convergence. These include requiring a decrease in the norm $||F||$ on each step or using a trust region to avoid steps which fall outside the linear regime.

To perform a weighted least-squares fit of a nonlinear model $Y(x,t)$ to data (t_i, y_i) with independent gaussian errors σ_i, use function components of the following form,

$$f_i = \frac{(Y(x, t_i) - y_i)}{\sigma_i}$$

Note that the model parameters are denoted by x in this chapter since the non-linear least-squares algorithms are described geometrically (i.e. finding the minimum of a surface). The independent variable of any data to be fitted is denoted by t.

With the definition above the Jacobian is $J_{ij} = (1/\sigma_i)\partial Y_i/\partial x_j$, where $Y_i = Y(x, t_i)$.

37.2 Initializing the Solver

gsl_multifit_fsolver * gsl_multifit_fsolver_alloc (const   Function
        gsl_multifit_fsolver_type * T, size_t n, size_t p)

This function returns a pointer to a newly allocated instance of a solver of type T for n observations and p parameters. The number of observations n must be greater than or equal to parameters p.

If there is insufficient memory to create the solver then the function returns a null pointer and the error handler is invoked with an error code of GSL_ENOMEM.

gsl_multifit_fdfsolver * gsl_multifit_fdfsolver_alloc (const   Function
        gsl_multifit_fdfsolver_type * T, size_t n, size_t p)

This function returns a pointer to a newly allocated instance of a derivative solver of type T for n observations and p parameters. For example, the following code creates an instance of a Levenberg-Marquardt solver for 100 data points and 3 parameters,

```
const gsl_multifit_fdfsolver_type * T
    = gsl_multifit_fdfsolver_lmder;
gsl_multifit_fdfsolver * s
    = gsl_multifit_fdfsolver_alloc (T, 100, 3);
```

The number of observations n must be greater than or equal to parameters p.

If there is insufficient memory to create the solver then the function returns a null pointer and the error handler is invoked with an error code of GSL_ENOMEM.

int gsl_multifit_fsolver_set (gsl_multifit_fsolver * s,   Function
        gsl_multifit_function * f, gsl_vector * x)

This function initializes, or reinitializes, an existing solver s to use the function f and the initial guess x.

int gsl_multifit_fdfsolver_set (gsl_multifit_fdfsolver * s,   Function
        gsl_multifit_function_fdf * fdf, gsl_vector * x)

This function initializes, or reinitializes, an existing solver s to use the function and derivative fdf and the initial guess x.

void gsl_multifit_fsolver_free (gsl_multifit_fsolver * s)   Function
void gsl_multifit_fdfsolver_free (gsl_multifit_fdfsolver * s)   Function

These functions free all the memory associated with the solver s.

const char * gsl_multifit_fsolver_name (const   Function
        gsl_multifit_fsolver * s)
const char * gsl_multifit_fdfsolver_name (const   Function
        gsl_multifit_fdfsolver * s)

These functions return a pointer to the name of the solver. For example,

```
printf ("s is a '%s' solver\n",
            gsl_multifit_fdfsolver_name (s));
```

would print something like s is a 'lmder' solver.

37.3 Providing the Function to be Minimized

You must provide n functions of p variables for the minimization algorithms to operate on. In order to allow for arbitrary parameters the functions are defined by the following data types:

`gsl_multifit_function`   Data Type
This data type defines a general system of functions with arbitrary parameters.

> `int (* f) (const gsl_vector * x, void * params, gsl_vector * f)`
> this function should store the vector result $f(x, params)$ in f for argument x and arbitrary parameters params, returning an appropriate error code if the function cannot be computed.

> `size_t n`
> the number of functions, i.e. the number of components of the vector f.

> `size_t p`
> the number of independent variables, i.e. the number of components of the vector x.

> `void * params`
> a pointer to the arbitrary parameters of the function.

`gsl_multifit_function_fdf`   Data Type
This data type defines a general system of functions with arbitrary parameters and the corresponding Jacobian matrix of derivatives,

> `int (* f) (const gsl_vector * x, void * params, gsl_vector * f)`
> this function should store the vector result $f(x, params)$ in f for argument x and arbitrary parameters params, returning an appropriate error code if the function cannot be computed.

> `int (* df) (const gsl_vector * x, void * params, gsl_matrix * J)`
> this function should store the n-by-p Jacobian matrix result $J_{ij} = \partial f_i(x, params)/\partial x_j$ in J for argument x and arbitrary parameters params, returning an appropriate error code if the function cannot be computed.

> `int (* fdf) (const gsl_vector * x, void * params, gsl_vector * f, gsl_matrix * J)`
> This function should set the values of the f and J as above, for arguments x and arbitrary parameters params. This function provides an optimization of the separate functions for $f(x)$ and $J(x)$—it is always faster to compute the function and its derivative at the same time.

> `size_t n`
> the number of functions, i.e. the number of components of the vector f.

> `size_t p`
> the number of independent variables, i.e. the number of components of the vector x.

```
void * params
```
a pointer to the arbitrary parameters of the function.

Note that when fitting a non-linear model against experimental data, the data is passed to the functions above using the *params* argument and the trial best-fit parameters through the *x* argument.

37.4 Iteration

The following functions drive the iteration of each algorithm. Each function performs one iteration to update the state of any solver of the corresponding type. The same functions work for all solvers so that different methods can be substituted at runtime without modifications to the code.

```
int gsl_multifit_fsolver_iterate (gsl_multifit_fsolver * s)       Function
int gsl_multifit_fdfsolver_iterate (gsl_multifit_fdfsolver *      Function
         s)
```
These functions perform a single iteration of the solver *s*. If the iteration encounters an unexpected problem then an error code will be returned. The solver maintains a current estimate of the best-fit parameters at all times.

The solver struct *s* contains the following entries, which can be used to track the progress of the solution:

```
gsl_vector * x
```
The current position.

```
gsl_vector * f
```
The function value at the current position.

```
gsl_vector * dx
```
The difference between the current position and the previous position, i.e. the last step, taken as a vector.

```
gsl_matrix * J
```
The Jacobian matrix at the current position (for the `gsl_multifit_fdfsolver` struct only)

The best-fit information also can be accessed with the following auxiliary functions,

```
gsl_vector * gsl_multifit_fsolver_position (const                 Function
         gsl_multifit_fsolver * s)
gsl_vector * gsl_multifit_fdfsolver_position (const               Function
         gsl_multifit_fdfsolver * s)
```
These functions return the current position (i.e. best-fit parameters) s->x of the solver *s*.

37.5 Search Stopping Parameters

A minimization procedure should stop when one of the following conditions is true:

- A minimum has been found to within the user-specified precision.
- A user-specified maximum number of iterations has been reached.
- An error has occurred.

The handling of these conditions is under user control. The functions below allow the user to test the current estimate of the best-fit parameters in several standard ways.

int gsl_multifit_test_delta (const gsl_vector * dx, const   Function
        gsl_vector * x, double *epsabs*, double *epsrel*)
    This function tests for the convergence of the sequence by comparing the last step *dx* with the absolute error *epsabs* and relative error *epsrel* to the current position *x*. The test returns GSL_SUCCESS if the following condition is achieved,

$$|dx_i| < epsabs + epsrel\,|x_i|$$

for each component of *x* and returns GSL_CONTINUE otherwise.

int gsl_multifit_test_gradient (const gsl_vector * g, double   Function
        epsabs)
    This function tests the residual gradient *g* against the absolute error bound *epsabs*. Mathematically, the gradient should be exactly zero at the minimum. The test returns GSL_SUCCESS if the following condition is achieved,

$$\sum_i |g_i| < epsabs$$

and returns GSL_CONTINUE otherwise. This criterion is suitable for situations where the precise location of the minimum, x, is unimportant provided a value can be found where the gradient is small enough.

int gsl_multifit_gradient (const gsl_matrix * J, const   Function
        gsl_vector * f, gsl_vector * g)
    This function computes the gradient *g* of $\Phi(x) = (1/2)\|F(x)\|^2$ from the Jacobian matrix J and the function values f, using the formula $g = J^T f$.

37.6 Minimization Algorithms using Derivatives

The minimization algorithms described in this section make use of both the function and its derivative. They require an initial guess for the location of the minimum. There is no absolute guarantee of convergence—the function must be suitable for this technique and the initial guess must be sufficiently close to the minimum for it to work.

`gsl_multifit_fdfsolver_lmsder`   Derivative Solver

This is a robust and efficient version of the Levenberg-Marquardt algorithm as implemented in the scaled LMDER routine in MINPACK. Minpack was written by Jorge J. Moré, Burton S. Garbow and Kenneth E. Hillstrom.

The algorithm uses a generalized trust region to keep each step under control. In order to be accepted a proposed new position x' must satisfy the condition $|D(x' - x)| < \delta$, where D is a diagonal scaling matrix and δ is the size of the trust region. The components of D are computed internally, using the column norms of the Jacobian to estimate the sensitivity of the residual to each component of x. This improves the behavior of the algorithm for badly scaled functions.

On each iteration the algorithm attempts to minimize the linear system $|F + Jp|$ subject to the constraint $|Dp| < \Delta$. The solution to this constrained linear system is found using the Levenberg-Marquardt method.

The proposed step is now tested by evaluating the function at the resulting point, x'. If the step reduces the norm of the function sufficiently, and follows the predicted behavior of the function within the trust region, then it is accepted and the size of the trust region is increased. If the proposed step fails to improve the solution, or differs significantly from the expected behavior within the trust region, then the size of the trust region is decreased and another trial step is computed.

The algorithm also monitors the progress of the solution and returns an error if the changes in the solution are smaller than the machine precision. The possible error codes are,

GSL_ETOLF

the decrease in the function falls below machine precision

GSL_ETOLX

the change in the position vector falls below machine precision

GSL_ETOLG

the norm of the gradient, relative to the norm of the function, falls below machine precision

These error codes indicate that further iterations will be unlikely to change the solution from its current value.

`gsl_multifit_fdfsolver_lmder`   Derivative Solver

This is an unscaled version of the LMDER algorithm. The elements of the diagonal scaling matrix D are set to 1. This algorithm may be useful in circumstances where the scaled version of LMDER converges too slowly, or the function is already scaled appropriately.

37.7 Minimization Algorithms without Derivatives

There are no algorithms implemented in this section at the moment.

37.8 Computing the covariance matrix of best fit parameters

int gsl_multifit_covar (const gsl_matrix * J, double *epsrel*,   Function
        gsl_matrix * *covar*)

This function uses the Jacobian matrix J to compute the covariance matrix of the best-fit parameters, *covar*. The parameter *epsrel* is used to remove linear-dependent columns when J is rank deficient.

The covariance matrix is given by,

$$C = (J^T J)^{-1}$$

and is computed by QR decomposition of J with column-pivoting. Any columns of R which satisfy

$$|R_{kk}| \leq epsrel|R_{11}|$$

are considered linearly-dependent and are excluded from the covariance matrix (the corresponding rows and columns of the covariance matrix are set to zero).

If the minimisation uses the weighted least-squares function $f_i = (Y(x, t_i) - y_i)/\sigma_i$ then the covariance matrix above gives the statistical error on the best-fit parameters resulting from the gaussian errors σ_i on the underlying data y_i. This can be verified from the relation $\delta f = J\delta c$ and the fact that the fluctuations in f from the data y_i are normalised by σ_i and so satisfy $\langle \delta f \delta f^T \rangle = I$.

For an unweighted least-squares function $f_i = (Y(x, t_i) - y_i)$ the covariance matrix above should be multiplied by the variance of the residuals about the best-fit $\sigma^2 = \sum(y_i - Y(x, t_i))^2/(n - p)$ to give the variance-covariance matrix $\sigma^2 C$. This estimates the statistical error on the best-fit parameters from the scatter of the underlying data.

For more information about covariance matrices see Section 36.1 [Fitting Overview], page 443.

37.9 Examples

The following example program fits a weighted exponential model with background to experimental data, $Y = A \exp(-\lambda t) + b$. The first part of the program sets up the functions expb_f and expb_df to calculate the model and its Jacobian. The appropriate fitting function is given by,

$$f_i = ((A \exp(-\lambda t_i) + b) - y_i)/\sigma_i$$

where we have chosen $t_i = i$. The Jacobian matrix J is the derivative of these functions with respect to the three parameters (A, λ, b). It is given by,

$$J_{ij} = \frac{\partial f_i}{\partial x_j}$$

where $x_0 = A$, $x_1 = \lambda$ and $x_2 = b$.

```
/* expfit.c -- model functions for exponential + background */

struct data {
  size_t n;
  double * y;
  double * sigma;
};

int
expb_f (const gsl_vector * x, void *data,
        gsl_vector * f)
{
  size_t n = ((struct data *)data)->n;
  double *y = ((struct data *)data)->y;
  double *sigma = ((struct data *) data)->sigma;

  double A = gsl_vector_get (x, 0);
  double lambda = gsl_vector_get (x, 1);
  double b = gsl_vector_get (x, 2);

  size_t i;

  for (i = 0; i < n; i++)
    {
      /* Model Yi = A * exp(-lambda * i) + b */
      double t = i;
      double Yi = A * exp (-lambda * t) + b;
      gsl_vector_set (f, i, (Yi - y[i])/sigma[i]);
    }

  return GSL_SUCCESS;
}

int
```

```
expb_df (const gsl_vector * x, void *data,
         gsl_matrix * J)
{
  size_t n = ((struct data *)data)->n;
  double *sigma = ((struct data *) data)->sigma;

  double A = gsl_vector_get (x, 0);
  double lambda = gsl_vector_get (x, 1);

  size_t i;

  for (i = 0; i < n; i++)
    {
      /* Jacobian matrix J(i,j) = dfi / dxj, */
      /* where fi = (Yi - yi)/sigma[i],      */
      /*          Yi = A * exp(-lambda * i) + b */
      /* and the xj are the parameters (A,lambda,b) */
      double t = i;
      double s = sigma[i];
      double e = exp(-lambda * t);
      gsl_matrix_set (J, i, 0, e/s);
      gsl_matrix_set (J, i, 1, -t * A * e/s);
      gsl_matrix_set (J, i, 2, 1/s);
    }
  return GSL_SUCCESS;
}

int
expb_fdf (const gsl_vector * x, void *data,
          gsl_vector * f, gsl_matrix * J)
{
  expb_f (x, data, f);
  expb_df (x, data, J);

  return GSL_SUCCESS;
}
```

The main part of the program sets up a Levenberg-Marquardt solver and some simulated random data. The data uses the known parameters (1.0,5.0,0.1) combined with gaussian noise (standard deviation = 0.1) over a range of 40 timesteps. The initial guess for the parameters is chosen as (0.0, 1.0, 0.0).

```
#include <stdlib.h>
#include <stdio.h>
#include <gsl/gsl_rng.h>
#include <gsl/gsl_randist.h>
#include <gsl/gsl_vector.h>
#include <gsl/gsl_blas.h>
#include <gsl/gsl_multifit_nlin.h>
```

```
#include "expfit.c"

#define N 40

void print_state (size_t iter, gsl_multifit_fdfsolver * s);

int
main (void)
{
  const gsl_multifit_fdfsolver_type *T;
  gsl_multifit_fdfsolver *s;
  int status;
  unsigned int i, iter = 0;
  const size_t n = N;
  const size_t p = 3;

  gsl_matrix *covar = gsl_matrix_alloc (p, p);
  double y[N], sigma[N];
  struct data d = { n, y, sigma};
  gsl_multifit_function_fdf f;
  double x_init[3] = { 1.0, 0.0, 0.0 };
  gsl_vector_view x = gsl_vector_view_array (x_init, p);
  const gsl_rng_type * type;
  gsl_rng * r;

  gsl_rng_env_setup();

  type = gsl_rng_default;
  r = gsl_rng_alloc (type);

  f.f = &expb_f;
  f.df = &expb_df;
  f.fdf = &expb_fdf;
  f.n = n;
  f.p = p;
  f.params = &d;

  /* This is the data to be fitted */

  for (i = 0; i < n; i++)
    {
      double t = i;
      y[i] = 1.0 + 5 * exp (-0.1 * t)
                 + gsl_ran_gaussian (r, 0.1);
      sigma[i] = 0.1;
      printf ("data: %u %g %g\n", i, y[i], sigma[i]);
    };
```

```
    T = gsl_multifit_fdfsolver_lmsder;
    s = gsl_multifit_fdfsolver_alloc (T, n, p);
    gsl_multifit_fdfsolver_set (s, &f, &x.vector);

    print_state (iter, s);

    do
      {
        iter++;
        status = gsl_multifit_fdfsolver_iterate (s);

        printf ("status = %s\n", gsl_strerror (status));

        print_state (iter, s);

        if (status)
          break;

        status = gsl_multifit_test_delta (s->dx, s->x,
                                          1e-4, 1e-4);
      }
    while (status == GSL_CONTINUE && iter < 500);

    gsl_multifit_covar (s->J, 0.0, covar);

#define FIT(i) gsl_vector_get(s->x, i)
#define ERR(i) sqrt(gsl_matrix_get(covar,i,i))

    {
      double chi = gsl_blas_dnrm2(s->f);
      double dof = n - p;
      double c = GSL_MAX_DBL(1, chi / sqrt(dof));

      printf("chisq/dof = %g\n",  pow(chi, 2.0) / dof);

      printf ("A      = %.5f +/- %.5f\n", FIT(0), c*ERR(0));
      printf ("lambda = %.5f +/- %.5f\n", FIT(1), c*ERR(1));
      printf ("b      = %.5f +/- %.5f\n", FIT(2), c*ERR(2));
    }

    printf ("status = %s\n", gsl_strerror (status));

    gsl_multifit_fdfsolver_free (s);
    gsl_matrix_free (covar);
    gsl_rng_free (r);
    return 0;
  }
```

```
void
print_state (size_t iter, gsl_multifit_fdfsolver * s)
{
  printf ("iter: %3u x = % 15.8f % 15.8f % 15.8f "
          "|f(x)| = %g\n",
          iter,
          gsl_vector_get (s->x, 0),
          gsl_vector_get (s->x, 1),
          gsl_vector_get (s->x, 2),
          gsl_blas_dnrm2 (s->f));
}
```

The iteration terminates when the change in x is smaller than 0.0001, as both an absolute and relative change. Here are the results of running the program:

```
iter: 0 x=1.00000000 0.00000000 0.00000000 |f(x)|=117.349
status=success
iter: 1 x=1.64659312 0.01814772 0.64659312 |f(x)|=76.4578
status=success
iter: 2 x=2.85876037 0.08092095 1.44796363 |f(x)|=37.6838
status=success
iter: 3 x=4.94899512 0.11942928 1.09457665 |f(x)|=9.58079
status=success
iter: 4 x=5.02175572 0.10287787 1.03388354 |f(x)|=5.63049
status=success
iter: 5 x=5.04520433 0.10405523 1.01941607 |f(x)|=5.44398
status=success
iter: 6 x=5.04535782 0.10404906 1.01924871 |f(x)|=5.44397
chisq/dof = 0.800996
A      = 5.04536 +/- 0.06028
lambda = 0.10405 +/- 0.00316
b      = 1.01925 +/- 0.03782
status = success
```

The approximate values of the parameters are found correctly, and the chi-squared value indicates a good fit (the chi-squared per degree of freedom is approximately 1). In this case the errors on the parameters can be estimated from the square roots of the diagonal elements of the covariance matrix.

If the chi-squared value shows a poor fit (i.e. $\chi^2/(n-p) \gg 1$) then the error estimates obtained from the covariance matrix will be too small. In the example program the error estimates are multiplied by $\sqrt{\chi^2/(n-p)}$ in this case, a common way of increasing the errors for a poor fit. Note that a poor fit will result from the use an inappropriate model, and the scaled error estimates may then be outside the range of validity for gaussian errors.

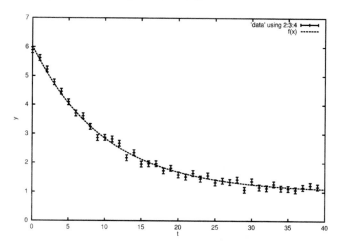

37.10 References and Further Reading

The MINPACK algorithm is described in the following article,

> J.J. Moré, *The Levenberg-Marquardt Algorithm: Implementation and Theory*, Lecture Notes in Mathematics, v630 (1978), ed G. Watson.

The following paper is also relevant to the algorithms described in this section,

> J.J. Moré, B.S. Garbow, K.E. Hillstrom, "Testing Unconstrained Optimization Software", ACM Transactions on Mathematical Software, Vol 7, No 1 (1981), p 17–41.

38 Basis Splines

This chapter describes functions for the computation of smoothing basis splines (B-splines). A smoothing spline differs from an interpolating spline in that the resulting curve is not required to pass through each datapoint. See Chapter 26 [Interpolation], page 359, for information about interpolating splines.

The header file 'gsl_bspline.h' contains the prototypes for the bspline functions and related declarations.

38.1 Overview

B-splines are commonly used as basis functions to fit smoothing curves to large data sets. To do this, the abscissa axis is broken up into some number of intervals, where the endpoints of each interval are called *breakpoints*. These breakpoints are then converted to *knots* by imposing various continuity and smoothness conditions at each interface. Given a nondecreasing knot vector $t = \{t_0, t_1, \ldots, t_{n+k-1}\}$, the n basis splines of order k are defined by

$$B_{i,1}(x) = \begin{cases} 1, & t_i \leq x < t_{i+1} \\ 0, & else \end{cases}$$

$$B_{i,k}(x) = \frac{(x - t_i)}{(t_{i+k-1} - t_i)} B_{i,k-1}(x) + \frac{(t_{i+k} - x)}{(t_{i+k} - t_{i+1})} B_{i+1,k-1}(x)$$

for $i = 0, \ldots, n-1$. The common case of cubic B-splines is given by $k = 4$. The above recurrence relation can be evaluated in a numerically stable way by the de Boor algorithm.

If we define appropriate knots on an interval $[a, b]$ then the B-spline basis functions form a complete set on that interval. Therefore we can expand a smoothing function as

$$f(x) = \sum_{i=0}^{n-1} c_i B_{i,k}(x)$$

given enough $(x_j, f(x_j))$ data pairs. The coefficients c_i can be readily obtained from a least-squares fit.

38.2 Initializing the B-splines solver

The computation of B-spline functions requires a preallocated workspace of type gsl_bspline_workspace. If B-spline derivatives are also required, an additional gsl_bspline_deriv_workspace is needed.

gsl_bspline_workspace * gsl_bspline_alloc (const size_t k,   *Function*
        const size_t nbreak)
    This function allocates a workspace for computing B-splines of order k. The
    number of breakpoints is given by nbreak. This leads to $n = nbreak + k - 2$
    basis functions. Cubic B-splines are specified by $k = 4$. The size of the
    workspace is $O(5k + nbreak)$.

void gsl_bspline_free (gsl_bspline_workspace * w)   *Function*
    This function frees the memory associated with the workspace w.

gsl_bspline_deriv_workspace * gsl_bspline_deriv_alloc (const   *Function*
        size_t k)
    This function allocates a workspace for computing the derivatives of a B-
    spline basis function of order k. The size of the workspace is $O(2k^2)$.

void gsl_bspline_deriv_free (gsl_bspline_deriv_workspace *   *Function*
        w)
    This function frees the memory associated with the derivative workspace w.

38.3 Constructing the knots vector

int gsl_bspline_knots (const gsl_vector * breakpts,   *Function*
        gsl_bspline_workspace * w)
    This function computes the knots associated with the given breakpoints and
    stores them internally in w->knots.

int gsl_bspline_knots_uniform (const double a, const double b,   *Function*
        gsl_bspline_workspace * w)
    This function assumes uniformly spaced breakpoints on $[a, b]$ and constructs
    the corresponding knot vector using the previously specified nbreak param-
    eter. The knots are stored in w->knots.

38.4 Evaluation of B-splines

int gsl_bspline_eval (const double x, gsl_vector * B,   *Function*
        gsl_bspline_workspace * w)
    This function evaluates all B-spline basis functions at the position x and
    stores them in the vector B, so that the i-th element is $B_i(x)$. The vector B
    must be of length $n = nbreak + k - 2$. This value may also be obtained by
    calling gsl_bspline_ncoeffs. Computing all the basis functions at once is
    more efficient than computing them individually, due to the nature of the
    defining recurrence relation.

int gsl_bspline_eval_nonzero (const double x, gsl_vector *   *Function*
        Bk, size_t * istart, size_t * iend, gsl_bspline_workspace *
        w)
    This function evaluates all potentially nonzero B-spline basis functions at
    the position x and stores them in the vector Bk, so that the i-th element
    is $B_{(istart+i)}(x)$. The last element of Bk is $B_{iend}(x)$. The vector Bk must
    be of length k. By returning only the nonzero basis functions, this function
    allows quantities involving linear combinations of the $B_i(x)$ to be computed
    without unnecessary terms (such linear combinations occur, for example,
    when evaluating an interpolated function).

size_t gsl_bspline_ncoeffs (gsl_bspline_workspace * w)   *Function*
    This function returns the number of B-spline coefficients given by $n = nbreak + k - 2$.

38.5 Evaluation of B-spline derivatives

int gsl_bspline_deriv_eval (const double x, const size_t   *Function*
        nderiv, gsl_matrix * dB, gsl_bspline_workspace * w,
        gsl_bspline_deriv_workspace * dw)
    This function evaluates all B-spline basis function derivatives of orders 0
    through nderiv (inclusive) at the position x and stores them in the matrix
    dB. The (i, j)-th element of dB is $d^j B_i(x)/dx^j$. The matrix dB must be
    of size $n = nbreak + k - 2$ by $nderiv + 1$. The value n may also be ob-
    tained by calling gsl_bspline_ncoeffs. Note that function evaluations are
    included as the zeroth order derivatives in dB. Computing all the basis func-
    tion derivatives at once is more efficient than computing them individually,
    due to the nature of the defining recurrence relation.

int gsl_bspline_deriv_eval_nonzero (const double x, const   *Function*
        size_t nderiv, gsl_matrix * dB, size_t * istart, size_t *
        iend, gsl_bspline_workspace * w,
        gsl_bspline_deriv_workspace * dw)
    This function evaluates all potentially nonzero B-spline basis function deriva-
    tives of orders 0 through nderiv (inclusive) at the position x and stores them
    in the matrix dB. The (i, j)-th element of dB is $d^j B_{(istart+i)}(x)/dx^j$. The
    last row of dB contains $d^j B_{iend}(x)/dx^j$. The matrix dB must be of size k by

at least $nderiv + 1$. Note that function evaluations are included as the zeroth order derivatives in dB. By returning only the nonzero basis functions, this function allows quantities involving linear combinations of the $B_i(x)$ and their derivatives to be computed without unnecessary terms.

38.6 Examples

The following program computes a linear least squares fit to data using cubic B-spline basis functions with uniform breakpoints. The data is generated from the curve $y(x) = \cos(x) \exp(-x/10)$ on the interval $[0, 15]$ with gaussian noise added.

```
#include <stdio.h>
#include <stdlib.h>
#include <math.h>
#include <gsl/gsl_bspline.h>
#include <gsl/gsl_multifit.h>
#include <gsl/gsl_rng.h>
#include <gsl/gsl_randist.h>
#include <gsl/gsl_statistics.h>

/* number of data points to fit */
#define N         200

/* number of fit coefficients */
#define NCOEFFS  12

/* nbreak = ncoeffs + 2 - k = ncoeffs - 2 since k = 4 */
#define NBREAK   (NCOEFFS - 2)

int
main (void)
{
  const size_t n = N;
  const size_t ncoeffs = NCOEFFS;
  const size_t nbreak = NBREAK;
  size_t i, j;
  gsl_bspline_workspace *bw;
  gsl_vector *B;
  double dy;
  gsl_rng *r;
  gsl_vector *c, *w;
  gsl_vector *x, *y;
  gsl_matrix *X, *cov;
  gsl_multifit_linear_workspace *mw;
  double chisq, Rsq, dof, tss;

  gsl_rng_env_setup();
  r = gsl_rng_alloc(gsl_rng_default);
```

```
/* allocate a cubic bspline workspace (k = 4) */
bw = gsl_bspline_alloc(4, nbreak);
B = gsl_vector_alloc(ncoeffs);

x = gsl_vector_alloc(n);
y = gsl_vector_alloc(n);
X = gsl_matrix_alloc(n, ncoeffs);
c = gsl_vector_alloc(ncoeffs);
w = gsl_vector_alloc(n);
cov = gsl_matrix_alloc(ncoeffs, ncoeffs);
mw = gsl_multifit_linear_alloc(n, ncoeffs);

printf("#m=0,S=0\n");
/* this is the data to be fitted */
for (i = 0; i < n; ++i)
  {
    double sigma;
    double xi = (15.0 / (N - 1)) * i;
    double yi = cos(xi) * exp(-0.1 * xi);

    sigma = 0.1 * yi;
    dy = gsl_ran_gaussian(r, sigma);
    yi += dy;

    gsl_vector_set(x, i, xi);
    gsl_vector_set(y, i, yi);
    gsl_vector_set(w, i, 1.0 / (sigma * sigma));

    printf("%f %f\n", xi, yi);
  }

/* use uniform breakpoints on [0, 15] */
gsl_bspline_knots_uniform(0.0, 15.0, bw);

/* construct the fit matrix X */
for (i = 0; i < n; ++i)
  {
    double xi = gsl_vector_get(x, i);

    /* compute B_j(xi) for all j */
    gsl_bspline_eval(xi, B, bw);

    /* fill in row i of X */
    for (j = 0; j < ncoeffs; ++j)
      {
        double Bj = gsl_vector_get(B, j);
        gsl_matrix_set(X, i, j, Bj);
```

```
        }
      }

    /* do the fit */
    gsl_multifit_wlinear(X, w, y, c, cov, &chisq, mw);

    dof = n - ncoeffs;
    tss = gsl_stats_wtss(w->data, 1, y->data, 1, y->size);
    Rsq = 1.0 - chisq / tss;

    fprintf(stderr, "chisq/dof = %e, Rsq = %f\n",
                    chisq / dof, Rsq);

    /* output the smoothed curve */
    {
      double xi, yi, yerr;

      printf("#m=1,S=0\n");
      for (xi = 0.0; xi < 15.0; xi += 0.1)
        {
          gsl_bspline_eval(xi, B, bw);
          gsl_multifit_linear_est(B, c, cov, &yi, &yerr);
          printf("%f %f\n", xi, yi);
        }
    }

    gsl_rng_free(r);
    gsl_bspline_free(bw);
    gsl_vector_free(B);
    gsl_vector_free(x);
    gsl_vector_free(y);
    gsl_matrix_free(X);
    gsl_vector_free(c);
    gsl_vector_free(w);
    gsl_matrix_free(cov);
    gsl_multifit_linear_free(mw);

    return 0;
  } /* main() */
```

The output can be plotted with GNU graph.

```
  $ ./a.out > bspline.dat
  chisq/dof = 1.118217e+00, Rsq = 0.989771
  $ graph -T ps -X x -Y y -x 0 15 -y -1 1.3 < bspline.dat
      > bspline.ps
```

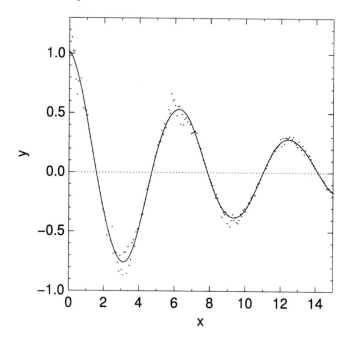

38.7 References and Further Reading

Further information on the algorithms described in this section can be found in the following book,

C. de Boor, *A Practical Guide to Splines* (1978), Springer-Verlag, ISBN 0-387-90356-9.

A large collection of B-spline routines is available in the PPPACK library available at http://www.netlib.org/pppack, which is also part of SLATEC.

39 Physical Constants

This chapter describes macros for the values of physical constants, such as the speed of light, c, and gravitational constant, G. The values are available in different unit systems, including the standard MKSA system (meters, kilograms, seconds, amperes) and the CGSM system (centimeters, grams, seconds, gauss), which is commonly used in Astronomy.

The definitions of constants in the MKSA system are available in the file 'gsl_const_mksa.h'. The constants in the CGSM system are defined in 'gsl_const_cgsm.h'. Dimensionless constants, such as the fine structure constant, which are pure numbers are defined in 'gsl_const_num.h'.

The full list of constants is described briefly below. Consult the header files themselves for the values of the constants used in the library.

39.1 Fundamental Constants

GSL_CONST_MKSA_SPEED_OF_LIGHT
    The speed of light in vacuum, c.

GSL_CONST_MKSA_VACUUM_PERMEABILITY
    The permeability of free space, μ_0. This constant is defined in the MKSA system only.

GSL_CONST_MKSA_VACUUM_PERMITTIVITY
    The permittivity of free space, ϵ_0. This constant is defined in the MKSA system only.

GSL_CONST_MKSA_PLANCKS_CONSTANT_H
    Planck's constant, h.

GSL_CONST_MKSA_PLANCKS_CONSTANT_HBAR
    Planck's constant divided by 2π, $\hbar$.

GSL_CONST_NUM_AVOGADRO
    Avogadro's number, N_a.

GSL_CONST_MKSA_FARADAY
    The molar charge of 1 Faraday.

GSL_CONST_MKSA_BOLTZMANN
    The Boltzmann constant, k.

GSL_CONST_MKSA_MOLAR_GAS
    The molar gas constant, R_0.

GSL_CONST_MKSA_STANDARD_GAS_VOLUME
    The standard gas volume, V_0.

GSL_CONST_MKSA_STEFAN_BOLTZMANN_CONSTANT
    The Stefan-Boltzmann radiation constant, σ.

GSL_CONST_MKSA_GAUSS
    The magnetic field of 1 Gauss.

39.2 Astronomy and Astrophysics

GSL_CONST_MKSA_ASTRONOMICAL_UNIT
   The length of 1 astronomical unit (mean earth-sun distance), au.

GSL_CONST_MKSA_GRAVITATIONAL_CONSTANT
   The gravitational constant, G.

GSL_CONST_MKSA_LIGHT_YEAR
   The distance of 1 light-year, ly.

GSL_CONST_MKSA_PARSEC
   The distance of 1 parsec, pc.

GSL_CONST_MKSA_GRAV_ACCEL
   The standard gravitational acceleration on Earth, g.

GSL_CONST_MKSA_SOLAR_MASS
   The mass of the Sun.

39.3 Atomic and Nuclear Physics

GSL_CONST_MKSA_ELECTRON_CHARGE
   The charge of the electron, e.

GSL_CONST_MKSA_ELECTRON_VOLT
   The energy of 1 electron volt, eV.

GSL_CONST_MKSA_UNIFIED_ATOMIC_MASS
   The unified atomic mass, amu.

GSL_CONST_MKSA_MASS_ELECTRON
   The mass of the electron, m_e.

GSL_CONST_MKSA_MASS_MUON
   The mass of the muon, m_μ.

GSL_CONST_MKSA_MASS_PROTON
   The mass of the proton, m_p.

GSL_CONST_MKSA_MASS_NEUTRON
   The mass of the neutron, m_n.

GSL_CONST_NUM_FINE_STRUCTURE
   The electromagnetic fine structure constant α.

GSL_CONST_MKSA_RYDBERG
   The Rydberg constant, Ry, in units of energy. This is related to the
   Rydberg inverse wavelength R by $Ry = hcR$.

GSL_CONST_MKSA_BOHR_RADIUS
   The Bohr radius, a_0.

GSL_CONST_MKSA_ANGSTROM
   The length of 1 angstrom.

GSL_CONST_MKSA_BARN
>   The area of 1 barn.

GSL_CONST_MKSA_BOHR_MAGNETON
>   The Bohr Magneton, μ_B.

GSL_CONST_MKSA_NUCLEAR_MAGNETON
>   The Nuclear Magneton, μ_N.

GSL_CONST_MKSA_ELECTRON_MAGNETIC_MOMENT
>   The absolute value of the magnetic moment of the electron, μ_e. The physical magnetic moment of the electron is negative.

GSL_CONST_MKSA_PROTON_MAGNETIC_MOMENT
>   The magnetic moment of the proton, μ_p.

GSL_CONST_MKSA_THOMSON_CROSS_SECTION
>   The Thomson cross section, σ_T.

GSL_CONST_MKSA_DEBYE
>   The electric dipole moment of 1 Debye, D.

39.4 Measurement of Time

GSL_CONST_MKSA_MINUTE
>   The number of seconds in 1 minute.

GSL_CONST_MKSA_HOUR
>   The number of seconds in 1 hour.

GSL_CONST_MKSA_DAY
>   The number of seconds in 1 day.

GSL_CONST_MKSA_WEEK
>   The number of seconds in 1 week.

39.5 Imperial Units

GSL_CONST_MKSA_INCH
>   The length of 1 inch.

GSL_CONST_MKSA_FOOT
>   The length of 1 foot.

GSL_CONST_MKSA_YARD
>   The length of 1 yard.

GSL_CONST_MKSA_MILE
>   The length of 1 mile.

GSL_CONST_MKSA_MIL
>   The length of 1 mil (1/1000th of an inch).

39.6 Speed and Nautical Units

GSL_CONST_MKSA_KILOMETERS_PER_HOUR
>   The speed of 1 kilometer per hour.

GSL_CONST_MKSA_MILES_PER_HOUR
>   The speed of 1 mile per hour.

GSL_CONST_MKSA_NAUTICAL_MILE
>   The length of 1 nautical mile.

GSL_CONST_MKSA_FATHOM
>   The length of 1 fathom.

GSL_CONST_MKSA_KNOT
>   The speed of 1 knot.

39.7 Printers Units

GSL_CONST_MKSA_POINT
>   The length of 1 printer's point (1/72 inch).

GSL_CONST_MKSA_TEXPOINT
>   The length of 1 TeX point (1/72.27 inch).

39.8 Volume, Area and Length

GSL_CONST_MKSA_MICRON
>   The length of 1 micron.

GSL_CONST_MKSA_HECTARE
>   The area of 1 hectare.

GSL_CONST_MKSA_ACRE
>   The area of 1 acre.

GSL_CONST_MKSA_LITER
>   The volume of 1 liter.

GSL_CONST_MKSA_US_GALLON
>   The volume of 1 US gallon.

GSL_CONST_MKSA_CANADIAN_GALLON
>   The volume of 1 Canadian gallon.

GSL_CONST_MKSA_UK_GALLON
>   The volume of 1 UK gallon.

GSL_CONST_MKSA_QUART
>   The volume of 1 quart.

GSL_CONST_MKSA_PINT
>   The volume of 1 pint.

39.9 Mass and Weight

GSL_CONST_MKSA_POUND_MASS
    The mass of 1 pound.

GSL_CONST_MKSA_OUNCE_MASS
    The mass of 1 ounce.

GSL_CONST_MKSA_TON
    The mass of 1 ton.

GSL_CONST_MKSA_METRIC_TON
    The mass of 1 metric ton (1000 kg).

GSL_CONST_MKSA_UK_TON
    The mass of 1 UK ton.

GSL_CONST_MKSA_TROY_OUNCE
    The mass of 1 troy ounce.

GSL_CONST_MKSA_CARAT
    The mass of 1 carat.

GSL_CONST_MKSA_GRAM_FORCE
    The force of 1 gram weight.

GSL_CONST_MKSA_POUND_FORCE
    The force of 1 pound weight.

GSL_CONST_MKSA_KILOPOUND_FORCE
    The force of 1 kilopound weight.

GSL_CONST_MKSA_POUNDAL
    The force of 1 poundal.

39.10 Thermal Energy and Power

GSL_CONST_MKSA_CALORIE
    The energy of 1 calorie.

GSL_CONST_MKSA_BTU
    The energy of 1 British Thermal Unit, btu.

GSL_CONST_MKSA_THERM
    The energy of 1 Therm.

GSL_CONST_MKSA_HORSEPOWER
    The power of 1 horsepower.

39.11 Pressure

GSL_CONST_MKSA_BAR
>   The pressure of 1 bar.

GSL_CONST_MKSA_STD_ATMOSPHERE
>   The pressure of 1 standard atmosphere.

GSL_CONST_MKSA_TORR
>   The pressure of 1 torr.

GSL_CONST_MKSA_METER_OF_MERCURY
>   The pressure of 1 meter of mercury.

GSL_CONST_MKSA_INCH_OF_MERCURY
>   The pressure of 1 inch of mercury.

GSL_CONST_MKSA_INCH_OF_WATER
>   The pressure of 1 inch of water.

GSL_CONST_MKSA_PSI
>   The pressure of 1 pound per square inch.

39.12 Viscosity

GSL_CONST_MKSA_POISE
>   The dynamic viscosity of 1 poise.

GSL_CONST_MKSA_STOKES
>   The kinematic viscosity of 1 stokes.

39.13 Light and Illumination

GSL_CONST_MKSA_STILB
>   The luminance of 1 stilb.

GSL_CONST_MKSA_LUMEN
>   The luminous flux of 1 lumen.

GSL_CONST_MKSA_LUX
>   The illuminance of 1 lux.

GSL_CONST_MKSA_PHOT
>   The illuminance of 1 phot.

GSL_CONST_MKSA_FOOTCANDLE
>   The illuminance of 1 footcandle.

GSL_CONST_MKSA_LAMBERT
>   The luminance of 1 lambert.

GSL_CONST_MKSA_FOOTLAMBERT
>   The luminance of 1 footlambert.

39.14 Radioactivity

GSL_CONST_MKSA_CURIE
> The activity of 1 curie.

GSL_CONST_MKSA_ROENTGEN
> The exposure of 1 roentgen.

GSL_CONST_MKSA_RAD
> The absorbed dose of 1 rad.

39.15 Force and Energy

GSL_CONST_MKSA_NEWTON
> The SI unit of force, 1 Newton.

GSL_CONST_MKSA_DYNE
> The force of 1 Dyne $= 10^{-5}$ Newton.

GSL_CONST_MKSA_JOULE
> The SI unit of energy, 1 Joule.

GSL_CONST_MKSA_ERG
> The energy 1 erg $= 10^{-7}$ Joule.

39.16 Prefixes

These constants are dimensionless scaling factors.

GSL_CONST_NUM_YOTTA
> 10^{24}

GSL_CONST_NUM_ZETTA
> 10^{21}

GSL_CONST_NUM_EXA
> 10^{18}

GSL_CONST_NUM_PETA
> 10^{15}

GSL_CONST_NUM_TERA
> 10^{12}

GSL_CONST_NUM_GIGA
> 10^{9}

GSL_CONST_NUM_MEGA
> 10^{6}

GSL_CONST_NUM_KILO
> 10^{3}

GSL_CONST_NUM_MILLI
> 10^{-3}

GSL_CONST_NUM_MICRO
10^{-6}

GSL_CONST_NUM_NANO
10^{-9}

GSL_CONST_NUM_PICO
10^{-12}

GSL_CONST_NUM_FEMTO
10^{-15}

GSL_CONST_NUM_ATTO
10^{-18}

GSL_CONST_NUM_ZEPTO
10^{-21}

GSL_CONST_NUM_YOCTO
10^{-24}

39.17 Examples

The following program demonstrates the use of the physical constants in a calculation. In this case, the goal is to calculate the range of light-travel times from Earth to Mars.

The required data is the average distance of each planet from the Sun in astronomical units (the eccentricities and inclinations of the orbits will be neglected for the purposes of this calculation). The average radius of the orbit of Mars is 1.52 astronomical units, and for the orbit of Earth it is 1 astronomical unit (by definition). These values are combined with the MKSA values of the constants for the speed of light and the length of an astronomical unit to produce a result for the shortest and longest light-travel times in seconds. The figures are converted into minutes before being displayed.

```
#include <stdio.h>
#include <gsl/gsl_const_mksa.h>

int
main (void)
{
  double c  = GSL_CONST_MKSA_SPEED_OF_LIGHT;
  double au = GSL_CONST_MKSA_ASTRONOMICAL_UNIT;
  double minutes = GSL_CONST_MKSA_MINUTE;

  /* distance stored in meters */
  double r_earth = 1.00 * au;
  double r_mars  = 1.52 * au;

  double t_min, t_max;

  t_min = (r_mars - r_earth) / c;
```

```
        t_max = (r_mars + r_earth) / c;

        printf ("light travel time from Earth to Mars:\n");
        printf ("minimum = %.1f minutes\n", t_min / minutes);
        printf ("maximum = %.1f minutes\n", t_max / minutes);

        return 0;
    }
```

Here is the output from the program,

```
    light travel time from Earth to Mars:
    minimum = 4.3 minutes
    maximum = 21.0 minutes
```

39.18 References and Further Reading

The authoritative sources for physical constants are the 2002 CODATA recommended values, published in the articles below. Further information on the values of physical constants is also available from the cited articles and the NIST website.

Journal of Physical and Chemical Reference Data, 28(6), 1713-1852, 1999

Reviews of Modern Physics, 72(2), 351-495, 2000

http://www.physics.nist.gov/cuu/Constants/index.html

http://physics.nist.gov/Pubs/SP811/appenB9.html

40 IEEE floating-point arithmetic

This chapter describes functions for examining the representation of floating point numbers and controlling the floating point environment of your program. The functions described in this chapter are declared in the header file 'gsl_ieee_utils.h'.

40.1 Representation of floating point numbers

The IEEE Standard for Binary Floating-Point Arithmetic defines binary formats for single and double precision numbers. Each number is composed of three parts: a *sign bit* (s), an *exponent* (E) and a *fraction* (f). The numerical value of the combination (s, E, f) is given by the following formula,

$$(-1)^s(1 \cdot fffff \ldots)2^E$$

The sign bit is either zero or one. The exponent ranges from a minimum value E_{min} to a maximum value E_{max} depending on the precision. The exponent is converted to an unsigned number e, known as the *biased exponent*, for storage by adding a *bias* parameter, $e = E + bias$. The sequence $fffff \ldots$ represents the digits of the binary fraction f. The binary digits are stored in *normalized form*, by adjusting the exponent to give a leading digit of 1. Since the leading digit is always 1 for normalized numbers it is assumed implicitly and does not have to be stored. Numbers smaller than $2^{E_{min}}$ are be stored in *denormalized form* with a leading zero,

$$(-1)^s(0 \cdot fffff \ldots)2^{E_{min}}$$

This allows gradual underflow down to $2^{E_{min}-p}$ for p bits of precision. A zero is encoded with the special exponent of $2^{E_{min}-1}$ and infinities with the exponent of $2^{E_{max}+1}$.

The format for single precision numbers uses 32 bits divided in the following way,

```
seeeeeeeefffffffffffffffffffffff

s = sign bit, 1 bit
e = exponent, 8 bits   (E_min=-126, E_max=127, bias=127)
f = fraction, 23 bits
```

The format for double precision numbers uses 64 bits divided in the following way,

```
seeeeeeeeeeefffffffffffffffffffffffffffffffffffffffffffffffffffff

s = sign bit, 1 bit
e = exponent, 11 bits  (E_min=-1022, E_max=1023, bias=1023)
f = fraction, 52 bits
```

It is often useful to be able to investigate the behavior of a calculation at the bit-level and the library provides functions for printing the IEEE representations in a human-readable form.

void gsl_ieee_fprintf_float (FILE * *stream*, const float * x)   Function
void gsl_ieee_fprintf_double (FILE * *stream*, const double * x)   Function
    These functions output a formatted version of the IEEE floating-point num-
    ber pointed to by x to the stream *stream*. A pointer is used to pass the
    number indirectly, to avoid any undesired promotion from float to double.
    The output takes one of the following forms,

    NaN
        the Not-a-Number symbol

    Inf, -Inf
        positive or negative infinity

    1.fffff...*2^E, -1.fffff...*2^E
        a normalized floating point number

    0.fffff...*2^E, -0.fffff...*2^E
        a denormalized floating point number

    0, -0
        positive or negative zero

    The output can be used directly in GNU Emacs Calc mode by preceding it
    with 2# to indicate binary.

void gsl_ieee_printf_float (const float * x)   Function
void gsl_ieee_printf_double (const double * x)   Function
    These functions output a formatted version of the IEEE floating-point num-
    ber pointed to by x to the stream stdout.

The following program demonstrates the use of the functions by printing the
single and double precision representations of the fraction 1/3. For comparison
the representation of the value promoted from single to double precision is also
printed.

```
#include <stdio.h>
#include <gsl/gsl_ieee_utils.h>

int
main (void)
{
  float f = 1.0/3.0;
  double d = 1.0/3.0;

  double fd = f; /* promote from float to double */

  printf (" f="); gsl_ieee_printf_float(&f);
  printf ("\n");

  printf ("fd="); gsl_ieee_printf_double(&fd);
  printf ("\n");
```

```
    printf (" d="); gsl_ieee_printf_double(&d);
    printf ("\n");

    return 0;
}
```

The binary representation of 1/3 is 0.01010101.... The output below shows that the IEEE format normalizes this fraction to give a leading digit of 1,

```
 f= 1.01010101010101010101011*2^-2
fd= 1.0101010101010101010101100000000000000000000000000000*2^-2
 d= 1.0101010101010101010101010101010101010101010101010101*2^-2
```

The output also shows that a single-precision number is promoted to double-precision by adding zeros in the binary representation.

40.2 Setting up your IEEE environment

The IEEE standard defines several *modes* for controlling the behavior of floating point operations. These modes specify the important properties of computer arithmetic: the direction used for rounding (e.g. whether numbers should be rounded up, down or to the nearest number), the rounding precision and how the program should handle arithmetic exceptions, such as division by zero.

Many of these features can now be controlled via standard functions such as fpsetround, which should be used whenever they are available. Unfortunately in the past there has been no universal API for controlling their behavior—each system has had its own low-level way of accessing them. To help you write portable programs GSL allows you to specify modes in a platform-independent way using the environment variable GSL_IEEE_MODE. The library then takes care of all the necessary machine-specific initializations for you when you call the function gsl_ieee_env_setup.

void gsl_ieee_env_setup ()   *Function*

This function reads the environment variable GSL_IEEE_MODE and attempts to set up the corresponding specified IEEE modes. The environment variable should be a list of keywords, separated by commas, like this,

GSL_IEEE_MODE = "*keyword,keyword,...*"

where *keyword* is one of the following mode-names,

single-precision

double-precision

extended-precision

round-to-nearest

round-down

round-up

round-to-zero

mask-all

mask-invalid

```
mask-denormalized
```

```
mask-division-by-zero
```

```
mask-overflow
```

```
mask-underflow
```

```
trap-inexact
```

```
trap-common
```

If `GSL_IEEE_MODE` is empty or undefined then the function returns immediately and no attempt is made to change the system's IEEE mode. When the modes from `GSL_IEEE_MODE` are turned on the function prints a short message showing the new settings to remind you that the results of the program will be affected.

If the requested modes are not supported by the platform being used then the function calls the error handler and returns an error code of `GSL_EUNSUP`.

When options are specified using this method, the resulting mode is based on a default setting of the highest available precision (double precision or extended precision, depending on the platform) in round-to-nearest mode, with all exceptions enabled apart from the INEXACT exception. The INEXACT exception is generated whenever rounding occurs, so it must generally be disabled in typical scientific calculations. All other floating-point exceptions are enabled by default, including underflows and the use of denormalized numbers, for safety. They can be disabled with the individual mask- settings or together using `mask-all`.

The following adjusted combination of modes is convenient for many purposes,

```
GSL_IEEE_MODE="double-precision,"\
              "mask-underflow,"\
                "mask-denormalized"
```

This choice ignores any errors relating to small numbers (either denormalized, or underflowing to zero) but traps overflows, division by zero and invalid operations.

Note that on the x86 series of processors this function sets both the original x87 mode and the newer MXCSR mode, which controls SSE floating-point operations. The SSE floating-point units do not have a precision-control bit, and always work in double-precision. The single-precision and extended-precision keywords have no effect in this case.

To demonstrate the effects of different rounding modes consider the following program which computes e, the base of natural logarithms, by summing a rapidly-decreasing series,

$$e = 1 + \frac{1}{2!} + \frac{1}{3!} + \frac{1}{4!} + \ldots = 2.71828182846\ldots$$

```
#include <stdio.h>
#include <gsl/gsl_math.h>
#include <gsl/gsl_ieee_utils.h>
```

```
int
main (void)
{
  double x = 1, oldsum = 0, sum = 0;
  int i = 0;

  gsl_ieee_env_setup (); /* read GSL_IEEE_MODE */

  do
    {
      i++;

      oldsum = sum;
      sum += x;
      x = x / i;

      printf ("i=%2d sum=%.18f error=%g\n",
              i, sum, sum - M_E);

      if (i > 30)
         break;
    }
  while (sum != oldsum);

  return 0;
}
```

Here are the results of running the program in round-to-nearest mode. This is the IEEE default so it isn't really necessary to specify it here,

```
$ GSL_IEEE_MODE="round-to-nearest" ./a.out
i= 1 sum=1.000000000000000000 error=-1.71828
i= 2 sum=2.000000000000000000 error=-0.718282
....
i=18 sum=2.718281828459045535 error=4.44089e-16
i=19 sum=2.718281828459045535 error=4.44089e-16
```

After nineteen terms the sum converges to within 4×10^{-16} of the correct value. If we now change the rounding mode to round-down the final result is less accurate,

```
$ GSL_IEEE_MODE="round-down" ./a.out
i= 1 sum=1.000000000000000000 error=-1.71828
....
i=19 sum=2.718281828459041094 error=-3.9968e-15
```

The result is about 4×10^{-15} below the correct value, an order of magnitude worse than the result obtained in the round-to-nearest mode.

If we change to rounding mode to round-up then the final result is higher than the correct value (when we add each term to the sum the final result is always rounded up, which increases the sum by at least one tick until the added term underflows to zero). To avoid this problem we would need to use a safer converge criterion, such as while (fabs(sum - oldsum) > epsilon), with a suitably chosen value of epsilon.

Finally we can see the effect of computing the sum using single-precision rounding, in the default round-to-nearest mode. In this case the program thinks it is still using double precision numbers but the CPU rounds the result of each floating point operation to single-precision accuracy. This simulates the effect of writing the program using single-precision float variables instead of double variables. The iteration stops after about half the number of iterations and the final result is much less accurate,

```
$ GSL_IEEE_MODE="single-precision" ./a.out
....
i=12 sum=2.718281984329223633 error=1.5587e-07
```

with an error of $O(10^{-7})$, which corresponds to single precision accuracy (about 1 part in 10^7). Continuing the iterations further does not decrease the error because all the subsequent results are rounded to the same value.

40.3 References and Further Reading

The reference for the IEEE standard is,

ANSI/IEEE Std 754-1985, IEEE Standard for Binary Floating-Point Arithmetic.

A more pedagogical introduction to the standard can be found in the following paper,

David Goldberg: What Every Computer Scientist Should Know About Floating-Point Arithmetic. *ACM Computing Surveys*, Vol. 23, No. 1 (March 1991), pages 5–48.

Corrigendum: *ACM Computing Surveys*, Vol. 23, No. 3 (September 1991), page 413. and see also the sections by B. A. Wichmann and Charles B. Dunham in Surveyor's Forum: "What Every Computer Scientist Should Know About Floating-Point Arithmetic". *ACM Computing Surveys*, Vol. 24, No. 3 (September 1992), page 319.

A detailed textbook on IEEE arithmetic and its practical use is available from SIAM Press,

Michael L. Overton, *Numerical Computing with IEEE Floating Point Arithmetic*, SIAM Press, ISBN 0898715717.

Appendix A Debugging Numerical Programs

This chapter describes some tips and tricks for debugging numerical programs which use GSL.

A.1 Using gdb

Any errors reported by the library are passed to the function `gsl_error`. By running your programs under gdb and setting a breakpoint in this function you can automatically catch any library errors. You can add a breakpoint for every session by putting

```
break gsl_error
```

into your '.gdbinit' file in the directory where your program is started.

If the breakpoint catches an error then you can use a backtrace (bt) to see the call-tree, and the arguments which possibly caused the error. By moving up into the calling function you can investigate the values of variables at that point. Here is an example from the program `fft/test_trap`, which contains the following line,

```
status = gsl_fft_complex_wavetable_alloc (0, &complex_wavetable);
```

The function `gsl_fft_complex_wavetable_alloc` takes the length of an FFT as its first argument. When this line is executed an error will be generated because the length of an FFT is not allowed to be zero.

To debug this problem we start gdb, using the file '.gdbinit' to define a breakpoint in `gsl_error`,

```
$ gdb test_trap
```

```
GDB is free software and you are welcome to distribute copies
of it under certain conditions; type "show copying" to see
the conditions.  There is absolutely no warranty for GDB;
type "show warranty" for details.  GDB 4.16 (i586-gnu-linux),
Copyright 1996 Free Software Foundation, Inc.

Breakpoint 1 at 0x8050b1e: file error.c, line 14.
```

When we run the program this breakpoint catches the error and shows the reason for it.

```
(gdb) run
Starting program: test_trap

Breakpoint 1, gsl_error (reason=0x8052b0d
    "length n must be positive integer",
    file=0x8052b04 "c_init.c", line=108, gsl_errno=1)
    at error.c:14
14          if (gsl_error_handler)
```

The first argument of gsl_error is always a string describing the error. Now we can look at the backtrace to see what caused the problem,

```
(gdb) bt
#0  gsl_error (reason=0x8052b0d
    "length n must be positive integer",
    file=0x8052b04 "c_init.c", line=108, gsl_errno=1)
    at error.c:14
#1  0x8049376 in gsl_fft_complex_wavetable_alloc (n=0,
    wavetable=0xbffff778) at c_init.c:108
#2  0x8048a00 in main (argc=1, argv=0xbffff9bc)
    at test_trap.c:94
#3  0x80488be in ___crt_dummy__ ()
```

We can see that the error was generated in the function gsl_fft_complex_wavetable_alloc when it was called with an argument of $n=0$. The original call came from line 94 in the file 'test_trap.c'.

By moving up to the level of the original call we can find the line that caused the error,

```
(gdb) up
#1  0x8049376 in gsl_fft_complex_wavetable_alloc (n=0,
    wavetable=0xbffff778) at c_init.c:108
108    GSL_ERROR ("length n must be positive integer", GSL_EDOM);
(gdb) up
#2  0x8048a00 in main (argc=1, argv=0xbffff9bc)
    at test_trap.c:94
94     status = gsl_fft_complex_wavetable_alloc (0,
           &complex_wavetable);
```

Thus we have found the line that caused the problem. From this point we could also print out the values of other variables such as complex_wavetable.

A.2 Examining floating point registers

The contents of floating point registers can be examined using the command info float (on supported platforms).

```
(gdb) info float
    st0: 0xc4018b895aa17a945000  Valid Normal -7.838871e+308
    st1: 0x3ff9ea3f50e4d7275000  Valid Normal 0.0285946
    st2: 0x3fe790c64ce27dad4800  Valid Normal 6.7415931e-08
    st3: 0x3ffaa3ef0df6607d7800  Spec  Normal 0.0400229
    st4: 0x3c028000000000000000  Valid Normal 4.4501477e-308
    st5: 0x3ffef5412c22219d9000  Zero  Normal 0.9580257
    st6: 0x3fff8000000000000000  Valid Normal 1
    st7: 0xc4028b65a1f6d243c800  Valid Normal -1.566206e+309
  fctrl: 0x0272 53 bit; NEAR; mask DENOR UNDER LOS;
  fstat: 0xb9ba flags 0001; top 7; excep DENOR OVERF UNDER LOS
   ftag: 0x3fff
    fip: 0x08048b5c
    fcs: 0x051a0023
```

```
fopoff: 0x08086820
fopsel: 0x002b
```

Individual registers can be examined using the variables $reg, where reg is the register name.

```
(gdb) p $st1
$1 = 0.028594644545426121021034719
```

A.3 Handling floating point exceptions

It is possible to stop the program whenever a SIGFPE floating point exception occurs. This can be useful for finding the cause of an unexpected infinity or NaN. The current handler settings can be shown with the command info signal SIGFPE.

```
(gdb) info signal SIGFPE
Signal  Stop  Print  Pass to program Description
SIGFPE  Yes   Yes    Yes             Arithmetic exception
```

Unless the program uses a signal handler the default setting should be changed so that SIGFPE is not passed to the program, as this would cause it to exit. The command handle SIGFPE stop nopass prevents this.

```
(gdb) handle SIGFPE stop nopass
Signal  Stop  Print  Pass to program Description
SIGFPE  Yes   Yes    No              Arithmetic exception
```

Depending on the platform it may be necessary to instruct the kernel to generate signals for floating point exceptions. For programs using GSL this can be achieved using the GSL_IEEE_MODE environment variable in conjunction with the function gsl_ieee_env_setup as described in see Chapter 40 [IEEE floating-point arithmetic], page 487.

```
(gdb) set env GSL_IEEE_MODE=double-precision
```

A.4 GCC warning options for numerical programs

Writing reliable numerical programs in C requires great care. The following GCC warning options are recommended when compiling numerical programs:

```
gcc -ansi -pedantic -Werror -Wall -W
    -Wmissing-prototypes -Wstrict-prototypes
    -Wconversion -Wshadow -Wpointer-arith
    -Wcast-qual -Wcast-align
    -Wwrite-strings -Wnested-externs
    -fshort-enums -fno-common -Dinline= -g -O2
```

For details of each option consult the manual *Using and Porting GCC*. The following table gives a brief explanation of what types of errors these options catch.

```
-ansi -pedantic
```
Use ANSI C, and reject any non-ANSI extensions. These flags help in writing portable programs that will compile on other systems.

-Werror

Consider warnings to be errors, so that compilation stops. This prevents warnings from scrolling off the top of the screen and being lost. You won't be able to compile the program until it is completely warning-free.

-Wall

This turns on a set of warnings for common programming problems. You need -Wall, but it is not enough on its own.

-O2

Turn on optimization. The warnings for uninitialized variables in -Wall rely on the optimizer to analyze the code. If there is no optimization then these warnings aren't generated.

-W   This turns on some extra warnings not included in -Wall, such as missing return values and comparisons between signed and unsigned integers.

-Wmissing-prototypes -Wstrict-prototypes

Warn if there are any missing or inconsistent prototypes. Without prototypes it is harder to detect problems with incorrect arguments.

-Wconversion

The main use of this option is to warn about conversions from signed to unsigned integers. For example, unsigned int x = -1. If you need to perform such a conversion you can use an explicit cast.

-Wshadow

This warns whenever a local variable shadows another local variable. If two variables have the same name then it is a potential source of confusion.

-Wpointer-arith -Wcast-qual -Wcast-align

These options warn if you try to do pointer arithmetic for types which don't have a size, such as void, if you remove a const cast from a pointer, or if you cast a pointer to a type which has a different size, causing an invalid alignment.

-Wwrite-strings

This option gives string constants a const qualifier so that it will be a compile-time error to attempt to overwrite them.

-fshort-enums

This option makes the type of enum as short as possible. Normally this makes an enum different from an int. Consequently any attempts to assign a pointer-to-int to a pointer-to-enum will generate a cast-alignment warning.

-fno-common

This option prevents global variables being simultaneously defined in different object files (you get an error at link time). Such a variable should be defined in one file and referred to in other files with an extern declaration.

-Wnested-externs

This warns if an extern declaration is encountered within a function.

`-Dinline=`

> The inline keyword is not part of ANSI C. Thus if you want to use -ansi with a program which uses inline functions you can use this preprocessor definition to remove the inline keywords.

`-g`   It always makes sense to put debugging symbols in the executable so that you can debug it using gdb. The only effect of debugging symbols is to increase the size of the file, and you can use the strip command to remove them later if necessary.

A.5 References and Further Reading

The following books are essential reading for anyone writing and debugging numerical programs with GCC and GDB.

> R.M. Stallman, *Using and Porting GNU CC*, Free Software Foundation, ISBN 1882114388

> R.M. Stallman, R.H. Pesch, *Debugging with GDB: The GNU Source-Level Debugger*, Free Software Foundation, ISBN 1882114779

For a tutorial introduction to the GNU C Compiler and related programs, see

> B.J. Gough, *An Introduction to GCC*, Network Theory Ltd, ISBN 0954161793

Appendix B Contributors to GSL

(See the AUTHORS file in the distribution for up-to-date information.)

Mark Galassi

Conceived GSL (with James Theiler) and wrote the design document. Wrote the simulated annealing package and the relevant chapter in the manual.

James Theiler

Conceived GSL (with Mark Galassi). Wrote the random number generators and the relevant chapter in this manual.

Jim Davies

Wrote the statistical routines and the relevant chapter in this manual.

Brian Gough

FFTs, numerical integration, random number generators and distributions, root finding, minimization and fitting, polynomial solvers, complex numbers, physical constants, permutations, vector and matrix functions, histograms, statistics, ieee-utils, revised CBLAS Level 2 & 3, matrix decompositions, eigensystems, cumulative distribution functions, testing, documentation and releases.

Reid Priedhorsky

Wrote and documented the initial version of the root finding routines while at Los Alamos National Laboratory, Mathematical Modeling and Analysis Group.

Gerard Jungman

Special Functions, Series acceleration, ODEs, BLAS, Linear Algebra, Eigensystems, Hankel Transforms.

Mike Booth

Wrote the Monte Carlo library.

Jorma Olavi Tähtinen

Wrote the initial complex arithmetic functions.

Thomas Walter

Wrote the initial heapsort routines and cholesky decomposition.

Fabrice Rossi

Multidimensional minimization.

Carlo Perassi

Implementation of the random number generators in Knuth's *Seminumerical Algorithms*, 3rd Ed.

Szymon Jaroszewicz

Wrote the routines for generating combinations.

Nicolas Darnis

Wrote the cyclic functions and the initial functions for canonical permutations.

Jason H. Stover

Wrote the major cumulative distribution functions.

Ivo Alxneit

Wrote the routines for wavelet transforms.

Tuomo Keskitalo

Improved the implementation of the ODE solvers.

Lowell Johnson

Implementation of the Mathieu functions.

Patrick Alken

Implementation of non-symmetric and generalized eigensystems and B-splines.

Thanks to Nigel Lowry for help in proofreading the manual.

The non-symmetric eigensystems routines contain code based on the LA-PACK linear algebra library. LAPACK is distributed under the following license:

Appendix C Autoconf Macros

For applications using autoconf the standard macro AC_CHECK_LIB can be used to link with GSL automatically from a configure script. The library itself depends on the presence of a CBLAS and math library as well, so these must also be located before linking with the main libgsl file. The following commands should be placed in the 'configure.ac' file to perform these tests,

```
AC_CHECK_LIB([m],[cos])
AC_CHECK_LIB([gslcblas],[cblas_dgemm])
AC_CHECK_LIB([gsl],[gsl_blas_dgemm])
```

It is important to check for libm and libgslcblas before libgsl, otherwise the tests will fail. Assuming the libraries are found the output during the configure stage looks like this,

```
checking for cos in -lm... yes
checking for cblas_dgemm in -lgslcblas... yes
checking for gsl_blas_dgemm in -lgsl... yes
```

If the library is found then the tests will define the macros HAVE_LIBGSL, HAVE_LIBGSLCBLAS, HAVE_LIBM and add the options -lgsl -lgslcblas -lm to the variable LIBS.

The tests above will find any version of the library. They are suitable for general use, where the versions of the functions are not important. An alternative macro is available in the file 'gsl.m4' to test for a specific version of the library. To use this macro simply add the following line to your 'configure.ac' file instead of the tests above:

```
AX_PATH_GSL(GSL_VERSION,
            [action-if-found],
            [action-if-not-found])
```

The argument GSL_VERSION should be the two or three digit MAJOR.MINOR or MAJOR.MINOR.MICRO version number of the release you require. A suitable choice for action-if-not-found is,

```
AC_MSG_ERROR(could not find required version of GSL)
```

Then you can add the variables GSL_LIBS and GSL_CFLAGS to your Makefile.am files to obtain the correct compiler flags. GSL_LIBS is equal to the output of the gsl-config --libs command and GSL_CFLAGS is equal to gsl-config --cflags command. For example,

```
libfoo_la_LDFLAGS = -lfoo $(GSL_LIBS) -lgslcblas
```

Note that the macro AX_PATH_GSL needs to use the C compiler so it should appear in the 'configure.ac' file before the macro AC_LANG_CPLUSPLUS for programs that use C++.

To test for inline the following test should be placed in your 'configure.ac' file,

```
AC_C_INLINE

if test "$ac_cv_c_inline" != no ; then
  AC_DEFINE(HAVE_INLINE,1)
  AC_SUBST(HAVE_INLINE)
fi
```

and the macro will then be defined in the compilation flags or by including the
file 'config.h' before any library headers.

The following autoconf test will check for extern inline,

```
dnl Check for "extern inline", using a modified version
dnl of the test for AC_C_INLINE from acspecific.mt
dnl
AC_CACHE_CHECK([for extern inline], ac_cv_c_extern_inline,
[ac_cv_c_extern_inline=no
AC_TRY_COMPILE([extern $ac_cv_c_inline double foo(double x);
extern $ac_cv_c_inline double foo(double x) { return x+1.0; };
double foo (double x) { return x + 1.0; };],
[ foo(1.0) ],
[ac_cv_c_extern_inline="yes"])
])

if test "$ac_cv_c_extern_inline" != no ; then
  AC_DEFINE(HAVE_INLINE,1)
  AC_SUBST(HAVE_INLINE)
fi
```

The substitution of portability functions can be made automatically if you
use autoconf. For example, to test whether the BSD function hypot is available
you can include the following line in the configure file 'configure.ac' for your
application,

```
AC_CHECK_FUNCS(hypot)
```

and place the following macro definitions in the file 'config.h.in',

```
/* Substitute gsl_hypot for missing system hypot */

#ifndef HAVE_HYPOT
#define hypot gsl_hypot
#endif
```

The application source files can then use the include command #include
<config.h> to substitute gsl_hypot for each occurrence of hypot when hypot
is not available.

Appendix D GSL CBLAS Library

The prototypes for the low-level CBLAS functions are declared in the file 'gsl_cblas.h'. For the definition of the functions consult the documentation available from Netlib (see Section 12.3 [BLAS References and Further Reading], page 151).

D.1 Level 1

float cblas_sdsdot (const int N, const float $alpha$, const   *Function*
        float * x, const int $incx$, const float * y, const int $incy$)

double cblas_dsdot (const int N, const float * x, const int   *Function*
        $incx$, const float * y, const int $incy$)

float cblas_sdot (const int N, const float * x, const int $incx$,   *Function*
        const float * y, const int $incy$)

double cblas_ddot (const int N, const double * x, const int   *Function*
        $incx$, const double * y, const int $incy$)

void cblas_cdotu_sub (const int N, const void * x, const int   *Function*
        $incx$, const void * y, const int $incy$, void * $dotu$)

void cblas_cdotc_sub (const int N, const void * x, const int   *Function*
        $incx$, const void * y, const int $incy$, void * $dotc$)

void cblas_zdotu_sub (const int N, const void * x, const int   *Function*
        $incx$, const void * y, const int $incy$, void * $dotu$)

void cblas_zdotc_sub (const int N, const void * x, const int   *Function*
        $incx$, const void * y, const int $incy$, void * $dotc$)

float cblas_snrm2 (const int N, const float * x, const int   *Function*
        $incx$)

float cblas_sasum (const int N, const float * x, const int   *Function*
        $incx$)

double cblas_dnrm2 (const int N, const double * x, const int   *Function*
        $incx$)

double cblas_dasum (const int N, const double * x, const int   *Function*
        $incx$)

float cblas_scnrm2 (const int N, const void * x, const int   *Function*
        $incx$)

float cblas_scasum (const int N, const void * x, const int   *Function*
        $incx$)

double cblas_dznrm2 (const int N, const void * x, const int   *Function*
        $incx$)

double cblas_dzasum (const int N, const void * x, const int   Function
      $incx$)

CBLAS_INDEX cblas_isamax (const int N, const float * x, const   Function
      int $incx$)

CBLAS_INDEX cblas_idamax (const int N, const double * x, const   Function
      int $incx$)

CBLAS_INDEX cblas_icamax (const int N, const void * x, const   Function
      int $incx$)

CBLAS_INDEX cblas_izamax (const int N, const void * x, const   Function
      int $incx$)

void cblas_sswap (const int N, float * x, const int $incx$, float   Function
      * y, const int $incy$)

void cblas_scopy (const int N, const float * x, const int $incx$,   Function
      float * y, const int $incy$)

void cblas_saxpy (const int N, const float $alpha$, const float *   Function
      x, const int $incx$, float * y, const int $incy$)

void cblas_dswap (const int N, double * x, const int $incx$,   Function
      double * y, const int $incy$)

void cblas_dcopy (const int N, const double * x, const int   Function
      $incx$, double * y, const int $incy$)

void cblas_daxpy (const int N, const double $alpha$, const   Function
      double * x, const int $incx$, double * y, const int $incy$)

void cblas_cswap (const int N, void * x, const int $incx$, void *   Function
      y, const int $incy$)

void cblas_ccopy (const int N, const void * x, const int $incx$,   Function
      void * y, const int $incy$)

void cblas_caxpy (const int N, const void * $alpha$, const void *   Function
      x, const int $incx$, void * y, const int $incy$)

void cblas_zswap (const int N, void * x, const int $incx$, void *   Function
      y, const int $incy$)

void cblas_zcopy (const int N, const void * x, const int $incx$,   Function
      void * y, const int $incy$)

void cblas_zaxpy (const int N, const void * $alpha$, const void *   Function
      x, const int $incx$, void * y, const int $incy$)

void cblas_srotg (float * a, float * b, float * c, float * s)   Function

void cblas_srotmg (float * $d1$, float * $d2$, float * $b1$, const   Function
      float $b2$, float * P)

void cblas_srot (const int N, float * x, const int $incx$, float   Function
    * y, const int $incy$, const float c, const float s)

void cblas_srotm (const int N, float * x, const int $incx$, float   Function
    * y, const int $incy$, const float * P)

void cblas_drotg (double * a, double * b, double * c, double *   Function
    s)

void cblas_drotmg (double * $d1$, double * $d2$, double * $b1$, const   Function
    double $b2$, double * P)

void cblas_drot (const int N, double * x, const int $incx$,   Function
    double * y, const int $incy$, const double c, const double s)

void cblas_drotm (const int N, double * x, const int $incx$,   Function
    double * y, const int $incy$, const double * P)

void cblas_sscal (const int N, const float $alpha$, float * x,   Function
    const int $incx$)

void cblas_dscal (const int N, const double $alpha$, double * x,   Function
    const int $incx$)

void cblas_cscal (const int N, const void * $alpha$, void * x,   Function
    const int $incx$)

void cblas_zscal (const int N, const void * $alpha$, void * x,   Function
    const int $incx$)

void cblas_csscal (const int N, const float $alpha$, void * x,   Function
    const int $incx$)

void cblas_zdscal (const int N, const double $alpha$, void * x,   Function
    const int $incx$)

D.2 Level 2

void cblas_sgemv (const enum CBLAS_ORDER *order*, const enum   Function
        CBLAS_TRANSPOSE *TransA*, const int *M*, const int *N*, const
        float *alpha*, const float * *A*, const int *lda*, const float * *x*,
        const int *incx*, const float *beta*, float * *y*, const int *incy*)

void cblas_sgbmv (const enum CBLAS_ORDER *order*, const enum   Function
        CBLAS_TRANSPOSE *TransA*, const int *M*, const int *N*, const
        int *KL*, const int *KU*, const float *alpha*, const float * *A*,
        const int *lda*, const float * *x*, const int *incx*, const float
        beta, float * *y*, const int *incy*)

void cblas_strmv (const enum CBLAS_ORDER *order*, const enum   Function
        CBLAS_UPLO *Uplo*, const enum CBLAS_TRANSPOSE *TransA*, const
        enum CBLAS_DIAG *Diag*, const int *N*, const float * *A*, const
        int *lda*, float * *x*, const int *incx*)

void cblas_stbmv (const enum CBLAS_ORDER *order*, const enum   Function
        CBLAS_UPLO *Uplo*, const enum CBLAS_TRANSPOSE *TransA*, const
        enum CBLAS_DIAG *Diag*, const int *N*, const int *K*, const float
        * *A*, const int *lda*, float * *x*, const int *incx*)

void cblas_stpmv (const enum CBLAS_ORDER *order*, const enum   Function
        CBLAS_UPLO *Uplo*, const enum CBLAS_TRANSPOSE *TransA*, const
        enum CBLAS_DIAG *Diag*, const int *N*, const float * *Ap*, float
        * *x*, const int *incx*)

void cblas_strsv (const enum CBLAS_ORDER *order*, const enum   Function
        CBLAS_UPLO *Uplo*, const enum CBLAS_TRANSPOSE *TransA*, const
        enum CBLAS_DIAG *Diag*, const int *N*, const float * *A*, const
        int *lda*, float * *x*, const int *incx*)

void cblas_stbsv (const enum CBLAS_ORDER *order*, const enum   Function
        CBLAS_UPLO *Uplo*, const enum CBLAS_TRANSPOSE *TransA*, const
        enum CBLAS_DIAG *Diag*, const int *N*, const int *K*, const float
        * *A*, const int *lda*, float * *x*, const int *incx*)

void cblas_stpsv (const enum CBLAS_ORDER *order*, const enum   Function
        CBLAS_UPLO *Uplo*, const enum CBLAS_TRANSPOSE *TransA*, const
        enum CBLAS_DIAG *Diag*, const int *N*, const float * *Ap*, float
        * *x*, const int *incx*)

void cblas_dgemv (const enum CBLAS_ORDER *order*, const enum   Function
        CBLAS_TRANSPOSE *TransA*, const int *M*, const int *N*, const
        double *alpha*, const double * *A*, const int *lda*, const double
        * *x*, const int *incx*, const double *beta*, double * *y*, const int
        incy)

void cblas_dgbmv (const enum CBLAS_ORDER *order*, const enum   Function
      CBLAS_TRANSPOSE *TransA*, const int M, const int N, const
      int KL, const int KU, const double *alpha*, const double *
      A, const int *lda*, const double * x, const int *incx*, const
      double *beta*, double * y, const int *incy*)

void cblas_dtrmv (const enum CBLAS_ORDER *order*, const enum   Function
      CBLAS_UPLO *Uplo*, const enum CBLAS_TRANSPOSE *TransA*, const
      enum CBLAS_DIAG *Diag*, const int N, const double * A, const
      int *lda*, double * x, const int *incx*)

void cblas_dtbmv (const enum CBLAS_ORDER *order*, const enum   Function
      CBLAS_UPLO *Uplo*, const enum CBLAS_TRANSPOSE *TransA*, const
      enum CBLAS_DIAG *Diag*, const int N, const int K, const
      double * A, const int *lda*, double * x, const int *incx*)

void cblas_dtpmv (const enum CBLAS_ORDER *order*, const enum   Function
      CBLAS_UPLO *Uplo*, const enum CBLAS_TRANSPOSE *TransA*, const
      enum CBLAS_DIAG *Diag*, const int N, const double * Ap,
      double * x, const int *incx*)

void cblas_dtrsv (const enum CBLAS_ORDER *order*, const enum   Function
      CBLAS_UPLO *Uplo*, const enum CBLAS_TRANSPOSE *TransA*, const
      enum CBLAS_DIAG *Diag*, const int N, const double * A, const
      int *lda*, double * x, const int *incx*)

void cblas_dtbsv (const enum CBLAS_ORDER *order*, const enum   Function
      CBLAS_UPLO *Uplo*, const enum CBLAS_TRANSPOSE *TransA*, const
      enum CBLAS_DIAG *Diag*, const int N, const int K, const
      double * A, const int *lda*, double * x, const int *incx*)

void cblas_dtpsv (const enum CBLAS_ORDER *order*, const enum   Function
      CBLAS_UPLO *Uplo*, const enum CBLAS_TRANSPOSE *TransA*, const
      enum CBLAS_DIAG *Diag*, const int N, const double * Ap,
      double * x, const int *incx*)

void cblas_cgemv (const enum CBLAS_ORDER *order*, const enum   Function
      CBLAS_TRANSPOSE *TransA*, const int M, const int N, const
      void * *alpha*, const void * A, const int *lda*, const void * x,
      const int *incx*, const void * *beta*, void * y, const int *incy*)

void cblas_cgbmv (const enum CBLAS_ORDER *order*, const enum   Function
      CBLAS_TRANSPOSE *TransA*, const int M, const int N, const
      int KL, const int KU, const void * *alpha*, const void * A,
      const int *lda*, const void * x, const int *incx*, const void *
      beta, void * y, const int *incy*)

void cblas_ctrmv (const enum CBLAS_ORDER *order*, const enum   Function
      CBLAS_UPLO *Uplo*, const enum CBLAS_TRANSPOSE *TransA*, const
      enum CBLAS_DIAG *Diag*, const int N, const void * A, const
      int *lda*, void * x, const int *incx*)

void cblas_ctbmv (const enum CBLAS_ORDER *order*, const enum   Function
    CBLAS_UPLO *Uplo*, const enum CBLAS_TRANSPOSE *TransA*, const
    enum CBLAS_DIAG *Diag*, const int *N*, const int *K*, const void
    * *A*, const int *lda*, void * *x*, const int *incx*)

void cblas_ctpmv (const enum CBLAS_ORDER *order*, const enum   Function
    CBLAS_UPLO *Uplo*, const enum CBLAS_TRANSPOSE *TransA*, const
    enum CBLAS_DIAG *Diag*, const int *N*, const void * *Ap*, void *
    x, const int *incx*)

void cblas_ctrsv (const enum CBLAS_ORDER *order*, const enum   Function
    CBLAS_UPLO *Uplo*, const enum CBLAS_TRANSPOSE *TransA*, const
    enum CBLAS_DIAG *Diag*, const int *N*, const void * *A*, const
    int *lda*, void * *x*, const int *incx*)

void cblas_ctbsv (const enum CBLAS_ORDER *order*, const enum   Function
    CBLAS_UPLO *Uplo*, const enum CBLAS_TRANSPOSE *TransA*, const
    enum CBLAS_DIAG *Diag*, const int *N*, const int *K*, const void
    * *A*, const int *lda*, void * *x*, const int *incx*)

void cblas_ctpsv (const enum CBLAS_ORDER *order*, const enum   Function
    CBLAS_UPLO *Uplo*, const enum CBLAS_TRANSPOSE *TransA*, const
    enum CBLAS_DIAG *Diag*, const int *N*, const void * *Ap*, void *
    x, const int *incx*)

void cblas_zgemv (const enum CBLAS_ORDER *order*, const enum   Function
    CBLAS_TRANSPOSE *TransA*, const int *M*, const int *N*, const
    void * *alpha*, const void * *A*, const int *lda*, const void * *x*,
    const int *incx*, const void * *beta*, void * *y*, const int *incy*)

void cblas_zgbmv (const enum CBLAS_ORDER *order*, const enum   Function
    CBLAS_TRANSPOSE *TransA*, const int *M*, const int *N*, const
    int *KL*, const int *KU*, const void * *alpha*, const void * *A*,
    const int *lda*, const void * *x*, const int *incx*, const void *
    beta, void * *y*, const int *incy*)

void cblas_ztrmv (const enum CBLAS_ORDER *order*, const enum   Function
    CBLAS_UPLO *Uplo*, const enum CBLAS_TRANSPOSE *TransA*, const
    enum CBLAS_DIAG *Diag*, const int *N*, const void * *A*, const
    int *lda*, void * *x*, const int *incx*)

void cblas_ztbmv (const enum CBLAS_ORDER *order*, const enum   Function
    CBLAS_UPLO *Uplo*, const enum CBLAS_TRANSPOSE *TransA*, const
    enum CBLAS_DIAG *Diag*, const int *N*, const int *K*, const void
    * *A*, const int *lda*, void * *x*, const int *incx*)

void cblas_ztpmv (const enum CBLAS_ORDER *order*, const enum   Function
    CBLAS_UPLO *Uplo*, const enum CBLAS_TRANSPOSE *TransA*, const
    enum CBLAS_DIAG *Diag*, const int *N*, const void * *Ap*, void *
    x, const int *incx*)

void cblas_ztrsv (const enum CBLAS_ORDER *order*, const enum   Function
        CBLAS_UPLO *Uplo*, const enum CBLAS_TRANSPOSE *TransA*, const
        enum CBLAS_DIAG *Diag*, const int N, const void * A, const
        int *lda*, void * *x*, const int *incx*)

void cblas_ztbsv (const enum CBLAS_ORDER *order*, const enum   Function
        CBLAS_UPLO *Uplo*, const enum CBLAS_TRANSPOSE *TransA*, const
        enum CBLAS_DIAG *Diag*, const int N, const int K, const void
        * A, const int *lda*, void * *x*, const int *incx*)

void cblas_ztpsv (const enum CBLAS_ORDER *order*, const enum   Function
        CBLAS_UPLO *Uplo*, const enum CBLAS_TRANSPOSE *TransA*, const
        enum CBLAS_DIAG *Diag*, const int N, const void * Ap, void *
        x, const int *incx*)

void cblas_ssymv (const enum CBLAS_ORDER *order*, const enum   Function
        CBLAS_UPLO *Uplo*, const int N, const float *alpha*, const
        float * A, const int *lda*, const float * *x*, const int *incx*,
        const float *beta*, float * *y*, const int *incy*)

void cblas_ssbmv (const enum CBLAS_ORDER *order*, const enum   Function
        CBLAS_UPLO *Uplo*, const int N, const int K, const float
        alpha, const float * A, const int *lda*, const float * *x*, const
        int *incx*, const float *beta*, float * *y*, const int *incy*)

void cblas_sspmv (const enum CBLAS_ORDER *order*, const enum   Function
        CBLAS_UPLO *Uplo*, const int N, const float *alpha*, const
        float * Ap, const float * *x*, const int *incx*, const float
        beta, float * *y*, const int *incy*)

void cblas_sger (const enum CBLAS_ORDER *order*, const int M,   Function
        const int N, const float *alpha*, const float * *x*, const int
        incx, const float * *y*, const int *incy*, float * A, const int
        lda)

void cblas_ssyr (const enum CBLAS_ORDER *order*, const enum   Function
        CBLAS_UPLO *Uplo*, const int N, const float *alpha*, const
        float * *x*, const int *incx*, float * A, const int *lda*)

void cblas_sspr (const enum CBLAS_ORDER *order*, const enum   Function
        CBLAS_UPLO *Uplo*, const int N, const float *alpha*, const
        float * *x*, const int *incx*, float * Ap)

void cblas_ssyr2 (const enum CBLAS_ORDER *order*, const enum   Function
        CBLAS_UPLO *Uplo*, const int N, const float *alpha*, const
        float * *x*, const int *incx*, const float * *y*, const int *incy*,
        float * A, const int *lda*)

void cblas_sspr2 (const enum CBLAS_ORDER *order*, const enum   Function
        CBLAS_UPLO *Uplo*, const int N, const float *alpha*, const
        float * *x*, const int *incx*, const float * *y*, const int *incy*,
        float * A)

void cblas_dsymv (const enum CBLAS_ORDER *order*, const enum   Function
         CBLAS_UPLO *Uplo*, const int *N*, const double *alpha*, const
         double * *A*, const int *lda*, const double * *x*, const int *incx*,
         const double *beta*, double * *y*, const int *incy*)

void cblas_dsbmv (const enum CBLAS_ORDER *order*, const enum   Function
         CBLAS_UPLO *Uplo*, const int *N*, const int *K*, const double
         alpha, const double * *A*, const int *lda*, const double * *x*,
         const int *incx*, const double *beta*, double * *y*, const int
         incy)

void cblas_dspmv (const enum CBLAS_ORDER *order*, const enum   Function
         CBLAS_UPLO *Uplo*, const int *N*, const double *alpha*, const
         double * *Ap*, const double * *x*, const int *incx*, const double
         beta, double * *y*, const int *incy*)

void cblas_dger (const enum CBLAS_ORDER *order*, const int *M*,   Function
         const int *N*, const double *alpha*, const double * *x*, const int
         incx, const double * *y*, const int *incy*, double * *A*, const int
         lda)

void cblas_dsyr (const enum CBLAS_ORDER *order*, const enum   Function
         CBLAS_UPLO *Uplo*, const int *N*, const double *alpha*, const
         double * *x*, const int *incx*, double * *A*, const int *lda*)

void cblas_dspr (const enum CBLAS_ORDER *order*, const enum   Function
         CBLAS_UPLO *Uplo*, const int *N*, const double *alpha*, const
         double * *x*, const int *incx*, double * *Ap*)

void cblas_dsyr2 (const enum CBLAS_ORDER *order*, const enum   Function
         CBLAS_UPLO *Uplo*, const int *N*, const double *alpha*, const
         double * *x*, const int *incx*, const double * *y*, const int *incy*,
         double * *A*, const int *lda*)

void cblas_dspr2 (const enum CBLAS_ORDER *order*, const enum   Function
         CBLAS_UPLO *Uplo*, const int *N*, const double *alpha*, const
         double * *x*, const int *incx*, const double * *y*, const int *incy*,
         double * *A*)

void cblas_chemv (const enum CBLAS_ORDER *order*, const enum   Function
         CBLAS_UPLO *Uplo*, const int *N*, const void * *alpha*, const void
         * *A*, const int *lda*, const void * *x*, const int *incx*, const
         void * *beta*, void * *y*, const int *incy*)

void cblas_chbmv (const enum CBLAS_ORDER *order*, const enum   Function
         CBLAS_UPLO *Uplo*, const int *N*, const int *K*, const void *
         alpha, const void * *A*, const int *lda*, const void * *x*, const
         int *incx*, const void * *beta*, void * *y*, const int *incy*)

void cblas_chpmv (const enum CBLAS_ORDER *order*, const enum   Function
         CBLAS_UPLO *Uplo*, const int *N*, const void * *alpha*, const void
         * *Ap*, const void * *x*, const int *incx*, const void * *beta*, void
         * *y*, const int *incy*)

void cblas_cgeru (const enum CBLAS_ORDER *order*, const int *M*,   Function
        const int *N*, const void * *alpha*, const void * *x*, const int
        incx, const void * *y*, const int *incy*, void * *A*, const int *lda*)

void cblas_cgerc (const enum CBLAS_ORDER *order*, const int *M*,   Function
        const int *N*, const void * *alpha*, const void * *x*, const int
        incx, const void * *y*, const int *incy*, void * *A*, const int *lda*)

void cblas_cher (const enum CBLAS_ORDER *order*, const enum   Function
        CBLAS_UPLO *Uplo*, const int *N*, const float *alpha*, const void
        * *x*, const int *incx*, void * *A*, const int *lda*)

void cblas_chpr (const enum CBLAS_ORDER *order*, const enum   Function
        CBLAS_UPLO *Uplo*, const int *N*, const float *alpha*, const void
        * *x*, const int *incx*, void * *A*)

void cblas_cher2 (const enum CBLAS_ORDER *order*, const enum   Function
        CBLAS_UPLO *Uplo*, const int *N*, const void * *alpha*, const void
        * *x*, const int *incx*, const void * *y*, const int *incy*, void * *A*,
        const int *lda*)

void cblas_chpr2 (const enum CBLAS_ORDER *order*, const enum   Function
        CBLAS_UPLO *Uplo*, const int *N*, const void * *alpha*, const void
        * *x*, const int *incx*, const void * *y*, const int *incy*, void *
        Ap)

void cblas_zhemv (const enum CBLAS_ORDER *order*, const enum   Function
        CBLAS_UPLO *Uplo*, const int *N*, const void * *alpha*, const void
        * *A*, const int *lda*, const void * *x*, const int *incx*, const
        void * *beta*, void * *y*, const int *incy*)

void cblas_zhbmv (const enum CBLAS_ORDER *order*, const enum   Function
        CBLAS_UPLO *Uplo*, const int *N*, const int *K*, const void *
        alpha, const void * *A*, const int *lda*, const void * *x*, const
        int *incx*, const void * *beta*, void * *y*, const int *incy*)

void cblas_zhpmv (const enum CBLAS_ORDER *order*, const enum   Function
        CBLAS_UPLO *Uplo*, const int *N*, const void * *alpha*, const void
        * *Ap*, const void * *x*, const int *incx*, const void * *beta*, void
        * *y*, const int *incy*)

void cblas_zgeru (const enum CBLAS_ORDER *order*, const int *M*,   Function
        const int *N*, const void * *alpha*, const void * *x*, const int
        incx, const void * *y*, const int *incy*, void * *A*, const int *lda*)

void cblas_zgerc (const enum CBLAS_ORDER *order*, const int *M*,   Function
        const int *N*, const void * *alpha*, const void * *x*, const int
        incx, const void * *y*, const int *incy*, void * *A*, const int *lda*)

void cblas_zher (const enum CBLAS_ORDER *order*, const enum   Function
        CBLAS_UPLO *Uplo*, const int *N*, const double *alpha*, const
        void * *x*, const int *incx*, void * *A*, const int *lda*)

void cblas_zhpr (const enum CBLAS_ORDER *order*, const enum   Function
    CBLAS_UPLO *Uplo*, const int N, const double *alpha*, const
    void * x, const int *incx*, void * A)

void cblas_zher2 (const enum CBLAS_ORDER *order*, const enum   Function
    CBLAS_UPLO *Uplo*, const int N, const void * *alpha*, const void
    * x, const int *incx*, const void * y, const int *incy*, void * A,
    const int *lda*)

void cblas_zhpr2 (const enum CBLAS_ORDER *order*, const enum   Function
    CBLAS_UPLO *Uplo*, const int N, const void * *alpha*, const void
    * x, const int *incx*, const void * y, const int *incy*, void *
    Ap)

D.3 Level 3

void cblas_sgemm (const enum CBLAS_ORDER *Order*, const enum   Function
    CBLAS_TRANSPOSE *TransA*, const enum CBLAS_TRANSPOSE *TransB*,
    const int M, const int N, const int K, const float *alpha*,
    const float * A, const int *lda*, const float * B, const int
    ldb, const float *beta*, float * C, const int *ldc*)

void cblas_ssymm (const enum CBLAS_ORDER *Order*, const enum   Function
    CBLAS_SIDE *Side*, const enum CBLAS_UPLO *Uplo*, const int M,
    const int N, const float *alpha*, const float * A, const int
    lda, const float * B, const int *ldb*, const float *beta*, float
    * C, const int *ldc*)

void cblas_ssyrk (const enum CBLAS_ORDER *Order*, const enum   Function
    CBLAS_UPLO *Uplo*, const enum CBLAS_TRANSPOSE *Trans*, const
    int N, const int K, const float *alpha*, const float * A,
    const int *lda*, const float *beta*, float * C, const int *ldc*)

void cblas_ssyr2k (const enum CBLAS_ORDER *Order*, const enum   Function
    CBLAS_UPLO *Uplo*, const enum CBLAS_TRANSPOSE *Trans*, const
    int N, const int K, const float *alpha*, const float * A,
    const int *lda*, const float * B, const int *ldb*, const float
    beta, float * C, const int *ldc*)

void cblas_strmm (const enum CBLAS_ORDER *Order*, const enum   Function
    CBLAS_SIDE *Side*, const enum CBLAS_UPLO *Uplo*, const enum
    CBLAS_TRANSPOSE *TransA*, const enum CBLAS_DIAG *Diag*, const
    int M, const int N, const float *alpha*, const float * A,
    const int *lda*, float * B, const int *ldb*)

void cblas_strsm (const enum CBLAS_ORDER *Order*, const enum   Function
    CBLAS_SIDE *Side*, const enum CBLAS_UPLO *Uplo*, const enum
    CBLAS_TRANSPOSE *TransA*, const enum CBLAS_DIAG *Diag*, const
    int M, const int N, const float *alpha*, const float * A,
    const int *lda*, float * B, const int *ldb*)

void cblas_dgemm (const enum CBLAS_ORDER *Order*, const enum   Function
      CBLAS_TRANSPOSE *TransA*, const enum CBLAS_TRANSPOSE *TransB*,
      const int M, const int N, const int K, const double *alpha*,
      const double * A, const int *lda*, const double * B, const int
      ldb, const double *beta*, double * C, const int *ldc*)

void cblas_dsymm (const enum CBLAS_ORDER *Order*, const enum   Function
      CBLAS_SIDE *Side*, const enum CBLAS_UPLO *Uplo*, const int M,
      const int N, const double *alpha*, const double * A, const int
      lda, const double * B, const int *ldb*, const double *beta*,
      double * C, const int *ldc*)

void cblas_dsyrk (const enum CBLAS_ORDER *Order*, const enum   Function
      CBLAS_UPLO *Uplo*, const enum CBLAS_TRANSPOSE *Trans*, const
      int N, const int K, const double *alpha*, const double * A,
      const int *lda*, const double *beta*, double * C, const int *ldc*)

void cblas_dsyr2k (const enum CBLAS_ORDER *Order*, const enum   Function
      CBLAS_UPLO *Uplo*, const enum CBLAS_TRANSPOSE *Trans*, const
      int N, const int K, const double *alpha*, const double * A,
      const int *lda*, const double * B, const int *ldb*, const double
      beta, double * C, const int *ldc*)

void cblas_dtrmm (const enum CBLAS_ORDER *Order*, const enum   Function
      CBLAS_SIDE *Side*, const enum CBLAS_UPLO *Uplo*, const enum
      CBLAS_TRANSPOSE *TransA*, const enum CBLAS_DIAG *Diag*, const
      int M, const int N, const double *alpha*, const double * A,
      const int *lda*, double * B, const int *ldb*)

void cblas_dtrsm (const enum CBLAS_ORDER *Order*, const enum   Function
      CBLAS_SIDE *Side*, const enum CBLAS_UPLO *Uplo*, const enum
      CBLAS_TRANSPOSE *TransA*, const enum CBLAS_DIAG *Diag*, const
      int M, const int N, const double *alpha*, const double * A,
      const int *lda*, double * B, const int *ldb*)

void cblas_cgemm (const enum CBLAS_ORDER *Order*, const enum   Function
      CBLAS_TRANSPOSE *TransA*, const enum CBLAS_TRANSPOSE *TransB*,
      const int M, const int N, const int K, const void * *alpha*,
      const void * A, const int *lda*, const void * B, const int *ldb*,
      const void * *beta*, void * C, const int *ldc*)

void cblas_csymm (const enum CBLAS_ORDER *Order*, const enum   Function
      CBLAS_SIDE *Side*, const enum CBLAS_UPLO *Uplo*, const int M,
      const int N, const void * *alpha*, const void * A, const int
      lda, const void * B, const int *ldb*, const void * *beta*, void *
      C, const int *ldc*)

void cblas_csyrk (const enum CBLAS_ORDER *Order*, const enum   Function
      CBLAS_UPLO *Uplo*, const enum CBLAS_TRANSPOSE *Trans*, const
      int N, const int K, const void * *alpha*, const void * A,
      const int *lda*, const void * *beta*, void * C, const int *ldc*)

void cblas_csyr2k (const enum CBLAS_ORDER *Order*, const enum   Function
    CBLAS_UPLO *Uplo*, const enum CBLAS_TRANSPOSE *Trans*, const
    int *N*, const int *K*, const void * *alpha*, const void * *A*,
    const int *lda*, const void * *B*, const int *ldb*, const void *
    beta, void * *C*, const int *ldc*)

void cblas_ctrmm (const enum CBLAS_ORDER *Order*, const enum   Function
    CBLAS_SIDE *Side*, const enum CBLAS_UPLO *Uplo*, const enum
    CBLAS_TRANSPOSE *TransA*, const enum CBLAS_DIAG *Diag*, const
    int *M*, const int *N*, const void * *alpha*, const void * *A*,
    const int *lda*, void * *B*, const int *ldb*)

void cblas_ctrsm (const enum CBLAS_ORDER *Order*, const enum   Function
    CBLAS_SIDE *Side*, const enum CBLAS_UPLO *Uplo*, const enum
    CBLAS_TRANSPOSE *TransA*, const enum CBLAS_DIAG *Diag*, const
    int *M*, const int *N*, const void * *alpha*, const void * *A*,
    const int *lda*, void * *B*, const int *ldb*)

void cblas_zgemm (const enum CBLAS_ORDER *Order*, const enum   Function
    CBLAS_TRANSPOSE *TransA*, const enum CBLAS_TRANSPOSE *TransB*,
    const int *M*, const int *N*, const int *K*, const void * *alpha*,
    const void * *A*, const int *lda*, const void * *B*, const int *ldb*,
    const void * *beta*, void * *C*, const int *ldc*)

void cblas_zsymm (const enum CBLAS_ORDER *Order*, const enum   Function
    CBLAS_SIDE *Side*, const enum CBLAS_UPLO *Uplo*, const int *M*,
    const int *N*, const void * *alpha*, const void * *A*, const int
    lda, const void * *B*, const int *ldb*, const void * *beta*, void *
    C, const int *ldc*)

void cblas_zsyrk (const enum CBLAS_ORDER *Order*, const enum   Function
    CBLAS_UPLO *Uplo*, const enum CBLAS_TRANSPOSE *Trans*, const
    int *N*, const int *K*, const void * *alpha*, const void * *A*,
    const int *lda*, const void * *beta*, void * *C*, const int *ldc*)

void cblas_zsyr2k (const enum CBLAS_ORDER *Order*, const enum   Function
    CBLAS_UPLO *Uplo*, const enum CBLAS_TRANSPOSE *Trans*, const
    int *N*, const int *K*, const void * *alpha*, const void * *A*,
    const int *lda*, const void * *B*, const int *ldb*, const void *
    beta, void * *C*, const int *ldc*)

void cblas_ztrmm (const enum CBLAS_ORDER *Order*, const enum   Function
    CBLAS_SIDE *Side*, const enum CBLAS_UPLO *Uplo*, const enum
    CBLAS_TRANSPOSE *TransA*, const enum CBLAS_DIAG *Diag*, const
    int *M*, const int *N*, const void * *alpha*, const void * *A*,
    const int *lda*, void * *B*, const int *ldb*)

void cblas_ztrsm (const enum CBLAS_ORDER *Order*, const enum   Function
    CBLAS_SIDE *Side*, const enum CBLAS_UPLO *Uplo*, const enum
    CBLAS_TRANSPOSE *TransA*, const enum CBLAS_DIAG *Diag*, const
    int *M*, const int *N*, const void * *alpha*, const void * *A*,
    const int *lda*, void * *B*, const int *ldb*)

void cblas_chemm (const enum CBLAS_ORDER *Order*, const enum   Function
    CBLAS_SIDE *Side*, const enum CBLAS_UPLO *Uplo*, const int M,
    const int N, const void * *alpha*, const void * A, const int
    lda, const void * B, const int *ldb*, const void * *beta*, void *
    C, const int *ldc*)

void cblas_cherk (const enum CBLAS_ORDER *Order*, const enum   Function
    CBLAS_UPLO *Uplo*, const enum CBLAS_TRANSPOSE *Trans*, const
    int N, const int K, const float *alpha*, const void * A,
    const int *lda*, const float *beta*, void * C, const int *ldc*)

void cblas_cher2k (const enum CBLAS_ORDER *Order*, const enum   Function
    CBLAS_UPLO *Uplo*, const enum CBLAS_TRANSPOSE *Trans*, const
    int N, const int K, const void * *alpha*, const void * A,
    const int *lda*, const void * B, const int *ldb*, const float
    beta, void * C, const int *ldc*)

void cblas_zhemm (const enum CBLAS_ORDER *Order*, const enum   Function
    CBLAS_SIDE *Side*, const enum CBLAS_UPLO *Uplo*, const int M,
    const int N, const void * *alpha*, const void * A, const int
    lda, const void * B, const int *ldb*, const void * *beta*, void *
    C, const int *ldc*)

void cblas_zherk (const enum CBLAS_ORDER *Order*, const enum   Function
    CBLAS_UPLO *Uplo*, const enum CBLAS_TRANSPOSE *Trans*, const
    int N, const int K, const double *alpha*, const void * A,
    const int *lda*, const double *beta*, void * C, const int *ldc*)

void cblas_zher2k (const enum CBLAS_ORDER *Order*, const enum   Function
    CBLAS_UPLO *Uplo*, const enum CBLAS_TRANSPOSE *Trans*, const
    int N, const int K, const void * *alpha*, const void * A,
    const int *lda*, const void * B, const int *ldb*, const double
    beta, void * C, const int *ldc*)

void cblas_xerbla (int p, const char * *rout*, const char * *form*,   Function
    ...)

D.4 Examples

The following program computes the product of two matrices using the Level-3 BLAS function SGEMM,

$$\begin{pmatrix} 0.11 & 0.12 & 0.13 \\ 0.21 & 0.22 & 0.23 \end{pmatrix} \begin{pmatrix} 1011 & 1012 \\ 1021 & 1022 \\ 1031 & 1031 \end{pmatrix} = \begin{pmatrix} 367.76 & 368.12 \\ 674.06 & 674.72 \end{pmatrix}$$

The matrices are stored in row major order but could be stored in column major order if the first argument of the call to cblas_sgemm was changed to CblasColMajor.

```
#include <stdio.h>
#include <gsl/gsl_cblas.h>

int
main (void)
{
  int lda = 3;

  float A[] = { 0.11, 0.12, 0.13,
                0.21, 0.22, 0.23 };

  int ldb = 2;

  float B[] = { 1011, 1012,
                1021, 1022,
                1031, 1032 };

  int ldc = 2;

  float C[] = { 0.00, 0.00,
                0.00, 0.00 };

  /* Compute C = A B */

  cblas_sgemm (CblasRowMajor,
               CblasNoTrans, CblasNoTrans, 2, 2, 3,
               1.0, A, lda, B, ldb, 0.0, C, ldc);

  printf ("[ %g, %g\n", C[0], C[1]);
  printf ("  %g, %g ]\n", C[2], C[3]);

  return 0;
}
```

To compile the program use the following command line,

```
$ gcc -Wall demo.c -lgslcblas
```

There is no need to link with the main library -lgsl in this case as the CBLAS library is an independent unit. Here is the output from the program,

```
$ ./a.out
[ 367.76, 368.12
  674.06, 674.72 ]
```

Appendix E GPG verification

The official source-code releases of the GNU Scientific Library on `ftp.gnu.org` are digitally signed with `gpg`, the GNU Project's cryptography tool. You can verify the integrity of the GSL source code by checking its signature against the maintainer's key.

For the benefit of owners of this printed edition of the manual, the official fingerprint of the maintainer's key is reproduced here:

D561 6F67 DF2F 5CCF 8A2B 6CA6 6E21 6FED 6406 9D5C

This fingerprint has been obtained directly from the GSL maintainer for inclusion in this printed manual. To check the GSL source code against the maintainer's key follow the procedure below, ensuring that exactly the same sequence of hexadecimal digits is shown as the fingerprint. In the event that the signing key changes, any new fingerprint will be printed in future editions of this manual.

Checking file signatures

To check a GSL release you will need the `gpg` and `gpgv` command-line tools installed, and the public key of the GSL maintainer, Brian Gough. This key is available from the Network Theory website—to download it use the GNU `wget` command like this,

```
$ wget http://www.network-theory.co.uk/download/gpg.txt
```

The key will be stored in the file 'gpg.txt'. You will then need to check the fingerprint of the downloaded key against the one printed in this manual—the security of this procedure depends on this step. The command to display the fingerprint of the downloaded key is,

```
$ gpg --with-fingerprint gpg.txt
pub   1024D/64069D5C 2002-03-19 Brian Gough
                                  <bjg@network-theory.co.uk>
      Key fingerprint = .... hexadecimal digits ....
sub   1024g/2E410647 2004-08-27
```

The hexadecimal digits of the key fingerprint in the output should match those printed in this manual. If there is any discrepancy, the downloaded key should not be used.[1]

---

[1] If the fingerprint does not match for any reason, please contact the publisher at info@network-theory.co.uk

Assuming the fingerprint is correct, import the key onto a keyring named 'gsl', ready for actual use:

```
$ gpg --no-default-keyring --keyring gsl --import gpg.txt
gpg: ~/.gnupg/gsl: keyring created
gpg: key 64069D5C: public key imported
gpg: Total number processed: 1
gpg:                 imported: 1
```

You can now check the source code. You need both the tar file, e.g. 'gsl-1.12.tar.gz', and its signature file, e.g. 'gsl-1.12.tar.gz.sig'[2] from the 'gnu/gsl' directory on ftp.gnu.org.

The gpgv command is used to verify the file,

```
$ gpgv --keyring gsl gsl-1.12.tar.gz.sig
gpgv: Signature made Mon 15 Dec 2008 18:26:47 GMT
   using DSA key ID 64069D5C
gpgv: Good signature from "Brian Gough
   <bjg@network-theory.co.uk>"
```

If you see the message Good signature you can be confident that the file has not been tampered with (assuming the fingerprint displayed earlier matches the one printed in this manual).

[2] The file extension '.asc' is sometimes used instead, for signatures which are ascii-encoded.

Free Software Needs Free Documentation

The following article was written by Richard Stallman, founder of the GNU Project.

The biggest deficiency in the free software community today is not in the software—it is the lack of good free documentation that we can include with the free software. Many of our most important programs do not come with free reference manuals and free introductory texts. Documentation is an essential part of any software package; when an important free software package does not come with a free manual and a free tutorial, that is a major gap. We have many such gaps today.

Consider Perl, for instance. The tutorial manuals that people normally use are non-free. How did this come about? Because the authors of those manuals published them with restrictive terms—no copying, no modification, source files not available—which exclude them from the free software world.

That wasn't the first time this sort of thing happened, and it was far from the last. Many times we have heard a GNU user eagerly describe a manual that he is writing, his intended contribution to the community, only to learn that he had ruined everything by signing a publication contract to make it non-free.

Free documentation, like free software, is a matter of freedom, not price. The problem with the non-free manual is not that publishers charge a price for printed copies—that in itself is fine. (The Free Software Foundation sells printed copies of manuals, too.) The problem is the restrictions on the use of the manual. Free manuals are available in source code form, and give you permission to copy and modify. Non-free manuals do not allow this.

The criteria of freedom for a free manual are roughly the same as for free software. Redistribution (including the normal kinds of commercial redistribution) must be permitted, so that the manual can accompany every copy of the program, both on-line and on paper.

Permission for modification of the technical content is crucial too. When people modify the software, adding or changing features, if they are conscientious they will change the manual too—so they can provide accurate and clear documentation for the modified program. A manual that leaves you no choice but to write a new manual to document a changed version of the program is not really available to our community.

Some kinds of limits on the way modification is handled are acceptable. For example, requirements to preserve the original author's copyright notice, the distribution terms, or the list of authors, are ok. It is also no problem to require modified versions to include notice that they were modified. Even entire sections that may not be deleted or changed are acceptable, as long as they deal with nontechnical topics (like this one). These kinds of restrictions are acceptable because they don't obstruct the community's normal use of the manual.

However, it must be possible to modify all the *technical* content of the manual, and then distribute the result in all the usual media, through all the usual channels. Otherwise, the restrictions obstruct the use of the manual, it is not free, and we need another manual to replace it.

Please spread the word about this issue. Our community continues to lose manuals to proprietary publishing. If we spread the word that free software needs free reference manuals and free tutorials, perhaps the next person who wants to contribute by writing documentation will realize, before it is too late, that only free manuals contribute to the free software community.

If you are writing documentation, please insist on publishing it under the GNU Free Documentation License or another free documentation license. Remember that this decision requires your approval—you don't have to let the publisher decide. Some commercial publishers will use a free license if you insist, but they will not propose the option; it is up to you to raise the issue and say firmly that this is what you want. If the publisher you are dealing with refuses, please try other publishers. If you're not sure whether a proposed license is free, write to licensing@gnu.org.

You can encourage commercial publishers to sell more free, copylefted manuals and tutorials by buying them, and particularly by buying copies from the publishers that paid for their writing or for major improvements. Meanwhile, try to avoid buying non-free documentation at all. Check the distribution terms of a manual before you buy it, and insist that whoever seeks your business must respect your freedom. Check the history of the book, and try reward the publishers that have paid or pay the authors to work on it.

The Free Software Foundation maintains a list of free documentation published by other publishers:

> http://www.fsf.org/doc/other-free-books.html

Other books from the publisher

Network Theory publishes books about free software under free documentation licenses. Our current catalogue includes the following titles:

- **GNU Octave Manual Version 3** by John W. Eaton et al. (ISBN 0-9546120-6-X) $39.95 (£24.95)

  This manual is the definitive guide to GNU Octave, an interactive environment for numerical computation with matrices and vectors. This edition covers version 3 of GNU Octave. For each copy sold $1 is donated to the GNU Octave Development Fund.

- **GNU Bash Reference Manual** by Chet Ramey and Brian Fox (ISBN 0-9541617-7-7) $29.95 (£19.95)

  This manual is the definitive reference for GNU Bash, the standard GNU command-line interpreter. GNU Bash is a complete implementation of the POSIX.2 Bourne shell specification, with additional features from the C-shell and Korn shell. For each copy of this manual sold, $1 is donated to the Free Software Foundation.

- **Version Management with CVS** by Per Cederqvist et al. (ISBN 0-9541617-1-8) $29.95 (£19.95)

  This manual describes how to use CVS, the concurrent versioning system— one of the most widely-used source-code management systems available today. The manual provides tutorial examples for new users of CVS, as well as the definitive reference documentation for every CVS command and configuration option.

- **Comparing and Merging Files with GNU diff and patch** by David MacKenzie, Paul Eggert, and Richard Stallman (ISBN 0-9541617-5-0) $19.95 (£12.95)

  This manual describes how to compare and merge files using GNU diff and patch. It includes an extensive tutorial that guides the reader through all the options of the diff and patch commands. For each copy of this manual sold, $1 is donated to the Free Software Foundation.

- **An Introduction to GCC** by Brian J. Gough, foreword by Richard M. Stallman. (ISBN 0-9541617-9-3) $19.95 (£12.95)

  This manual provides a tutorial introduction to the GNU C and C++ compilers, gcc and g++. Many books teach the C and C++ languages, but this book explains how to use the compiler itself. Based on years of observation of questions posted on mailing lists, it guides the reader straight to the important options of GCC.

- **Valgrind 3.3** by J. Seward, N. Nethercote, J. Weidendorfer et al. (ISBN 0-9546120-5-1) $19.95 (£12.95)

  This manual describes how to use Valgrind, an award-winning suite of tools for debugging and profiling GNU/Linux programs. Valgrind detects memory and threading bugs automatically, avoiding hours of frustrating bug-hunting and making programs more stable. For each copy of this manual sold, $1 is donated to the Valgrind developers.

- **An Introduction to Python** by Guido van Rossum and Fred L. Drake, Jr. (ISBN 0-9541617-6-9) $19.95 (£12.95)

  This tutorial provides an introduction to Python, an easy to learn object oriented programming language. For each copy of this manual sold, $1 is donated to the Python Software Foundation.

- **Python Language Reference Manual** by Guido van Rossum and Fred L. Drake, Jr. (ISBN 0-9541617-8-5) $19.95 (£12.95)

  This manual is the official reference for the Python language itself. It describes the syntax of Python and its built-in datatypes in depth, This manual is suitable for readers who need to be familiar with the details and rules of the Python language and its object system. For each copy of this manual sold, $1 is donated to the Python Software Foundation.

- **An Introduction to R** by W.N. Venables, D.M. Smith and the R Development Core Team (ISBN 0-9541617-4-2) $19.95 (£12.95)

  This tutorial manual provides a comprehensive introduction to GNU R, a free software package for statistical computing and graphics.

- **PostgreSQL Reference Manual: Volumes 1–3** (ISBN 0-9546120-2-7) $49.95 (£32.00), (ISBN 0-9546120-3-5) $34.95 (£19.95), (ISBN 0-9546120-4-3) $24.95 (£13.95)

  These manuals documents the SQL language and commands of PostgreSQL, its client and server programming interfaces, and the configuration and maintenance of PostgreSQL servers. For each copy of these manuals sold, $1 is donated to the PostgreSQL project.

All titles are available for order from bookstores worldwide.

Sales of the manuals fund the development of more free software and documentation.

For details, visit the website http://www.network-theory.co.uk/. For questions or comments, please contact sales@network-theory.co.uk.

GNU General Public License

Version 3, 29 June 2007

Copyright © 2007 Free Software Foundation, Inc. http://fsf.org/

Everyone is permitted to copy and distribute verbatim copies of this
license document, but changing it is not allowed.

Preamble

The GNU General Public License is a free, copyleft license for software and other kinds of works.

The licenses for most software and other practical works are designed to take away your freedom to share and change the works. By contrast, the GNU General Public License is intended to guarantee your freedom to share and change all versions of a program–to make sure it remains free software for all its users. We, the Free Software Foundation, use the GNU General Public License for most of our software; it applies also to any other work released this way by its authors. You can apply it to your programs, too.

When we speak of free software, we are referring to freedom, not price. Our General Public Licenses are designed to make sure that you have the freedom to distribute copies of free software (and charge for them if you wish), that you receive source code or can get it if you want it, that you can change the software or use pieces of it in new free programs, and that you know you can do these things.

To protect your rights, we need to prevent others from denying you these rights or asking you to surrender the rights. Therefore, you have certain responsibilities if you distribute copies of the software, or if you modify it: responsibilities to respect the freedom of others.

For example, if you distribute copies of such a program, whether gratis or for a fee, you must pass on to the recipients the same freedoms that you received. You must make sure that they, too, receive or can get the source code. And you must show them these terms so they know their rights.

Developers that use the GNU GPL protect your rights with two steps: (1) assert copyright on the software, and (2) offer you this License giving you legal permission to copy, distribute and/or modify it.

For the developers' and authors' protection, the GPL clearly explains that there is no warranty for this free software. For both users' and authors' sake, the GPL requires that modified versions be marked as changed, so that their problems will not be attributed erroneously to authors of previous versions.

Some devices are designed to deny users access to install or run modified versions of the software inside them, although the manufacturer can do so. This is fundamentally incompatible with the aim of protecting users' freedom to change the software. The systematic pattern of such abuse occurs in the area of products for individuals to use, which is precisely where it is most unacceptable. Therefore, we have designed this version of the GPL to prohibit the practice for those products. If such problems arise substantially in other domains, we stand ready to extend this provision to those domains in future versions of the GPL, as needed to protect the freedom of users.

Finally, every program is threatened constantly by software patents. States should not allow patents to restrict development and use of software on general-purpose computers, but in those that do, we wish to avoid the special danger that patents applied to a free program could make it effectively proprietary. To prevent this, the GPL assures that patents cannot be used to render the program non-free.

The precise terms and conditions for copying, distribution and modification follow.

TERMS AND CONDITIONS

0. Definitions.

"This License" refers to version 3 of the GNU General Public License.

"Copyright" also means copyright-like laws that apply to other kinds of works, such as semiconductor masks.

"The Program" refers to any copyrightable work licensed under this License. Each licensee is addressed as "you". "Licensees" and "recipients" may be individuals or organizations.

To "modify" a work means to copy from or adapt all or part of the work in a fashion requiring copyright permission, other than the making of an exact copy. The resulting work is called a "modified version" of the earlier work or a work "based on" the earlier work.

A "covered work" means either the unmodified Program or a work based on the Program.

To "propagate" a work means to do anything with it that, without permission, would make you directly or secondarily liable for infringement under applicable copyright law, except executing it on a computer or modifying a private copy. Propagation includes copying, distribution (with or without modification), making available to the public, and in some countries other activities as well.

To "convey" a work means any kind of propagation that enables other parties to make or receive copies. Mere interaction with a user through a computer network, with no transfer of a copy, is not conveying.

An interactive user interface displays "Appropriate Legal Notices" to the extent that it includes a convenient and prominently visible feature that (1) displays an appropriate copyright notice, and (2) tells the user that there is no warranty for the work (except to the extent that warranties are provided), that licensees may convey the work under this License, and how to view a copy of this License. If the interface presents a list of user commands or options, such as a menu, a prominent item in the list meets this criterion.

1. Source Code.

The "source code" for a work means the preferred form of the work for making modifications to it. "Object code" means any non-source form of a work.

A "Standard Interface" means an interface that either is an official standard defined by a recognized standards body, or, in the case of interfaces specified for a particular programming language, one that is widely used among developers working in that language.

The "System Libraries" of an executable work include anything, other than the work as a whole, that (a) is included in the normal form of packaging a Major Component, but which is not part of that Major Component, and (b) serves only to enable use of the work with that Major Component, or to implement a Standard Interface for which an implementation is available to the public in source code form. A "Major Component", in this context, means a major essential component (kernel, window system, and so on) of the specific operating system (if any) on which the executable work runs, or a compiler used to produce the work, or an object code interpreter used to run it.

The "Corresponding Source" for a work in object code form means all the source code needed to generate, install, and (for an executable work) run the object code and to modify the work, including scripts to control those activities. However, it does not include the work's System Libraries, or general-purpose tools or generally available free programs which are used unmodified in performing those activities but which are not part of the work. For example, Corresponding Source includes interface definition files associated with source files for the work, and the source code for shared libraries and dynamically linked subprograms that the work is specifically designed to require, such as by intimate data communication or control flow between those subprograms and other parts of the work.

The Corresponding Source need not include anything that users can regenerate automatically from other parts of the Corresponding Source.

The Corresponding Source for a work in source code form is that same work.

2. Basic Permissions.

All rights granted under this License are granted for the term of copyright on the Program, and are irrevocable provided the stated conditions are met. This License explicitly affirms your unlimited permission to run the unmodified Program. The output from running a covered work is covered by this License only if the output, given its content, constitutes a covered work. This License acknowledges your rights of fair use or other equivalent, as provided by copyright law.

You may make, run and propagate covered works that you do not convey, without conditions so long as your license otherwise remains in force. You may convey covered works to others for the sole purpose of having them make modifications exclusively for you, or provide you with facilities for running those works, provided that you comply with the terms of this License in conveying all material for which you do not control copyright. Those thus making or running the covered works for you must do so exclusively on your behalf, under your direction and control, on terms that prohibit them from making any copies of your copyrighted material outside their relationship with you.

Conveying under any other circumstances is permitted solely under the conditions stated below. Sublicensing is not allowed; section 10 makes it unnecessary.

3. Protecting Users' Legal Rights From Anti-Circumvention Law.

No covered work shall be deemed part of an effective technological measure under any applicable law fulfilling obligations under article 11 of the WIPO copyright treaty adopted on 20 December 1996, or similar laws prohibiting or restricting circumvention of such measures.

When you convey a covered work, you waive any legal power to forbid circumvention of technological measures to the extent such circumvention is effected by exercising rights under this License with respect to the covered work, and you disclaim any intention to limit operation or modification of the work as a means of enforcing, against the work's users, your or third parties' legal rights to forbid circumvention of technological measures.

4. Conveying Verbatim Copies.

You may convey verbatim copies of the Program's source code as you receive it, in any medium, provided that you conspicuously and appropriately publish on each copy an appropriate copyright notice; keep intact all notices stating that this License and any non-permissive terms added in accord with section 7 apply to the code; keep intact all notices of the absence of any warranty; and give all recipients a copy of this License along with the Program.

You may charge any price or no price for each copy that you convey, and you may offer support or warranty protection for a fee.

5. Conveying Modified Source Versions.

You may convey a work based on the Program, or the modifications to produce it from the Program, in the form of source code under the terms of section 4, provided that you also meet all of these conditions:

    a. The work must carry prominent notices stating that you modified it, and giving a relevant date.

    b. The work must carry prominent notices stating that it is released under this License and any conditions added under section 7. This requirement modifies the requirement in section 4 to "keep intact all notices".

    c. You must license the entire work, as a whole, under this License to anyone who comes into possession of a copy. This License will therefore apply, along with any applicable section 7 additional terms, to the whole of the work, and all its parts, regardless of how they are packaged. This License gives no permission to license the work in any other way, but it does not invalidate such permission if you have separately received it.

d. If the work has interactive user interfaces, each must display Appropriate Legal Notices; however, if the Program has interactive interfaces that do not display Appropriate Legal Notices, your work need not make them do so.

A compilation of a covered work with other separate and independent works, which are not by their nature extensions of the covered work, and which are not combined with it such as to form a larger program, in or on a volume of a storage or distribution medium, is called an "aggregate" if the compilation and its resulting copyright are not used to limit the access or legal rights of the compilation's users beyond what the individual works permit. Inclusion of a covered work in an aggregate does not cause this License to apply to the other parts of the aggregate.

6. Conveying Non-Source Forms.

You may convey a covered work in object code form under the terms of sections 4 and 5, provided that you also convey the machine-readable Corresponding Source under the terms of this License, in one of these ways:

a. Convey the object code in, or embodied in, a physical product (including a physical distribution medium), accompanied by the Corresponding Source fixed on a durable physical medium customarily used for software interchange.

b. Convey the object code in, or embodied in, a physical product (including a physical distribution medium), accompanied by a written offer, valid for at least three years and valid for as long as you offer spare parts or customer support for that product model, to give anyone who possesses the object code either (1) a copy of the Corresponding Source for all the software in the product that is covered by this License, on a durable physical medium customarily used for software interchange, for a price no more than your reasonable cost of physically performing this conveying of source, or (2) access to copy the Corresponding Source from a network server at no charge.

c. Convey individual copies of the object code with a copy of the written offer to provide the Corresponding Source. This alternative is allowed only occasionally and noncommercially, and only if you received the object code with such an offer, in accord with subsection 6b.

d. Convey the object code by offering access from a designated place (gratis or for a charge), and offer equivalent access to the Corresponding Source in the same way through the same place at no further charge. You need not require recipients to copy the Corresponding Source along with the object code. If the place to copy the object code is a network server, the Corresponding Source may be on a different server (operated by you or a third party) that supports equivalent copying facilities, provided you maintain clear directions next to the object code saying where to find the Corresponding Source. Regardless of what server hosts the Corresponding Source, you remain obligated to ensure that it is available for as long as needed to satisfy these requirements.

e. Convey the object code using peer-to-peer transmission, provided you inform other peers where the object code and Corresponding Source of the work are being offered to the general public at no charge under subsection 6d.

A separable portion of the object code, whose source code is excluded from the Corresponding Source as a System Library, need not be included in conveying the object code work.

A "User Product" is either (1) a "consumer product", which means any tangible personal property which is normally used for personal, family, or household purposes, or (2) anything designed or sold for incorporation into a dwelling. In determining whether a product is a consumer product, doubtful cases shall be resolved in favor of coverage. For a particular product received by a particular user, "normally used" refers to a typical or common use of that class of product, regardless of the status of the particular user or of the way in which the particular user actually uses, or expects or is expected to use, the product. A product is a consumer product regardless of whether the product has substantial commercial, industrial or non-consumer uses, unless such uses represent the only significant mode of use of the product.

"Installation Information" for a User Product means any methods, procedures, authorization keys, or other information required to install and execute modified versions of a covered work in that User Product from a modified version of its Corresponding Source. The information must suffice to ensure that the continued functioning of the modified object code is in no case prevented or interfered with solely because modification has been made.

If you convey an object code work under this section in, or with, or specifically for use in, a User Product, and the conveying occurs as part of a transaction in which the right of possession and use of the User Product is transferred to the recipient in perpetuity or for a fixed term (regardless of how the transaction is characterized), the Corresponding Source conveyed under this section must be accompanied by the Installation Information. But this requirement does not apply if neither you nor any third party retains the ability to install modified object code on the User Product (for example, the work has been installed in ROM).

The requirement to provide Installation Information does not include a requirement to continue to provide support service, warranty, or updates for a work that has been modified or installed by the recipient, or for the User Product in which it has been modified or installed. Access to a network may be denied when the modification itself materially and adversely affects the operation of the network or violates the rules and protocols for communication across the network.

Corresponding Source conveyed, and Installation Information provided, in accord with this section must be in a format that is publicly documented (and with an implementation available to the public in source code form), and must require no special password or key for unpacking, reading or copying.

7. Additional Terms.

"Additional permissions" are terms that supplement the terms of this License by making exceptions from one or more of its conditions. Additional permissions that are applicable to the entire Program shall be treated as though they were included in this License, to the extent that they are valid under applicable law. If additional permissions apply only to part of the Program, that part may be used separately under those permissions, but the entire Program remains governed by this License without regard to the additional permissions.

When you convey a copy of a covered work, you may at your option remove any additional permissions from that copy, or from any part of it. (Additional permissions may be written to require their own removal in certain cases when you modify the work.) You may place additional permissions on material, added by you to a covered work, for which you have or can give appropriate copyright permission.

Notwithstanding any other provision of this License, for material you add to a covered work, you may (if authorized by the copyright holders of that material) supplement the terms of this License with terms:

a. Disclaiming warranty or limiting liability differently from the terms of sections 15 and 16 of this License; or

b. Requiring preservation of specified reasonable legal notices or author attributions in that material or in the Appropriate Legal Notices displayed by works containing it; or

c. Prohibiting misrepresentation of the origin of that material, or requiring that modified versions of such material be marked in reasonable ways as different from the original version; or

d. Limiting the use for publicity purposes of names of licensors or authors of the material; or

e. Declining to grant rights under trademark law for use of some trade names, trademarks, or service marks; or

f. Requiring indemnification of licensors and authors of that material by anyone who conveys the material (or modified versions of it) with contractual assumptions of liability to the recipient, for any liability that these contractual assumptions directly impose on those licensors and authors.

All other non-permissive additional terms are considered "further restrictions" within the meaning of section 10. If the Program as you received it, or any part of it, contains a notice stating that it is governed by this License along with a term that is a further restriction, you may remove that term. If a license document contains a further restriction but permits relicensing or conveying under this License, you may add to a covered work material governed by the terms of that license document, provided that the further restriction does not survive such relicensing or conveying.

If you add terms to a covered work in accord with this section, you must place, in the relevant source files, a statement of the additional terms that apply to those files, or a notice indicating where to find the applicable terms.

Additional terms, permissive or non-permissive, may be stated in the form of a separately written license, or stated as exceptions; the above requirements apply either way.

8. Termination.

You may not propagate or modify a covered work except as expressly provided under this License. Any attempt otherwise to propagate or modify it is void, and will automatically terminate your rights under this License (including any patent licenses granted under the third paragraph of section 11).

However, if you cease all violation of this License, then your license from a particular copyright holder is reinstated (a) provisionally, unless and until the copyright holder explicitly and finally terminates your license, and (b) permanently, if the copyright holder fails to notify you of the violation by some reasonable means prior to 60 days after the cessation.

Moreover, your license from a particular copyright holder is reinstated permanently if the copyright holder notifies you of the violation by some reasonable means, this is the first time you have received notice of violation of this License (for any work) from that copyright holder, and you cure the violation prior to 30 days after your receipt of the notice.

Termination of your rights under this section does not terminate the licenses of parties who have received copies or rights from you under this License. If your rights have been terminated and not permanently reinstated, you do not qualify to receive new licenses for the same material under section 10.

9. Acceptance Not Required for Having Copies.

You are not required to accept this License in order to receive or run a copy of the Program. Ancillary propagation of a covered work occurring solely as a consequence of using peer-to-peer transmission to receive a copy likewise does not require acceptance. However, nothing other than this License grants you permission to propagate or modify any covered work. These actions infringe copyright if you do not accept this License. Therefore, by modifying or propagating a covered work, you indicate your acceptance of this License to do so.

10. Automatic Licensing of Downstream Recipients.

Each time you convey a covered work, the recipient automatically receives a license from the original licensors, to run, modify and propagate that work, subject to this License. You are not responsible for enforcing compliance by third parties with this License.

An "entity transaction" is a transaction transferring control of an organization, or substantially all assets of one, or subdividing an organization, or merging organizations. If propagation of a covered work results from an entity transaction, each party to that transaction who receives a copy of the work also receives whatever licenses to the work the party's predecessor in interest had or could give under the previous paragraph, plus a right to possession of the Corresponding Source of the work from the predecessor in interest, if the predecessor has it or can get it with reasonable efforts.

You may not impose any further restrictions on the exercise of the rights granted or affirmed under this License. For example, you may not impose a license fee, royalty, or other charge for exercise of rights granted under this License, and you may not initiate litigation (including a cross-claim or counterclaim in a lawsuit) alleging that any patent claim is infringed by making, using, selling, offering for sale, or importing the Program or any portion of it.

11. Patents.

A "contributor" is a copyright holder who authorizes use under this License of the Program or a work on which the Program is based. The work thus licensed is called the contributor's "contributor version".

A contributor's "essential patent claims" are all patent claims owned or controlled by the contributor, whether already acquired or hereafter acquired, that would be infringed by some manner, permitted by this License, of making, using, or selling its contributor version, but do not include claims that would be infringed only as a consequence of further modification of the contributor version. For purposes of this definition, "control" includes the right to grant patent sublicenses in a manner consistent with the requirements of this License.

Each contributor grants you a non-exclusive, worldwide, royalty-free patent license under the contributor's essential patent claims, to make, use, sell, offer for sale, import and otherwise run, modify and propagate the contents of its contributor version.

In the following three paragraphs, a "patent license" is any express agreement or commitment, however denominated, not to enforce a patent (such as an express permission to practice a patent or covenant not to sue for patent infringement). To "grant" such a patent license to a party means to make such an agreement or commitment not to enforce a patent against the party.

If you convey a covered work, knowingly relying on a patent license, and the Corresponding Source of the work is not available for anyone to copy, free of charge and under the terms of this License, through a publicly available network server or other readily accessible means, then you must either (1) cause the Corresponding Source to be so available, or (2) arrange to deprive yourself of the benefit of the patent license for this particular work, or (3) arrange, in a manner consistent with the requirements of this License, to extend the patent license to downstream recipients. "Knowingly relying" means you have actual knowledge that, but for the patent license, your conveying the covered work in a country, or your recipient's use of the covered work in a country, would infringe one or more identifiable patents in that country that you have reason to believe are valid.

If, pursuant to or in connection with a single transaction or arrangement, you convey, or propagate by procuring conveyance of, a covered work, and grant a patent license to some of the parties receiving the covered work authorizing them to use, propagate, modify or convey a specific copy of the covered work, then the patent license you grant is automatically extended to all recipients of the covered work and works based on it.

A patent license is "discriminatory" if it does not include within the scope of its coverage, prohibits the exercise of, or is conditioned on the non-exercise of one or more of the rights that are specifically granted under this License. You may not convey a covered work if you are a party to an arrangement with a third party that is in the business of distributing software, under which you make payment to the third party based on the extent of your activity of conveying the work, and under which the third party grants, to any of the parties who would receive the covered work from you, a discriminatory patent license (a) in connection with copies of the covered work conveyed by you (or copies made from those copies), or (b) primarily for and in connection with specific products or compilations that contain the covered work, unless you entered into that arrangement, or that patent license was granted, prior to 28 March 2007.

Nothing in this License shall be construed as excluding or limiting any implied license or other defenses to infringement that may otherwise be available to you under applicable patent law.

12. No Surrender of Others' Freedom.

If conditions are imposed on you (whether by court order, agreement or otherwise) that contradict the conditions of this License, they do not excuse you from the conditions of this License. If you cannot convey a covered work so as to satisfy simultaneously your obligations under this License and any other pertinent obligations, then as a consequence you may not convey it at all. For example, if you agree to terms that obligate you to collect a royalty for further conveying from those to whom you convey

the Program, the only way you could satisfy both those terms and this License would
be to refrain entirely from conveying the Program.

13. Use with the GNU Affero General Public License.

Notwithstanding any other provision of this License, you have permission to link or
combine any covered work with a work licensed under version 3 of the GNU Affero
General Public License into a single combined work, and to convey the resulting work.
The terms of this License will continue to apply to the part which is the covered work,
but the special requirements of the GNU Affero General Public License, section 13,
concerning interaction through a network will apply to the combination as such.

14. Revised Versions of this License.

The Free Software Foundation may publish revised and/or new versions of the GNU
General Public License from time to time. Such new versions will be similar in spirit
to the present version, but may differ in detail to address new problems or concerns.

Each version is given a distinguishing version number. If the Program specifies that
a certain numbered version of the GNU General Public License "or any later version"
applies to it, you have the option of following the terms and conditions either of that
numbered version or of any later version published by the Free Software Foundation.
If the Program does not specify a version number of the GNU General Public License,
you may choose any version ever published by the Free Software Foundation.

If the Program specifies that a proxy can decide which future versions of the GNU
General Public License can be used, that proxy's public statement of acceptance of a
version permanently authorizes you to choose that version for the Program.

Later license versions may give you additional or different permissions. However, no
additional obligations are imposed on any author or copyright holder as a result of
your choosing to follow a later version.

15. Disclaimer of Warranty.

THERE IS NO WARRANTY FOR THE PROGRAM, TO THE EXTENT PER-
MITTED BY APPLICABLE LAW. EXCEPT WHEN OTHERWISE STATED IN
WRITING THE COPYRIGHT HOLDERS AND/OR OTHER PARTIES PROVIDE
THE PROGRAM "AS IS" WITHOUT WARRANTY OF ANY KIND, EITHER EX-
PRESSED OR IMPLIED, INCLUDING, BUT NOT LIMITED TO, THE IMPLIED
WARRANTIES OF MERCHANTABILITY AND FITNESS FOR A PARTICULAR
PURPOSE. THE ENTIRE RISK AS TO THE QUALITY AND PERFORMANCE
OF THE PROGRAM IS WITH YOU. SHOULD THE PROGRAM PROVE DEFEC-
TIVE, YOU ASSUME THE COST OF ALL NECESSARY SERVICING, REPAIR
OR CORRECTION.

16. Limitation of Liability.

IN NO EVENT UNLESS REQUIRED BY APPLICABLE LAW OR AGREED TO IN
WRITING WILL ANY COPYRIGHT HOLDER, OR ANY OTHER PARTY WHO
MODIFIES AND/OR CONVEYS THE PROGRAM AS PERMITTED ABOVE, BE
LIABLE TO YOU FOR DAMAGES, INCLUDING ANY GENERAL, SPECIAL, IN-
CIDENTAL OR CONSEQUENTIAL DAMAGES ARISING OUT OF THE USE OR
INABILITY TO USE THE PROGRAM (INCLUDING BUT NOT LIMITED TO
LOSS OF DATA OR DATA BEING RENDERED INACCURATE OR LOSSES SUS-
TAINED BY YOU OR THIRD PARTIES OR A FAILURE OF THE PROGRAM
TO OPERATE WITH ANY OTHER PROGRAMS), EVEN IF SUCH HOLDER OR
OTHER PARTY HAS BEEN ADVISED OF THE POSSIBILITY OF SUCH DAM-
AGES.

17. Interpretation of Sections 15 and 16.

If the disclaimer of warranty and limitation of liability provided above cannot be given
local legal effect according to their terms, reviewing courts shall apply local law that
most closely approximates an absolute waiver of all civil liability in connection with
the Program, unless a warranty or assumption of liability accompanies a copy of the
Program in return for a fee.

END OF TERMS AND CONDITIONS

How to Apply These Terms to Your New Programs

If you develop a new program, and you want it to be of the greatest possible use to the public, the best way to achieve this is to make it free software which everyone can redistribute and change under these terms.

To do so, attach the following notices to the program. It is safest to attach them to the start of each source file to most effectively state the exclusion of warranty; and each file should have at least the "copyright" line and a pointer to where the full notice is found.

```
one line to give the program's name and a brief idea
of what it does.
Copyright (C) year name of author

This program is free software: you can redistribute it and/or modify
it under the terms of the GNU General Public License as published by
the Free Software Foundation, either version 3 of the License, or (at
your option) any later version.

This program is distributed in the hope that it will be useful, but
WITHOUT ANY WARRANTY; without even the implied warranty of
MERCHANTABILITY or FITNESS FOR A PARTICULAR PURPOSE.  See the GNU
General Public License for more details.

You should have received a copy of the GNU General Public License
along with this program.  If not, see http://www.gnu.org/licenses/.
```

Also add information on how to contact you by electronic and paper mail.

If the program does terminal interaction, make it output a short notice like this when it starts in an interactive mode:

```
program Copyright (C) year name of author
This program comes with ABSOLUTELY NO WARRANTY; for details type 'show w'.
This is free software, and you are welcome to redistribute it
under certain conditions; type 'show c' for details.
```

The hypothetical commands 'show w' and 'show c' should show the appropriate parts of the General Public License. Of course, your program's commands might be different; for a GUI interface, you would use an "about box".

You should also get your employer (if you work as a programmer) or school, if any, to sign a "copyright disclaimer" for the program, if necessary. For more information on this, and how to apply and follow the GNU GPL, see http://www.gnu.org/licenses/.

The GNU General Public License does not permit incorporating your program into proprietary programs. If your program is a subroutine library, you may consider it more useful to permit linking proprietary applications with the library. If this is what you want to do, use the GNU Lesser General Public License instead of this License. But first, please read http://www.gnu.org/philosophy/why-not-lgpl.html.

GNU Free Documentation License

Version 1.3, 3 November 2008

Copyright © 2000, 2001, 2002, 2007, 2008 Free Software Foundation, Inc.
http://fsf.org/

Everyone is permitted to copy and distribute verbatim copies of this license
document, but changing it is not allowed.

0. PREAMBLE

The purpose of this License is to make a manual, textbook, or other functional and
useful document *free* in the sense of freedom: to assure everyone the effective freedom
to copy and redistribute it, with or without modifying it, either commercially or noncommercially. Secondarily, this License preserves for the author and publisher a way
to get credit for their work, while not being considered responsible for modifications
made by others.

This License is a kind of "copyleft", which means that derivative works of the document must themselves be free in the same sense. It complements the GNU General
Public License, which is a copyleft license designed for free software.

We have designed this License in order to use it for manuals for free software, because
free software needs free documentation: a free program should come with manuals
providing the same freedoms that the software does. But this License is not limited
to software manuals; it can be used for any textual work, regardless of subject matter
or whether it is published as a printed book. We recommend this License principally
for works whose purpose is instruction or reference.

1. APPLICABILITY AND DEFINITIONS

This License applies to any manual or other work, in any medium, that contains a
notice placed by the copyright holder saying it can be distributed under the terms
of this License. Such a notice grants a world-wide, royalty-free license, unlimited in
duration, to use that work under the conditions stated herein. The "Document",
below, refers to any such manual or work. Any member of the public is a licensee,
and is addressed as "you". You accept the license if you copy, modify or distribute
the work in a way requiring permission under copyright law.

A "Modified Version" of the Document means any work containing the Document or
a portion of it, either copied verbatim, or with modifications and/or translated into
another language.

A "Secondary Section" is a named appendix or a front-matter section of the Document
that deals exclusively with the relationship of the publishers or authors of the Document to the Document's overall subject (or to related matters) and contains nothing
that could fall directly within that overall subject. (Thus, if the Document is in part
a textbook of mathematics, a Secondary Section may not explain any mathematics.)
The relationship could be a matter of historical connection with the subject or with
related matters, or of legal, commercial, philosophical, ethical or political position
regarding them.

The "Invariant Sections" are certain Secondary Sections whose titles are designated,
as being those of Invariant Sections, in the notice that says that the Document is
released under this License. If a section does not fit the above definition of Secondary
then it is not allowed to be designated as Invariant. The Document may contain zero
Invariant Sections. If the Document does not identify any Invariant Sections then
there are none.

The "Cover Texts" are certain short passages of text that are listed, as Front-Cover
Texts or Back-Cover Texts, in the notice that says that the Document is released
under this License. A Front-Cover Text may be at most 5 words, and a Back-Cover
Text may be at most 25 words.

A "Transparent" copy of the Document means a machine-readable copy, represented
in a format whose specification is available to the general public, that is suitable

for revising the document straightforwardly with generic text editors or (for images composed of pixels) generic paint programs or (for drawings) some widely available drawing editor, and that is suitable for input to text formatters or for automatic translation to a variety of formats suitable for input to text formatters. A copy made in an otherwise Transparent file format whose markup, or absence of markup, has been arranged to thwart or discourage subsequent modification by readers is not Transparent. An image format is not Transparent if used for any substantial amount of text. A copy that is not "Transparent" is called "Opaque".

Examples of suitable formats for Transparent copies include plain ASCII without markup, Texinfo input format, LaTeX input format, SGML or XML using a publicly available DTD, and standard-conforming simple HTML, PostScript or PDF designed for human modification. Examples of transparent image formats include PNG, XCF and JPG. Opaque formats include proprietary formats that can be read and edited only by proprietary word processors, SGML or XML for which the DTD and/or processing tools are not generally available, and the machine-generated HTML, PostScript or PDF produced by some word processors for output purposes only.

The "Title Page" means, for a printed book, the title page itself, plus such following pages as are needed to hold, legibly, the material this License requires to appear in the title page. For works in formats which do not have any title page as such, "Title Page" means the text near the most prominent appearance of the work's title, preceding the beginning of the body of the text.

The "publisher" means any person or entity that distributes copies of the Document to the public.

A section "Entitled XYZ" means a named subunit of the Document whose title either is precisely XYZ or contains XYZ in parentheses following text that translates XYZ in another language. (Here XYZ stands for a specific section name mentioned below, such as "Acknowledgements", "Dedications", "Endorsements", or "History".) To "Preserve the Title" of such a section when you modify the Document means that it remains a section "Entitled XYZ" according to this definition.

The Document may include Warranty Disclaimers next to the notice which states that this License applies to the Document. These Warranty Disclaimers are considered to be included by reference in this License, but only as regards disclaiming warranties: any other implication that these Warranty Disclaimers may have is void and has no effect on the meaning of this License.

2. VERBATIM COPYING

You may copy and distribute the Document in any medium, either commercially or noncommercially, provided that this License, the copyright notices, and the license notice saying this License applies to the Document are reproduced in all copies, and that you add no other conditions whatsoever to those of this License. You may not use technical measures to obstruct or control the reading or further copying of the copies you make or distribute. However, you may accept compensation in exchange for copies. If you distribute a large enough number of copies you must also follow the conditions in section 3.

You may also lend copies, under the same conditions stated above, and you may publicly display copies.

3. COPYING IN QUANTITY

If you publish printed copies (or copies in media that commonly have printed covers) of the Document, numbering more than 100, and the Document's license notice requires Cover Texts, you must enclose the copies in covers that carry, clearly and legibly, all these Cover Texts: Front-Cover Texts on the front cover, and Back-Cover Texts on the back cover. Both covers must also clearly and legibly identify you as the publisher of these copies. The front cover must present the full title with all words of the title equally prominent and visible. You may add other material on the covers in addition. Copying with changes limited to the covers, as long as they preserve the title of the Document and satisfy these conditions, can be treated as verbatim copying in other respects.

If the required texts for either cover are too voluminous to fit legibly, you should put the first ones listed (as many as fit reasonably) on the actual cover, and continue the rest onto adjacent pages.

If you publish or distribute Opaque copies of the Document numbering more than 100, you must either include a machine-readable Transparent copy along with each Opaque copy, or state in or with each Opaque copy a computer-network location from which the general network-using public has access to download using public-standard network protocols a complete Transparent copy of the Document, free of added material. If you use the latter option, you must take reasonably prudent steps, when you begin distribution of Opaque copies in quantity, to ensure that this Transparent copy will remain thus accessible at the stated location until at least one year after the last time you distribute an Opaque copy (directly or through your agents or retailers) of that edition to the public.

It is requested, but not required, that you contact the authors of the Document well before redistributing any large number of copies, to give them a chance to provide you with an updated version of the Document.

4. MODIFICATIONS

You may copy and distribute a Modified Version of the Document under the conditions of sections 2 and 3 above, provided that you release the Modified Version under precisely this License, with the Modified Version filling the role of the Document, thus licensing distribution and modification of the Modified Version to whoever possesses a copy of it. In addition, you must do these things in the Modified Version:

A. Use in the Title Page (and on the covers, if any) a title distinct from that of the Document, and from those of previous versions (which should, if there were any, be listed in the History section of the Document). You may use the same title as a previous version if the original publisher of that version gives permission.

B. List on the Title Page, as authors, one or more persons or entities responsible for authorship of the modifications in the Modified Version, together with at least five of the principal authors of the Document (all of its principal authors, if it has fewer than five), unless they release you from this requirement.

C. State on the Title page the name of the publisher of the Modified Version, as the publisher.

D. Preserve all the copyright notices of the Document.

E. Add an appropriate copyright notice for your modifications adjacent to the other copyright notices.

F. Include, immediately after the copyright notices, a license notice giving the public permission to use the Modified Version under the terms of this License, in the form shown in the Addendum below.

G. Preserve in that license notice the full lists of Invariant Sections and required Cover Texts given in the Document's license notice.

H. Include an unaltered copy of this License.

I. Preserve the section Entitled "History", Preserve its Title, and add to it an item stating at least the title, year, new authors, and publisher of the Modified Version as given on the Title Page. If there is no section Entitled "History" in the Document, create one stating the title, year, authors, and publisher of the Document as given on its Title Page, then add an item describing the Modified Version as stated in the previous sentence.

J. Preserve the network location, if any, given in the Document for public access to a Transparent copy of the Document, and likewise the network locations given in the Document for previous versions it was based on. These may be placed in the "History" section. You may omit a network location for a work that was published at least four years before the Document itself, or if the original publisher of the version it refers to gives permission.

K. For any section Entitled "Acknowledgements" or "Dedications", Preserve the Title of the section, and preserve in the section all the substance and tone of each of the contributor acknowledgements and/or dedications given therein.

L. Preserve all the Invariant Sections of the Document, unaltered in their text and in their titles. Section numbers or the equivalent are not considered part of the section titles.

M. Delete any section Entitled "Endorsements". Such a section may not be included in the Modified Version.

N. Do not retitle any existing section to be Entitled "Endorsements" or to conflict in title with any Invariant Section.

O. Preserve any Warranty Disclaimers.

If the Modified Version includes new front-matter sections or appendices that qualify as Secondary Sections and contain no material copied from the Document, you may at your option designate some or all of these sections as invariant. To do this, add their titles to the list of Invariant Sections in the Modified Version's license notice. These titles must be distinct from any other section titles.

You may add a section Entitled "Endorsements", provided it contains nothing but endorsements of your Modified Version by various parties—for example, statements of peer review or that the text has been approved by an organization as the authoritative definition of a standard.

You may add a passage of up to five words as a Front-Cover Text, and a passage of up to 25 words as a Back-Cover Text, to the end of the list of Cover Texts in the Modified Version. Only one passage of Front-Cover Text and one of Back-Cover Text may be added by (or through arrangements made by) any one entity. If the Document already includes a cover text for the same cover, previously added by you or by arrangement made by the same entity you are acting on behalf of, you may not add another; but you may replace the old one, on explicit permission from the previous publisher that added the old one.

The author(s) and publisher(s) of the Document do not by this License give permission to use their names for publicity for or to assert or imply endorsement of any Modified Version.

5. COMBINING DOCUMENTS

You may combine the Document with other documents released under this License, under the terms defined in section 4 above for modified versions, provided that you include in the combination all of the Invariant Sections of all of the original documents, unmodified, and list them all as Invariant Sections of your combined work in its license notice, and that you preserve all their Warranty Disclaimers.

The combined work need only contain one copy of this License, and multiple identical Invariant Sections may be replaced with a single copy. If there are multiple Invariant Sections with the same name but different contents, make the title of each such section unique by adding at the end of it, in parentheses, the name of the original author or publisher of that section if known, or else a unique number. Make the same adjustment to the section titles in the list of Invariant Sections in the license notice of the combined work.

In the combination, you must combine any sections Entitled "History" in the various original documents, forming one section Entitled "History"; likewise combine any sections Entitled "Acknowledgements", and any sections Entitled "Dedications". You must delete all sections Entitled "Endorsements."

6. COLLECTIONS OF DOCUMENTS

You may make a collection consisting of the Document and other documents released under this License, and replace the individual copies of this License in the various documents with a single copy that is included in the collection, provided that you follow the rules of this License for verbatim copying of each of the documents in all other respects.

You may extract a single document from such a collection, and distribute it individually under this License, provided you insert a copy of this License into the extracted document, and follow this License in all other respects regarding verbatim copying of that document.

7. AGGREGATION WITH INDEPENDENT WORKS

A compilation of the Document or its derivatives with other separate and independent documents or works, in or on a volume of a storage or distribution medium, is called an "aggregate" if the copyright resulting from the compilation is not used to limit the legal rights of the compilation's users beyond what the individual works permit. When the Document is included in an aggregate, this License does not apply to the other works in the aggregate which are not themselves derivative works of the Document.

If the Cover Text requirement of section 3 is applicable to these copies of the Document, then if the Document is less than one half of the entire aggregate, the Document's Cover Texts may be placed on covers that bracket the Document within the aggregate, or the electronic equivalent of covers if the Document is in electronic form. Otherwise they must appear on printed covers that bracket the whole aggregate.

8. TRANSLATION

Translation is considered a kind of modification, so you may distribute translations of the Document under the terms of section 4. Replacing Invariant Sections with translations requires special permission from their copyright holders, but you may include translations of some or all Invariant Sections in addition to the original versions of these Invariant Sections. You may include a translation of this License, and all the license notices in the Document, and any Warranty Disclaimers, provided that you also include the original English version of this License and the original versions of those notices and disclaimers. In case of a disagreement between the translation and the original version of this License or a notice or disclaimer, the original version will prevail.

If a section in the Document is Entitled "Acknowledgements", "Dedications", or "History", the requirement (section 4) to Preserve its Title (section 1) will typically require changing the actual title.

9. TERMINATION

You may not copy, modify, sublicense, or distribute the Document except as expressly provided under this License. Any attempt otherwise to copy, modify, sublicense, or distribute it is void, and will automatically terminate your rights under this License.

However, if you cease all violation of this License, then your license from a particular copyright holder is reinstated (a) provisionally, unless and until the copyright holder explicitly and finally terminates your license, and (b) permanently, if the copyright holder fails to notify you of the violation by some reasonable means prior to 60 days after the cessation.

Moreover, your license from a particular copyright holder is reinstated permanently if the copyright holder notifies you of the violation by some reasonable means, this is the first time you have received notice of violation of this License (for any work) from that copyright holder, and you cure the violation prior to 30 days after your receipt of the notice.

Termination of your rights under this section does not terminate the licenses of parties who have received copies or rights from you under this License. If your rights have been terminated and not permanently reinstated, receipt of a copy of some or all of the same material does not give you any rights to use it.

10. FUTURE REVISIONS OF THIS LICENSE

The Free Software Foundation may publish new, revised versions of the GNU Free Documentation License from time to time. Such new versions will be similar in spirit to the present version, but may differ in detail to address new problems or concerns. See http://www.gnu.org/copyleft/.

Each version of the License is given a distinguishing version number. If the Document specifies that a particular numbered version of this License "or any later version" applies to it, you have the option of following the terms and conditions either of that specified version or of any later version that has been published (not as a draft) by the Free Software Foundation. If the Document does not specify a version number of this License, you may choose any version ever published (not as a draft) by the Free Software Foundation. If the Document specifies that a proxy can decide which future versions of this License can be used, that proxy's public statement of acceptance of a version permanently authorizes you to choose that version for the Document.

11. RELICENSING

"Massive Multiauthor Collaboration Site" (or "MMC Site") means any World Wide Web server that publishes copyrightable works and also provides prominent facilities for anybody to edit those works. A public wiki that anybody can edit is an example of such a server. A "Massive Multiauthor Collaboration" (or "MMC") contained in the site means any set of copyrightable works thus published on the MMC site.

"CC-BY-SA" means the Creative Commons Attribution-Share Alike 3.0 license published by Creative Commons Corporation, a not-for-profit corporation with a principal place of business in San Francisco, California, as well as future copyleft versions of that license published by that same organization.

"Incorporate" means to publish or republish a Document, in whole or in part, as part of another Document.

An MMC is "eligible for relicensing" if it is licensed under this License, and if all works that were first published under this License somewhere other than this MMC, and subsequently incorporated in whole or in part into the MMC, (1) had no cover texts or invariant sections, and (2) were thus incorporated prior to November 1, 2008.

The operator of an MMC Site may republish an MMC contained in the site under CC-BY-SA on the same site at any time before August 1, 2009, provided the MMC is eligible for relicensing.

ADDENDUM: How to use this License for your documents

To use this License in a document you have written, include a copy of the License in the document and put the following copyright and license notices just after the title page:

```
Copyright (C)  year  your name.
Permission is granted to copy, distribute and/or modify
this document under the terms of the GNU Free
Documentation License, Version 1.3 or any later version
published by the Free Software Foundation; with no
Invariant Sections, no Front-Cover Texts, and no
Back-Cover Texts.  A copy of the license is included in
the section entitled ''GNU Free Documentation License''.
```

If you have Invariant Sections, Front-Cover Texts and Back-Cover Texts, replace the "with...Texts." line with this:

```
with the Invariant Sections being list their
titles, with the Front-Cover Texts being list, and
with the Back-Cover Texts being list.
```

If you have Invariant Sections without Cover Texts, or some other combination of the three, merge those two alternatives to suit the situation.

If your document contains nontrivial examples of program code, we recommend releasing these examples in parallel under your choice of free software license, such as the GNU General Public License, to permit their use in free software.

History

This section gives the history of the modifications made to the manual by the publisher, as required by the GNU Free Documentation License.

12/2008 "GNU Scientific Library Reference Manual"
Original release. M. Galassi, J. Davies, J. Theiler, B. Gough, G. Jungman, P. Alken, M. Booth, F. Rossi. Publisher: GSL Team.

1/2009 "GNU Scientific Library Reference Manual"
Edited for publication by Brian Gough. Publisher: Network Theory Ltd. Title used with permission. Added publisher's preface. Minor modifications for publication as a printed book: reformatted examples to fit smaller page width, other minor changes to improve line-breaking. Added this "History" section.

The source code for the original version of this document is available from ftp.gnu.org/gnu/gsl/ in the file 'gsl-1.12.tar.gz'.

The source code for this version is available from http://www.network-theory .co.uk/gsl/manual/src/. A complete set of differences can be obtained from the same location.

Function Index

C

G

Type and Variable Index

Concept Index

E

F

CPSIA information can be obtained at www.ICGtesting.com
Printed in the USA
BVOW071051180613

323625BV00001B/98/P